AF078416

The Evans Equations of Unified Field Theory

Laurence G. Felker

This book is dedicated to my mother, Clara Jane Wenzel Felker.
There is too much to say to put here.

Published 2007 by arima publishing

www.arimapublishing.com

ISBN 978-1-84549-214-4

© Laurence G. Felker 2007

All rights reserved

This book is copyright. Subject to statutory exception and to provisions of relevant collective licensing agreements, no part of this publication may be reproduced, stored in a retrieval system, or transmitted in any form or by any means, without the prior written permission of the author.

Printed and bound in the United Kingdom

Typeset in Arial

This book is sold subject to the conditions that it shall not, by way of trade or otherwise, be lent, re-sold, hired out, or otherwise circulated without the publisher's prior consent in any form of binding or cover other than that which it is published and without a similar condition including this condition being imposed on the subsequent purchaser.

abramis is an imprint of arima publishing

arima publishing
ASK House, Northgate Avenue
Bury St Edmunds, Suffolk IP32 6BB
t: (+44) 01284 700321

www.arimapublishing.com

Contents

Introduction
- Don't Panic 1
- General Relativity and Quantum Theory 2
- Einstein-Cartan-Evans Unified Field Theory 4
- Evans' Results 5
- The Nature of Spacetime 6
- The Four Forces 12
- The Particles 14
- What We Will See 15

Chapter 1 Special Relativity 16
- Relativity and Quantum Theory 16
- Special Relativity 19
- Invariant distance 22
- Correspondence Principle 24
- Vectors 25
- The Metric 31
- Summary 32

Chapter 2 General Relativity 34
- Introduction 34
- Curved Spacetime 41
- Curvature 44
- Vacuum 46
- Manifolds and Mathematical Spaces 47
- The Metric and the Tetrad 51
- Tensors and Differential Geometry 53
- Equivalence Principles 54

Chapter 3 Quantum Theory 56
- Quantum Theory 56
- Schrodinger's Equation 61
- Vector Space 62
- Dirac and Klein-Gordon Equations 63
- The Quantum Hypothesis 64
- Heisenberg Uncertainty 65
- Quantum Numbers 66
- Quantum Electrodynamics and Chromodynamics 67
- Quantum Gravity and other theories 68
- Planck's Constant 69
- Geometricized and Planck units 73
- Quantum Mechanics 74
- Summary 76

Chapter 4 Geometry 77
- Introduction 77
- Eigenvalues 78
- Riemann Curved Spacetime 79
- Torsion 81

Derivatives	81
∂ The Partial Differential	82
Vectors	83
Scalar, or Inner Product	84
Cross or Vector Product	85
Curl	86
Divergence	87
Wedge Product	87
Outer, Tensor, or Exterior Product	89
4-Vectors and the Scalar Product	90
Tensors	90
Some Other Tensors	93
Matrix Algebra	95
The Tetrad, $q^a{}_\mu$	96
Contravariant and covariant vectors and one-forms	102
Wave Equations	104
The Gradient Vector and Directional Derivative, $\wedge$	104
Exterior Derivative	107
Covariant Exterior Derivative $D \wedge$	108
Vector Multiplication	108
Summary	109

Chapter 5 Well Known Equations 110

Introduction	110
Newton's laws of motion	110
Electrical Equations	111
Maxwell's Equations	114
Newton's law of gravitation	120
Poisson's Equations	122
The d'Alembertian $\square$	123
Einstein's Equations	125
Wave equations	126
Compton and de Broglie wavelengths	127
Schrodinger's Equation	129
Dirac Equation	130
Mathematics and Physics	131
Summary	132

Chapter 6 The Evans Field Equation 134

Introduction	134
Einstein Field Equation	135
Curvature and Torsion	136
Torsion	141
Classical and Quantum	145
The Tetrad	145
Field Descriptions	151
Summary	152
Evans Field Equation Extensions	155

Chapter 7 The Evans Wave Equation 156
Introduction 156
The Wave Equation 159
Basic Description of the Evans Wave Equation 160
Electromagnetism 168
Weak force 169
Strong force 170
Equations of Physics 172
Summary 174

Chapter 8 Implications of the Evans Equations 177
Introduction 177
Geometry 177
Very Strong Equivalence Principle 180
A Mechanical Example 183
Implications of the Matrix Symmetries 184
Electrodynamics 186
Particle physics 189
Charge 190
The Symmetries of the Evans Wave Equation 191
$R = -kT$ 192
Summary 194

Chapter 9 The Dirac, Klein-Gordon, and Evans Equations 197
Introduction 197
Dirac and Klein-Gordon Equations 201
Compton Wavelength and Rest Curvature 204
Relationship between r and λ 206
Particles 210
The Shape of the Electron 210
Summary 211

Chapter 10 Replacement of the Heisenberg Uncertainty Principle 214
Basic Concepts 214
Replacement of the Heisenberg Uncertainty Principle 217
The Klein Gordon Equation 220
Summary 221

Chapter 11 The Evans $B^{(3)}$ Spin Field 222
Introduction 222
The Evans $B^{(3)}$ Spin Field 223
Basic geometric description 225
The Metric 228
Summary 231

Chapter 12 Electro-Weak Theory 233
Introduction 233
Derivation of the Boson Masses 237
Particle Scattering 239
The Neutrino Oscillation Mass 241
Standard Model with Higgs versus the Evans method 244

    Generally Covariant Description 244

Chapter 13 The Aharonov-Bohm (AB) Effect 245
    Phase Effects 245
    The Aharonov-Bohm Effect 248
    The Helix versus the Circle 251
    Summary 253

Chapter 14 Geometric Concepts 254
    Introduction 254
    The Electrogravitic Equation 254
    Principle of Least Curvature 256
    Non-local Effects 257
    EMAB and RFR 258
    Differential Geometry 259
    Fundamental Invariants of the Evans Field Theory 260
    Origin of Wave Number 262
    Summary 263

Chapter 15 A Unified Viewpoint 264
    Introduction 264
    Review 265
    Curvature and Torsion 269
    Mathematics = Physics 269
    The Tetrad and Causality 271
    Heisenberg Uncertainty 271
    Non-locality (entanglement) 272
    Principle of Least Curvature 272
    The Nature of Spacetime 273
    The Particles 276
    The Electromagnetic Field – The photon 279
    The Neutrino 281
    The Electron 281
    The Neutron 282
    Unified Wave Theory 284
    Oscillatory Universe 286
    Generally Covariant Physical Optics 286
    Charge and Antiparticles 287
    The Electrogravitic Field 288
    A Summary of Einstein-Cartan-Evans Field Theory 289

Glossary 292

References 367

Index 368

Introduction

> Genius is 1% inspiration and 99% perspiration.
> Thomas Edison

Don't Panic

Don't panic if the technical terms are intimidating at this point. While this is not a fundamental book, the basic terms are all defined in the glossary and the first five chapters. These chapters are a review of quantum mechanics and relativity.

This is a book about equations. However, the attempt is made to describe them in such a way that the reader does not have to do any math calculations nor even understand the full meaning of the equations. To just give verbal explanations and pictures and ignore the equations is to lie by omission and would hide the fuller beauty that the equations expose. However, the non-physicist needs verbal and pictorial explanations. Where possible, two or three approaches are taken on any subject. We describe phenomena in words, pictures, and mathematics. Most readers are most interested in the *concepts* not the deep math; know that the author is well aware of it. So…don't panic if the technical terms are intimidating at first. When you have finished this book and you reread this introduction, you will say, "Oh, that's obvious," about a lot of the material that confused you at first. There is a glossary with additional explanations of terms and a number of web sites are given in the references for those wishing to get either more basic or more thorough explanations.

Do keep in mind Edison's statement that genius is 99% hard work. The more you put into this subject, the more you get out.

General Relativity and Quantum Theory

It is well recognized in physics that the two basic physics theories, general relativity and quantum theory, are each lacking in complete descriptions of reality. Each is correct and makes precise predictions when restricted to its own realm. Neither explains interactions between gravitation and electromagnetism (radiated fields) nor the complete inner construction of particles (matter fields). We could say that general relativity and quantum theory do not commute.

General relativity has shown that gravitation is curved spacetime and has shown that large collections of particles can become black holes – masses so large that spacetime curves within itself and nothing can leave the region.

Electric charge and magnetic fields were unified into electromagnetism in the 19th century... Quantum theory has correctly explained or predicted many features of particle interactions. The standard model of the forces in physics is primarily quantum theory and indicates that electromagnetism and radioactivity can be combined into a single electroweak theory and are the same thing at high energies. However, gravitation and the strong force are still unconnected in the theory. Quantum theory uses special relativity, which is just an approximation to relativistic effects; it cannot describe nor determine the effects on reactions due to gravitational effects. Quantum mechanics uses discrete energy levels in its formulations. Changes always occur in small jumps and in between there is no existence. It leaves the vague and unproven impression that spacetime itself may be quantized - that the vacuum is composed of little dots of something with nothing in between.

Gravitation is not quantized. Einstein's general relativity is an "analog" theory; it envisions and operates on a spacetime that is divisible down to the point. Since points are zero, they cannot exist and some minimum size should exist. The quantum of energy is not defined within it.

Thus, our understanding of our existence is incomplete. We do not know what spacetime is. We do not know the basic composition of particles. We have mysteries still to solve. Back in the 1890's and again in the 1990's, some physicists believed that we were on the border of knowing everything in physics. Don't believe it. Thousands of years from now, we will still be asking new questions and searching for answers.

Unified field theory is the combination of general relativity and quantum theory. We hope that it will answer some of the open questions.

In addition to open questions, we have a number of erroneous, forced concepts that have crept into physics. Some explanations are wrong due to attempts to explain experimental results without the correct basic understanding. Among these are the Aharonov-Bohm effect (Chapter 13) and maybe the quark description of the particle. We also have entanglement and apparently non-local effects that are not adequately understood.

For example, the origin of charge is explained in the standard model by symmetry in Minkowski spacetime. It requires that there is a scalar field with two complex components. These would indicate positive and negative components. However, the existence of the two types of charge, positive and negative, is used to conclude that the scalar field must be complex. This is circular reasoning (and it borders on gobbledygook. We want to explain things and sometimes we force that explanation.)

There are four fields in physics – gravitation, electromagnetism, strong (particle), and weak (radioactivity). However only the gravitational field is generally covariant – that is, objectively the same regardless of the observer's coordinate system (reference frame) due to gravitational fields or differing velocity. The other three fields exist inside gravitational fields but neither general relativity nor quantum theory can describe the interactions due to gravitation.

The Evans equations show how general relativity and quantum theory, previously separate areas of physics, can both be derived from Einstein's postulate of general relativity. In the early decades of the 20th century, these two theories were developed and a totally new understanding of physics resulted.

Relativity deals with the geometric nature of spacetime and gravitation.[1] In the past, it was applied more to large-scale processes, like black holes. Quantum theory deals primarily with the nature of individual particles, energy, and the vacuum. It has been applied very successfully at micro scales.

Einstein-Cartan-Evans Unified Field Theory

The combination of general relativity and quantum theory into one unified theory was Einstein's goal for the last 30 years of his life.[2] A number of other physicists have also worked on developing unified field theories. String theory has been one such effort. While it has developed into excellent mathematical studies, it has not gotten beyond mathematics and is unphysical; it makes no predictions and remains untested.

Unified field theory is then the combination of general relativity and quantum theory into one theory that describes both using the same fundamental equations. The mutual effects of all four fields must be explained and calculations of those effects must be possible.

This has been achieved by Professor Myron Wyn Evans using Einstein's general relativity as the foundation, Cartan's differential geometry to define the spacetime, and his own wave equation to describe both relativity and quantum mechanics. Evans provides the derivation of electrodynamics from geometry. The geometric nature of spacetime and gravitation is reaffirmed. Interestingly, while Evans starts with general relativity, most of his papers apply it to electromagnetic theory and processes that have been traditionally subjects in quantum theory. This is absolutely explicable given the need to unify subjects. The strength of unified field theory is just that – unification of concepts.

---

[1] Einstein developed relativity using ideas drawn from Lorentz, Mach, and Riemann. Minkowski added to the theory shortly after Einstein's initial publication.
[2] For example see *Unified Field Theory based on Riemannian Metrics and distant Parallelism*, Albert Einstein, Mathematische Annalen 102 (1930) pp 685-697. Translated by A. Unzicker in www.alexander-unzicker.de/einst.html. (Accessed July, 2007.) The term distant Parallelism refers to the Cartan tetrad.

Evans' Results

Professor Myron Wyn Evans has written over 700 published technical papers and books primarily in physical chemistry and in the last few years on unification of quantum and relativistic physics. He has worked at several universities in the US and now resides in Wales in his ancestral home. The five volume Enigmatic Photon series and his research on the $B^{(3)}$ field and O(3) electrodynamics have been past major contributions to physics.

As of this writing Evans has produced 85 papers discussing various aspects of unified field theory. They can be briefly summarized:

1) The unified field theory is primarily a theory of general relativity. The nature of spacetime is again redefined just as Einstein twice reestablished it. The change is simple and the mathematics was well established by Cartan and used in part by Einstein.
2) Quantum theory emerges from, and is part of, general relativity. Together they have power to explain more about basic reality than either had alone. For example, a minimum volume of a particle always exists and can be calculated from basic constants. This implies that black holes are not singularities (zero points); the universe had a minimum volume and the big bang was likely a re-expansion from a previous universe.
3) The unified field theory allows study of the mutual influence of gravitation. electromagnetism, strong atomic force, and the weak radioactive force. Equations can be solved in a variety of gravitational fields.
4) From the unified field theory all the equations of physics can be derived. Reinterpretation of some of quantum theory is required.
5) The Heisenberg uncertainty principle has been rejected in favor of causal quantum mechanics. In the unified field theory explanations exist for new experiments showing Heisenberg and the Copenhagen probabilistic school of thought were incorrect. Einstein was right saying that quantum physics was incomplete. The Higgs mechanism is rejected as an unnecessary complication. The existence of quarks as anything more than temporary curvature states or mathematical constructs is questioned.

6) The origins of various phenomena are found in general relativity. Among these are optical phase laws, the origins of electromagnetism and the Evans $B^{(3)}$ spin field, and the equivalence of wave number and curvature.

We will explain what the meaning of some of these things as we go along in this book. Not many people have heard of some of them (e.g., AB effect, $B^{(3)}$ field) since they are in more uncommon areas of electrodynamics.

Professor Evans frequently gives credit to the members of the Alpha Institute for Advanced Study (aias) group. The group members have made suggestions, encouraged and supported him, acted as a sounding board, and criticized and proofed his writings. In particular, Professor Emeritus John B. Hart of Xavier University has strongly supported development of the unified field and was "The Father of the House" for aias until his death. A number of members of aias have helped with funding and considerable time. Among them are The Ted Annis Foundation, Craddock, Inc., Franklin Amador and David Feustel.

However, it is Myron Evans' hard work that has developed the theory.

The Nature of Spacetime

The nature of spacetime is the starting point of unified field theory, just as it was essential to the other three main theories of physics.

The metric of the spacetime is key in each of the accepted physics theories in its historical period. Classical Newtonian, Einstein's two theories – special and general relativity, and Evans' (Cartan differential geometry) are the four. Quantum theory uses special relativity's spacetime. See Figures I-1 and I-2 for a depiction of spacetimes.

A metric is a formal map from one point in a spacetime to another. Newton's space was nothing; it was an empty 3-dimensional background upon which matter moved and one could measure distances. Our everyday idea of spacetime is much like Newton's and unless one deals in the very small or large, or in high velocities, it is a good enough. It is a 3-dimensional concept and time is outside its definition. The speed of electromagnetism and gravitation were assumed to be instantaneous in Newtonian physics.

Einstein's special relativistic metric is called Minkowski spacetime. It is flat without gravitational curvature; however it has 4-dimensional distance defined. It is a combination of space and time into one inseparable 4-dimensional spacetime. In addition, that spacetime can compress due to velocity. The distinction between spacetime and the matter within it starts to blur. We will deal with it in a bit of detail in Chapter 2.

Einstein's general relativistic metric is that of Riemann geometry. Spacetime has four dimensions and gravity is geometry with curvature. Special relativity can be derived from general relativity when there is no curvature. Time can be treated like a fourth spatial dimension.

Historically, there is a progression in the concept of spacetime as understanding has increased.

Newton's flat, empty, 3-D Euclidean space with instantaneous time is our everyday experience and is usually adequate.

Einstein-Minkowski spacetime of special relativity and quantum mechanics has well known limitations, but answers more questions. It is 4-D and invariant, but still flat Euclidean in nature.

In General Relativity, Riemann 4-D Curved geometry defines gravitational spacetime. However, electromagnetism remains unexplained.

Cartan differential geometry is sufficient to explain both gravitation and electromagnetism.

Each geometric-spacetime includes those that preceded it while adding new insights.

Figure I-1   Spacetime

Newton's flat empty space with time a separate concept.

Euclidean geometry.

Einstein-Minkowski spacetime of special relativity.

Flat non-Euclidean geometry with invariant 4-dimensional distances.

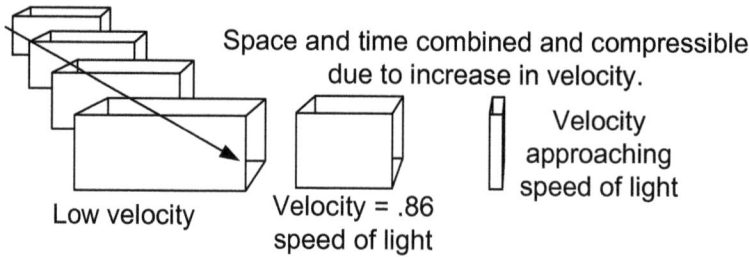

Space and time combined and compressible due to increase in velocity.

Low velocity   Velocity = .86 speed of light   Velocity approaching speed of light

General relativity with curved distance, but no turning .

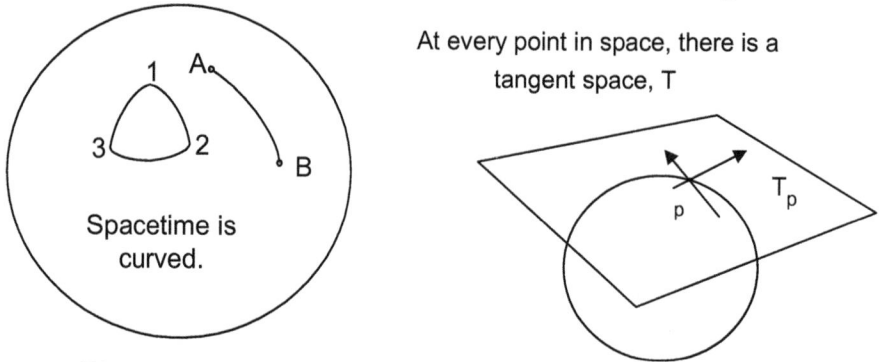

Spacetime is curved.

At every point in space, there is a tangent space, T

Riemann curved 4-dimensional geometry .

The development of general relativity indicated that spacetime is a mathematical analytical, fully differentiable manifold. That is, it was an active, real existence that had points that connected smoothly. The physical description starts to become mathematical. We do not have the necessary experience or the mechanical concepts to describe spacetime any other way than mathematically. One will hear physicists say physics is mathematics or general relativity is geometry. There is no doubt truth in this and until our engineering skills can look at the vacuum itself, it must be the way we think.

Quantum theory uses special relativity's spacetime. It does not have the ability to describe processes in different gravitational fields nor to calculate the gravitational affects very close to the particle. It does have the ability to make very precise predictions of electrical and nuclear processes in flat spacetime.

Newton used standard flat three dimensional Euclidean geometry. The metric was simple distances between points.

Special relativity used flat four dimensional geometry. Time was included in the metric.

General relativity used the curved four dimensional geometry developed by Riemann. It had four dimensional distances however specifically had no spinning of the metric.

Evans spacetime uses Cartan differential geometry. It is differential geometry which contains Riemann geometry and more. In order to unify general relativity and quantum mechanics, in particular electromagnetism, the spacetime must be able to curve and spin. Einstein's Riemann spacetime has curvature but expressly sets spin to zero. Einstein and others took this spacetime and placed equations (tensors) describing spin on top of it. This approach has not worked and Evans simply uses a spacetime that has curvature, R, and spin, more formally called torsion, T^a. Others have attempted to use Cartan geometry also so the approach is not new. However Evans developed a wave equation which solves the problem.

Figure I-2   Evans Spacetime

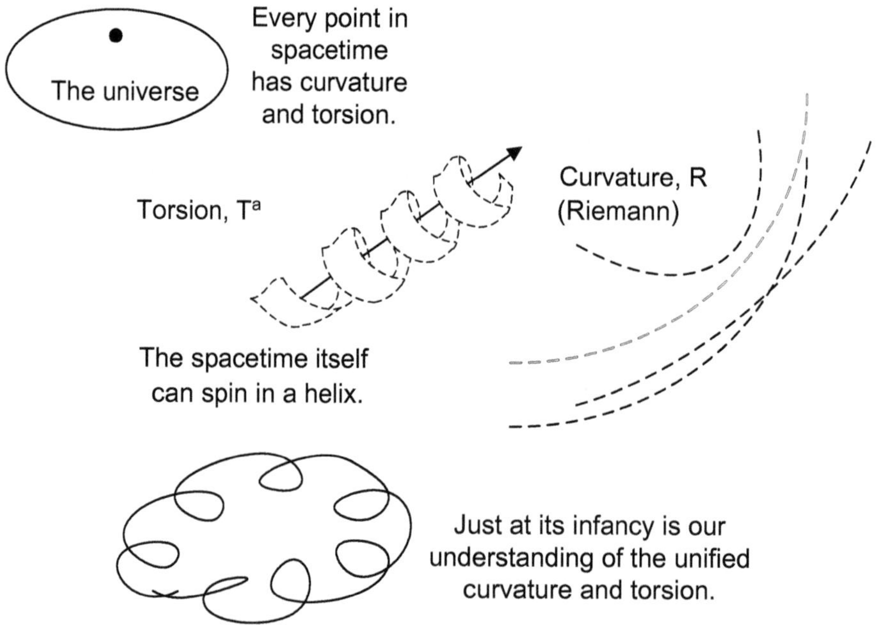

The universe

Every point in spacetime has curvature and torsion.

Torsion, T^a

Curvature, R (Riemann)

The spacetime itself can spin in a helix.

Just at its infancy is our understanding of the unified curvature and torsion.

In order to do calculations in the curved spacetimes, physicists have to use vector spaces. While the calculations may be involved, the idea is simple. At every point in a spacetime there are mathematical boxes that can be used to hold information.[3] In general relativity the box is called "tangent space." In quantum theory the box is called "the vector space."

General relativists consider the tangent space to be real. That it is a geometrical space and the information has a solid connection to reality. One looks at the numbers that define the mass, velocity, etc. at the point and can move them to some other point. Quantum physicists consider the vector space to be imaginary. It is a pure abstract mathematical space for calculation. One goes into it, adds and divides, and brings the answers out. Most of the vectors

---

[3] These are a lot like escrow accounts

and answers are mysteries. Why they work is mostly unknown. Trial and error experiment shows they are just simply correct. The ideas behind the tangent space and the vector space are shown in Figure I-3.

Figure I-3   Geometrical Tangent Space and Abstract Fiber Bundle

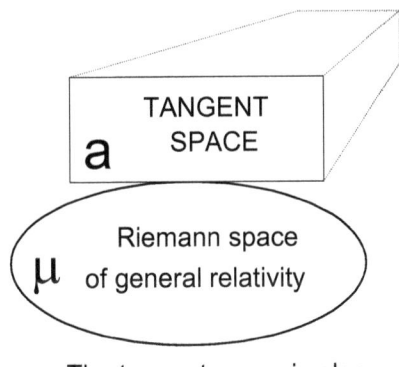
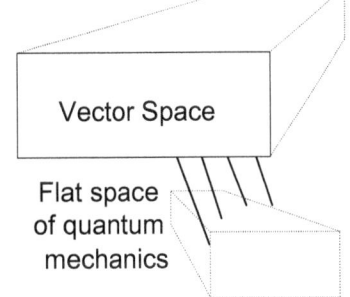

The tangent space is also called the orthonormal tangent space. It is considered physical

The vector space is also referred to as the abstract fiber bundle space or Hilbert Space. It is considered to be an abstract mathematical space..

Just what is spacetime? A classical description would be like an everyday experience definition. The reader experiences spacetime as distances between objects and the passing of time. That is not the worst description. The physicist must be a bit more detailed since he wants to perform calculations, explain experiments, and predict the results of new experiments.

Spacetime is four dimensional. It is curved - there are no straight lines since gravity is present everywhere in the universe. It can twist – this is the new point that using Cartan geometry in Evans spacetime adds. It has distances from one point to another. A variety of actions can occur within it.

"The vacuum" is a related concept. In quantum mechanics the vacuum refers to the mostly empty space surrounding particles. The Minkowski spacetime is a mathematical construct and Evans defines it as the vacuum. We also use the word vacuum (without "the") to mean the very low density space between planets and suns.

The use of mathematics to define the spacetime of our everyday universe for calculations leads to defining spacetime itself as that mathematics. (One of these may be a chicken or an egg.) Einstein clearly stated that physics is geometry. Evans has echoed him. We can see the logic in the idea. Lacking any concrete definition of spacetime, we use that which is most appropriate - geometry. Until such time comes that someone finds a better underlying concept, we must tentatively accept the idea that spacetime is geometry.

Unified field theory again requires revision of the existing concept of spacetime just as Einstein twice revised it. Einstein added a dimension for special relativity; then he added curvature for general relativity. Evans adds spin-torsion for unified field theory. We will see how Evans uncovers how the forces are related to the spacetime. This starts on the very small scale of the particle and vacuum and can go all the way to black holes and galaxies.

The Four Forces

Physics recognizes four forces which are depicted in Figure I-4.

<u>Gravitation is curved geometric spacetime</u>. Any formula or theory retains its form under coordinate transformations in general relativity. With acceleration or a gravitational field change, the objects within a reference frame change in a linear, predictable and organized fashion. Within the frame, all objects retain their relationships to one another. From the viewpoint of an external observer things may compress or twist, but the changes are seen as organized and can be moved to yet another reference frame in an organized fashion and later moved back to their original form.

Einstein's Principle of General Relativity states that every theory of physics should be generally covariant – that it should retain its form under general coordinate transformations. A unified field theory should also be generally covariant. At present only gravitation is described by a generally covariant field equation - Einstein's field equation of gravitation. This is written in tensor geometry.

Figure I-4   The Four Forces

Gravitation

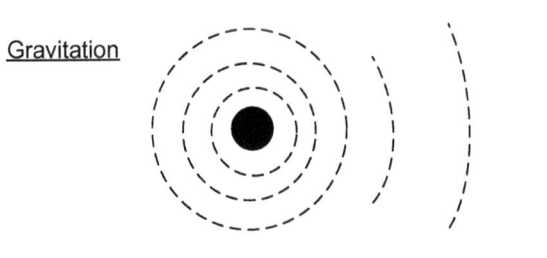

Curved spacetime due to mass

Electromagnetism
  Magnetic (B) and Electric (E) Fields

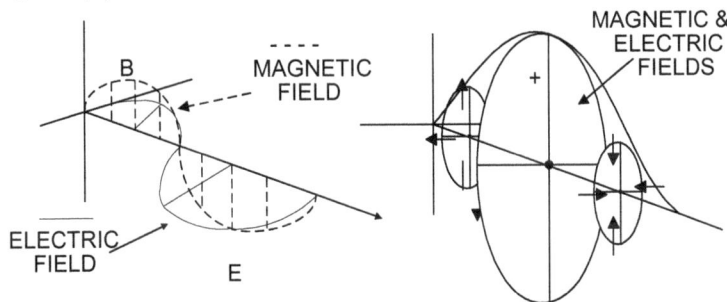

Strong nuclear force

Presently described by gluons, quarks and pions.

Weak force

Radioactivity = W&Z boson momentum exchange

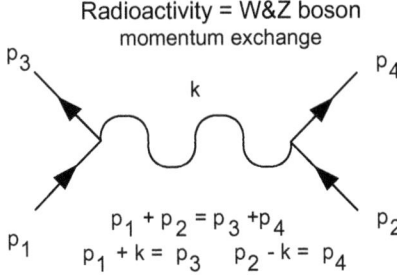

k is the W or Z boson
p indicates the momentum of the particles

$p_1 + p_2 = p_3 + p_4$
$p_1 + k = p_3 \quad p_2 - k = p_4$

Electromagnetism has historically been seen as a force superimposed upon spacetime. It is some spinning energy but otherwise is unclear. The "electro" part is charge – electrons and protons and their antiparticles.

Charge polarizes spacetime into regions we see as positive and negative. The "magnetic" part can be seen as a field or invisible lines of force which reach out and affect particles and other fields. Charge and magnetism are combined into the electromagnetic field. This was the first unification made in the 19th century and described in the Maxwell equations.

The weak force is seen in radioactivity. Its mediators are the W and Z bosons. Weak force holds the lone neutron together for about 10 minutes before it changes into an electron, proton, and antineutrino. It has been shown that the weak force and electromagnetism are the same thing at high energies – this unifies them into the electro-weak force. Gravitation and the strong force remain unconnected to the electro-weak force.

The strong force holds atoms together. It has been described mathematically using three "quarks" which have never been seen physically but are surmised from particle scattering statistics. "Gluons" hold the quarks together. Although the quark-gluon description is usually given as gospel, the conceptual description is incomplete.

The Particles

There are stable, long-lived particles that are recognized as the basic building blocks of the universe. There are also fleeting "particles" that exist in accelerator collisions for very short durations. The most organized sense that the particles can be put within is yet to be found.

The interrelations among the types of existence are not fully understood. The electron weighs 1/1826 of a proton, but has the exact same charge. The neutron decomposes into an electron, proton, and antineutrino in 10 minutes if left outside an atom. A photon can hit an atom and cause the electron surrounding the proton to spread out a bit – it gives the electron a spatial component.

We will touch upon the particle puzzles during the course of this book and return to it in the last chapter.

What We Will See

The first five chapters are introductory review material on relativity, quantum mechanics, and equations that concern both. The next three chapters introduce the Evans equations. The next six cover implications of the unified field equations. Finally, the last chapter is a review with some speculation about further ramifications. Evans has published extensive electromagnetic material that is not covered in this book.[4] The overall subject is too large for a single book.

Mathematically there are two ways to describe general relativity and differential geometry. The first is in abstract form – for example, "space is curved". This is just like a verbal language if one learns the meaning of terms – for example, R means curvature. Once demystified to a certain degree, the mathematics is understandable. The second way is in coordinates. This is very hard with detailed calculations necessary to say just how much spacetime curvature exists. In general, it is not necessary to state what the precise curvature is near a black hole; it is sufficient to say it is curved a lot.

Any mistakes in this book are the responsibility of this author. Professor Evans helped guide and allowed free use of his writings, but he has not corrected the book.

Laurence G. Felker, Reno NV

---

[4] As this book was being written, it was hard to keep up with developments. For example, explanation of the Faraday homopolar generator which has been a mystery for 130 years has been published.

Chapter 1 Special Relativity

> The theory of relativity is intimately connected with the theory of space and time. I shall therefore begin with a brief investigation of the origin of our ideas of space and time, although in doing so I know that I introduce a controversial subject. The object of all science, whether natural science or psychology, is to co-ordinate our experiences and to bring them into a logical system.
>
> Albert Einstein[5]

Relativity and Quantum Theory

Einstein published two papers in 1905 that eventually changed all physics. One paper used Planck's quantum hypothesis[6] to explain the photoelectric effect and was an important step in the development of quantum theory. This is discussed in Chapter 3. Another paper established special relativity. Special relativity in its initial stages was primarily a theory of electrodynamics – moving electric and magnetic fields.

The basic postulate of special relativity is that there are no special reference frames and certain physical quantities are invariant. Regardless of the velocity or direction of travel of any observer, the laws of physics are the same and certain measurements should always give the same number.

Measurements of the speed of light (electromagnetic waves) in the vacuum will always be the same. This was a departure from Newtonian physics. In order for measurements of the speed of light to be frame independent, the

---

[5] Most quotations from Einstein are from his book, The Meaning of Relativity, Princeton University Press, 1921, 1945. This is on page 1.
[6] There are many terms used here without definition in the text. The Glossary has more information on many; one can see the references at the back of the book; and web search engines are helpful.

nature of space and time had to be redefined to recognize them as spacetime – a single entity.

A reference frame is a system, like a particle, spaceship, or a laboratory on earth that can be clearly distinguished due to its velocity or gravitational field. A single particle, a photon, or a dot on a curve can also be a reference frame.

Figure 1-1 Reference Frames

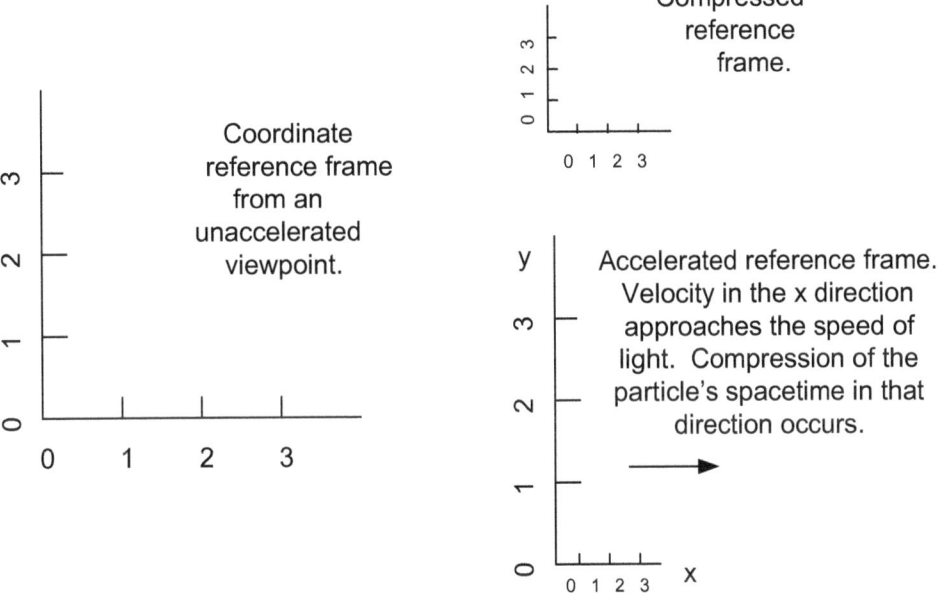

Figure 1-1 shows the basic concept of reference frames. Spacetime changes with the energy density – velocity or gravitation. In special relativity, only velocity is considered. From within a reference frame, no change in the spacetime can be observed since all measuring devices also change with the spacetime. From outside in a higher or lower energy density reference frame, the spacetime of another frame can be seen to be different as it goes through compression or expansion.

The Newtonian idea is that velocities sum linearly. If one is walking 2 km/hr on a train in the same direction as the train moving 50 km/hr with respect to the tracks, then the velocity is 52 km/hr. If one walks in the opposite direction, then the velocity is 48 km/hr. This is true, or at least any difference is unnoticeable, for low velocities, but not true for velocities near that of light. In the case of light, it will always be measured to be the same regardless of the velocity of the observer-experimenter. This is expressed as V1 + V2 = V3 for Newtonian physics, but V3 can never be greater than c in actuality as special relativity shows us.

Regardless of whether one is in a spaceship traveling at high velocity or in a high gravitational field, the laws of physics are the same. One of the first phenomena Einstein explained was the invariance of the speed of light which is a constant in the vacuum from the viewpoint within any given reference frame.

In order for the speed of light to be constant as observed in experiments, the lengths of objects and the passage of time must change for the observers within different reference frames.

Spacetime is a mathematical construct telling us that space and time are not separate entities as was thought before special relativity. Relativity tells us about the shape of spacetime. The equations quite clearly describe compression of spacetime due to high energy density and experiments have confirmed this. Physicists and mathematicians tend to speak of "curvature" rather than "compression"; they are the same thing with only a connotation difference. In special relativity, "contraction" is a more common term.

Among the implications of special relativity is that mass = energy. Although they are not identical, particles and energy are interconvertible when the proper action is taken. $E = mc^2$ is probably the most famous equation on Earth (but not necessarily the most important). It means that energy and mass are interconvertible, not that they are identical, for clearly they are not, at least at the energy levels of everyday existence.

As a rough first definition, particles = compressed energy or very high frequency standing waves and charge = electrons. In special relativity, these are seen to move inside spacetime.[7] They are also interconvertible. If we accept the Big Bang and the laws of conservation and the idea that the entire universe was

---

[7] After exposure to the Evans equations and their implications, one will start to see that the seemingly individual entities are all versions of spacetime.

once compressed into a homogeneous Planck size region (1.6 x 10^{-35}m), then it is clear that all the different aspects of existence that we now see were identical. The present universe of different particles, energy, and spacetime vacuum all originated from the same primordial nut. They were presumably initially identical.

The different aspects of existence have characteristics that define them. Spin, mass, and polarization are among them. Classically (non-quantum) a particle with spin will behave like a tiny bar magnet. The photon's polarization is the direction of its electric field. See Spin in Glossary.

Where special relativity is primarily involved with constant velocity, general relativity extends the concepts to acceleration and gravitation. Mathematics is necessary to explain experiments and make predictions. We cannot see spacetime or the vacuum. However, we can calculate orbits and gravitational forces and then watch when mass or photons travel in order to prove the math was correct.

Quantum theory is a special relativistic theory. It cannot deal with the effects of gravitation. It assumes that spacetime is flat - that is, contracted without the effects of gravitation. This indicates that quantum theory has a problem – the effects of gravitation on the electromagnetic, weak and strong forces are unknown.

Special Relativity

As stated, the basic postulate of special relativity is that the laws of physics are the same in all reference frames. Regardless of the velocity at which a particle or spaceship is moving, certain processes are invariant. The speed of light is one of these. Mass and charge are constant – these are basic existence and the laws of conservation result since they must be conserved. Spacetime changes within the reference frame to keep those constants the same. As the energy within a reference frame increases, the spacetime distance decreases.

The nature of spacetime is the cause of invariance. Special relativity gives us the results of what happens when we accelerate a particle or spaceship (reference frame) to near the velocity of light.

$$\gamma = 1/\sqrt{(1-(v/c)^2)} \tag{1}$$

This is the Lorentz–Fitzgerald contraction, a simple Pythagorean formula. See Figure 1-2. It was devised for electrodynamics and special relativity draws upon it for explanations.

γ is the Greek letter gamma, v is the velocity of the reference frame, say a particle, and c is the speed of light, about 300,000 kilometers per second.

Let the hypotenuse of a right triangle be X. Let one side be X times v/c; and let the third side be X'. Then $X' = X$ times $\sqrt{(1-(v/c)^2)}$.

If we divide 1 by $\sqrt{(1-(v/c)^2)}$ we arrive at gamma.

How we use gamma is to find the change in length of a distance or of a length of time by multiplying or dividing by gamma.

If v = .87 c, then

$(v/c)^2 = .76$ $1 - .76 = .24$ and $\sqrt{.24}$ = about .5

$X' = X$ times $\sqrt{(1-(v/c)^2)} = X$ times .5

The distance as viewed from outside the reference frame will shrink to .5 of its original.

Time is treated the same way. t', the time experienced by the person traveling at .87c, = t, time experienced by the person standing almost still, times $\sqrt{(1-(v/c)^2)}$. The time that passes for the accelerated observer is ½ that of the observer moving at low velocity.

If a measuring rod is 1 meter long and we accelerate it to 87% of the speed of light, it will become shorter as viewed from our low energy density reference frame. It will appear to be .5 meter long. If we were traveling alongside the rod (a "co-moving reference frame") and we used the rod to measure the speed of light we would get a value that is the same as if we were standing still. We would be accelerated and our bodies, ship, etc. would be foreshortened the same amount as any measuring instrument.

Figure 1-2   Lorentz-Fitzgerald contraction

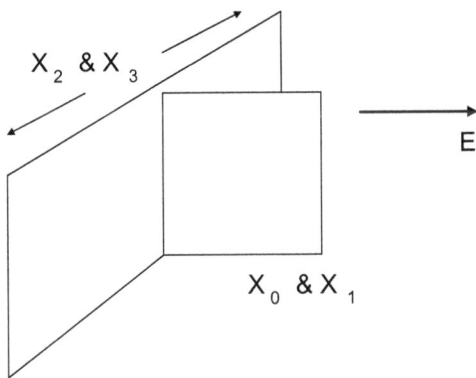

As a particle is accelerated, its kinetic energy increases. It becomes "shorter" in the direction of travel - spatial or timewise.

X' refers to X_0 or X_1
X' is in the high energy density reference frame

The math to find the "size" of a dimension - a length or the time - is simply that of a triangle as shown to the right.

$$X' = X(1-(v/c)^2)^{1/2}$$

X in the low energy density reference frame

$$X^2 = (v/c\ X)^2 + X'^2$$

The High Energy Density Reference Frame is the accelerated frame. The low energy density reference frame is the "normal" at rest, unaccelerated reference frame.

See Figure 1-3 for a graph of the decrease in distance with respect to velocity as seen from a low energy density reference frame. The high energy density reference frame is compressed; this is common to both special relativity and general relativity.

> By the compression of the spacetime of a high energy density system, the speed of light will be measured to be the same by any observer. The effects of a gravitational field are similar. Gravitation changes the geometry of local spacetime and thus produces relativistic effects.

Figure 1-3   Graph of Stress Energy in Relation to Compression due to velocity

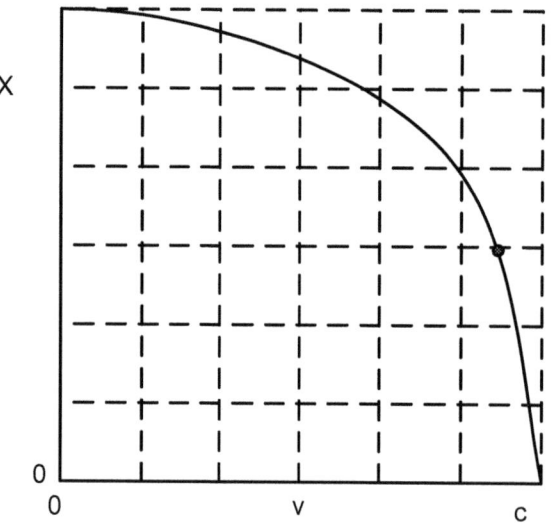

X will have compressed to .5 of its original length when v = .87 of the speed of light.

Invariant distance

These changes cannot be observed from within the observer's reference frame and therefore we use mathematical models and experiments to uncover the true nature of spacetime. By using the concept of energy density reference frames, we avoid confusion. The high energy density frame (high velocity particle or space near a black hole) experiences contraction in space and time.

The spacetime or *manifold* that is assumed in special relativity is that of Minkowski. The invariant distance exists in Minkowski spacetime. The distance between two events is separated by time and space. One observer far away may see a flash of light long after another. One observer traveling at a high speed may have his time dilated. Regardless of the distances or relative velocities, the invariant distance will always be measured as the same in any reference frame. Because of this realization, the concept of spacetime was defined. Time and space are both part of the same four-dimensional manifold; it is flat space (no gravitational fields), but it is not Euclidean.

Chapter 1 – Special Relativity

Figure 1-4 Invariant distance

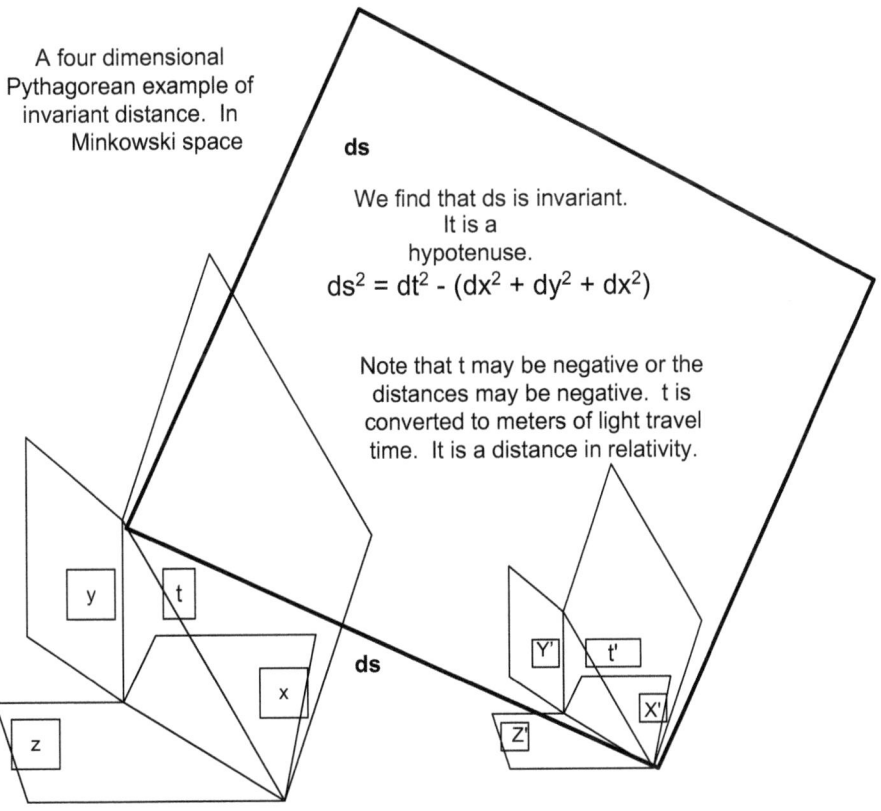

This is an example of an invariant distance, ds, in a simple Pythagorean situation.

The distance ds is here depicted as a 2 dimensional distance with dy and dz suppressed. In 4 dimensions, the distance between two events, A and B, is a constant. While dt and dx vary, the sum of their squares does not.

A four dimensional Pythagorean example of invariant distance. In Minkowski space

We find that ds is invariant. It is a hypotenuse.

$$ds^2 = dt^2 - (dx^2 + dy^2 + dx^2)$$

Note that t may be negative or the distances may be negative. t is converted to meters of light travel time. It is a distance in relativity.

Tensors produce the same effect for multiple dimensional objects.

23

An example is shown in Figure 1-4. The hypotenuse of a triangle is invariant when various right triangles are drawn. The invariant distance in special relativity subtracts time from the spatial distances (or vice versa). The spacetime of special relativity is called Minkowski spacetime or the Minkowski metric; it is flat, but it has a metric unlike that of Newtonian space.

The time variable in Figure 1-4 can be written two ways, t or ct. ct means that the time in seconds is multiplied by the speed of light in order for the equation to be correct. If two seconds of light travel separate two events, then t in the formula is actually 2 seconds x 300,000 kilometers/second = 600,000 kilometers. Time in relativity is meters of photon travel. All times are changed into distances. When one sees t, the ct is understood.

Proper time is the time (distance of light travel during a duration) measured from inside the moving reference frame; it will be different for the observer on earth and the cosmic ray approaching the earth's atmosphere. In special relativity, the Lorentz contraction is applied. In general relativity it is more common to use τ, tau, meaning proper time. Proper time is the time (distance) as measured in the reference frame of the moving particle or the gravitational field – the high energy density reference frame. The time measured from a relatively stationary position is not applicable to the high velocity particle.

The transformation using the Lorentz-Fitzgerald contraction formula can be applied to position, momentum, time, energy, or angular momentum.

The Newton formula for momentum was **p** = m**v**. Einstein's special relativity formula is **p** = γm**v** with $\gamma = 1/\sqrt{(1-v^2/c^2)}$. Since $\gamma = 1$ when v = 0, this results in Newton's formula in the "weak limit" or in flat Euclidean space at low velocities.

Correspondence Principle

The correspondence principle says that any advanced generalized theory must produce the same results as the older more specialized theory. In particular, Einstein's relativity had to produce Newton's well-known and established theories as well as explain new phenomena. While the contraction formula worked for high velocities, it had to predict the same results as Newton for low velocities. When Einstein developed general relativity, special relativity had to be derivable from it.

The Evans equations are new to general relativity. To be judged valid, they must result in all the known equations of physics. To make such a strong claim to be a unified theory, both general relativity and quantum mechanics must be derived clearly. To be of spectacular value, they must explain more and even change understanding of the standard theories.

A full appreciation of physics cannot be had without mathematics. Many of the explanations in physics are most difficult in words, but are quite simple in mathematics. Indeed, in Evans' equations, physics is differential geometry. This completes Einstein's vision.

Vectors

Arrow vectors are lines that point from one event to another. They have a numerical value and a direction. In relativity four-dimensional vectors are used. They give us the distance between points. "Δt" means "delta t" which equals the difference in time.

A car's movement is an example. The car travels at 100 km per hour. Δt is one hour, Δx is 100 km; these are scalars. By also giving the direction, say north, we have a velocity which is a vector. Therefore, a scalar provides only magnitude and the vector provides magnitude and direction.

We also see "dt" meaning "difference between time one and time two" and "∂t", which means the same thing when two variables are involved. A vector is not a scalar because it has direction as well as magnitude. A scalar is a simple number.

There are many short form abbreviations used in physics, but once the definition is understood, a lot is demystified.

A is the symbol for a vector named A. It could be the velocity of car A and another vector, **B**, stands for the velocity of car B. A vector is typically shown as a **bold lower case** letter. A tensor may use bold upper or lower case so one needs to be aware of context.

> For a summary of the types of products of vectors and matrices, see the Glossary under "Products."

a • b is the dot product and geometrically means we multiply the value of the projection of **a** on **b** times **b** to get a scalar number. Typically it is an invariant distance and usually the dot is not shown.

The dot product is called an inner product in four dimensions.

a x b is called the cross product. If **a** and **b** are 2-dimensional vectors, the cross product gives a resulting vector in a third dimension. Generalizing to 4 dimensions, if **A** and **B** are vectors in the xyz-volume, then vector **C = A x B** will be perpendicular or "orthogonal" to the xyz-volume. In special relativity this is time; in general relativity this is a spatial dimension. While special relativity treats the 4th dimension as time, general relativity is as comfortable treating it as a spatial dimension.

The wedge product is the four dimensional cross product. It is covered in more detail in Chapter 4 along with more about vectors.

Torque calculations are an example of cross products.[8] τ,[9] torque is the turning force. It is a vector pointing straight up. Figure 1-5 shows the torque vector. It points in the 3rd dimension although the actions appear to be only in the other two dimensions.

At first this seems strange since the vector of the momentum is in the circle's two dimensions. However, for the torque to be able to be translated into

---

[8] For a good moving demonstration of the vector cross product in three dimensions see the JAVA interactive tutorial at http://www.phy.syr.edu/courses/java-suite/crosspro.html (Accessed July, 2007.)
[9] τ is used for proper time and torque. The context distinguishes them.

Chapter 1 – Special Relativity

other quantities in the plane of the turning, its vector must be outside those dimensions.

Figure 1-5   Torque as an example of a cross product

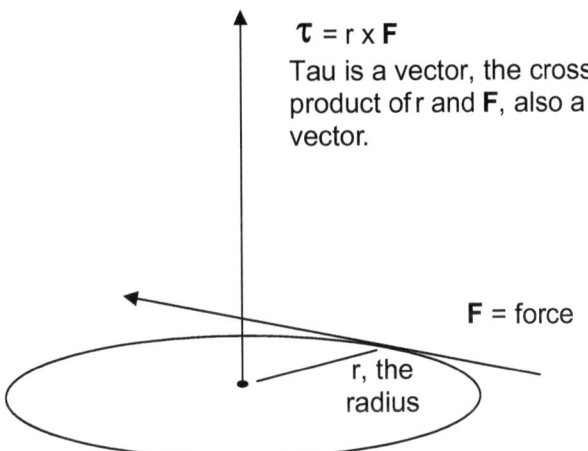

Imagine a top spinning on a table. There is angular momentum from the mass spinning around in a certain direction. The value of the momentum would be calculated using a cross product.

In relativity, Greek indices indicate the 4 dimensions – 0, 1, 2, 3 as they are often labeled with t, x, y, and z understood. A Roman letter indicates the three spatial indices or dimensions. These are the conventions.

The cross product of r and F is the torque. It is in a vector space and can be moved to another location to see the results. Tensors are similar. When a tensor is moved from one reference frame to another, the distances (and other values) calculated will remain invariant.

The cross product is not defined in four dimensions; the wedge product is essentially the same thing.

Tensors and the tetrad use mathematics similar to that used with vectors.

Evans Equations of Unified Field Theory

We will discuss the tetrad in later chapters. The tetrad, which we will see is fundamental to the Evans equations, has a Latin and a Greek index. The "a" of the tetrad refers to the Euclidean tangent space and is t, x, y, z of our invariant distance. The "μ" refers to the four dimensions of the universe. The tetrad is a vector valued matrix.

Figure 1-6 shows the basic idea of vectors and Figure 1-7 shows more about basis vectors, e.

Figure 1-6   Vectors

Vector and Basis Vectors, e

Vector **A** with its components, **b** and **c**, which are right angled vectors which add up to **A**. The two components have a unit length, **e**, marked off for measuring. e_1 and e_2 are not necessarily the same length.

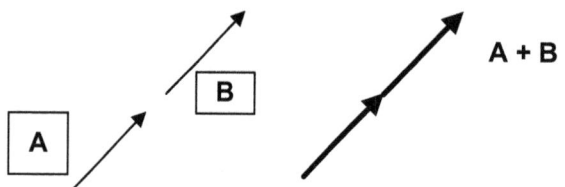

Adding **A** plus **B** gives a vector that is twice as long and has the same direction.

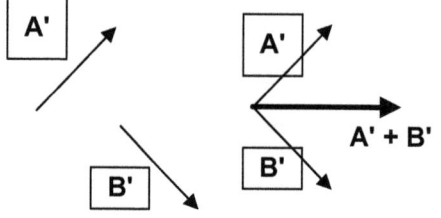

Adding **A'** plus **B'** gives a vector that is shown to the left. It changes direction and its length is a function of the original vectors.

28

Chapter 1 – Special Relativity

Figure 1-7   Basis Vectors, e

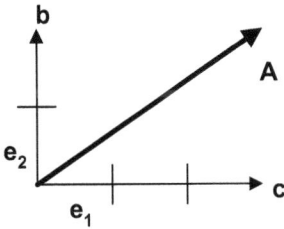

Reference Frame One - High Energy Density

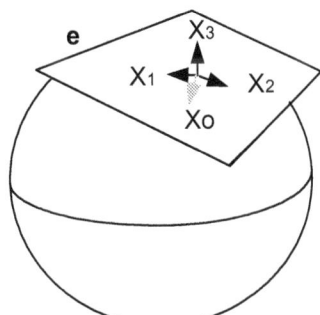

Curved space with a 4-vector
point and a tangent plane.

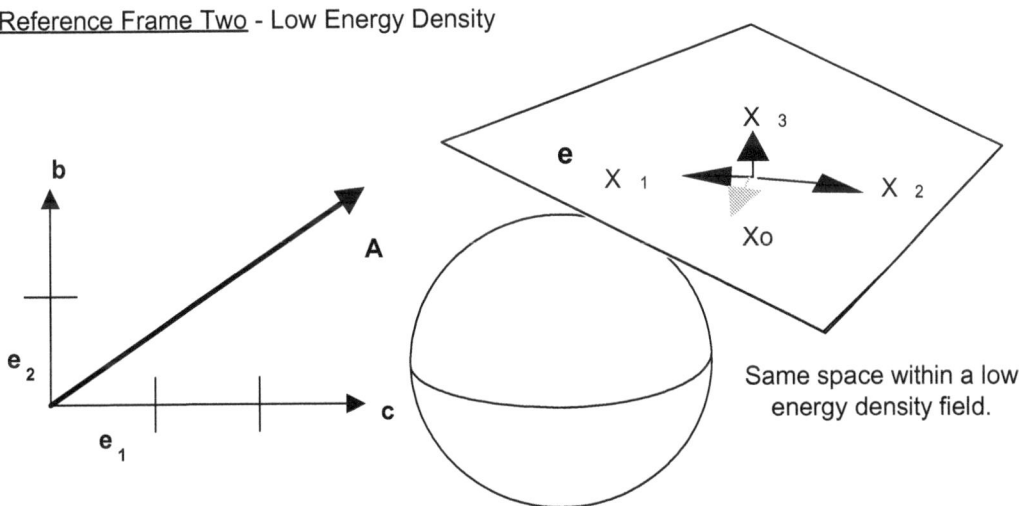

Basis vectors are yardsticks (metersticks?). They can be moved from one
reference frame to another. Regular vectors cannot be moved without losing
their proportions. The vectors in the new reference frame are formed from
the basis. All proportions will be correct.

Another type of vector is used extensively in general relativity – *the one form*. It is introduced in Figure 1-8 and the Glossary has more information. All the vectors in a complicated pattern that have the same value are like a one-form. Given that they are discrete values, they are also like eigen or quantized values. More later.

29

Evans Equations of Unified Field Theory

In relativity and quantum theory, vectors are used to analyze properties of the spacetime, the masses, events, and the state of any particle. They are used and interpreted differently, but the math is the same.

Figure 1-8   One-form example

A one-form defines the constant values of a function. The same elevation bands in a topological map are one-forms.

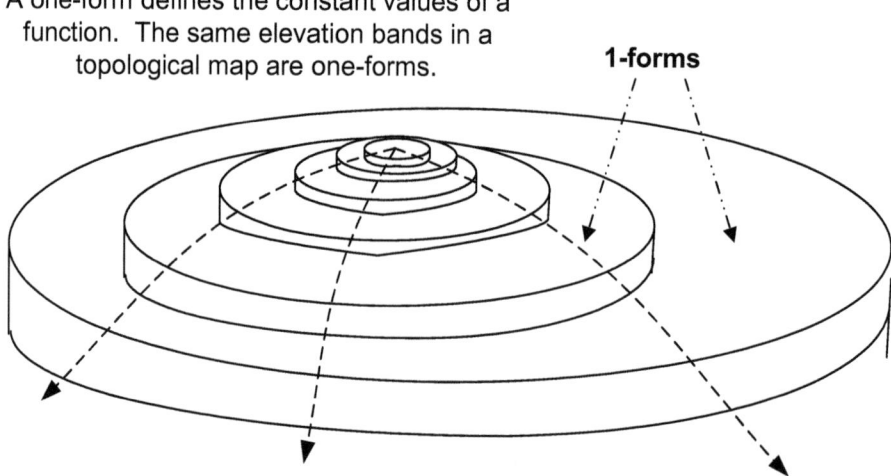

A vector can exist in itself. This makes it a "geometrical object." It does not have to refer to any space in particular or it can be put into any and all spaces. We assume that there are mathematical spaces in our imaginations. Coordinates on the other hand always refer to a specific space, like the region near a black hole or an atom. Vectors can define those coordinates.

The Metric

The metric of special relativity is the Minkowski metric. A metric is a map in a spacetime that defines the distance and also some mathematical functions of the spacetime, like simple addition and multiplication. If one looks around the region in which they find themselves, a metric exists. It is the reality within which they find themselves and is a simple mapping. It establishes a distance from every point to every other. In relativity the metric is typically designated as $\eta_{\mu\nu}$ = (-1, 1, 1, 1) and is multiplied against distances. At times it appears as (+1, -1, -1, -1). For example, $(dt^2, dx^2, dy^2, $ and $ dz^2)$ give the distances between two four dimensional points. Then $\eta_{\alpha\beta} (dt^2, dx^2, dy^2,$ and $ dz^2) = -dt^2 + dx^2 + dy^2 + dz^2$

The invariant distance is ds where
$$ds^2 = -dt^2 + dx^2 + dy^2 + dz^2 \qquad (2)$$
Alternatively, letting dx stand for dx, dy, and dz combined, $ds^2 = dx^2 - dt^2$.

This distance is sometimes called the "line element." In relativity – both special and general – the spacetime metric is defined as above. When calculating various quantities, distance or momentum or energy, etc., the metric must be considered. Movement in time does change quantities just as movement in space does.

We will see the equation g(**a**, **b**) or simply g used. It is the metric tensor and is a function of two vectors. All it means is the dot, more precisely the inner product of the two vectors combined with the spacetime metric. The result is a 4-dimensional distance as shown in Figure 1-9 using curved Riemann geometry instead of the Euclidean in the figure. (See the Glossary under *Metric Tensor*.)

The nature of the metric becomes very important as one moves on to general relativity and unified field theory. The metric of quantum theory is the same as that of special relativity. It is a good mathematical model for approximations, but it is not the metric of our real universe. Gravitation cannot be described. In the same manner, Riemann geometry describes gravitational curvature, but does not allow spinning of the metric. For that, Cartan geometry is needed.

Figure 1-9   Metric and invariant distance, ds

$ds^2 = dx^2 - dt^2$ dx stands for the dx, dy, dz three dimensional distance.

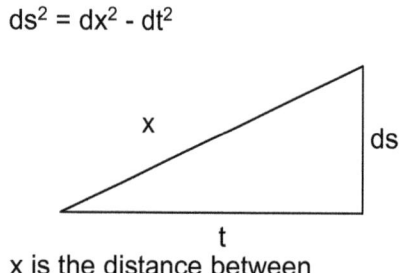

ds, the invariant distance

x is the distance between two events, say flashes of light

dt is the time between the two events

Regardless of the spacetime compression of any observer's reference frame, ds is calculated to be the same. The spatial and temporal distances will be different in different energy density reference frames, however the difference is always the same.

Summary

Special relativity showed us that space and time are part of the same physical construct, spacetime. Spacetime is a real physical construct; reference frames are the mathematical description.

The spacetime of a particle or any reference frame expands and compresses with its energy density as velocity varies. The velocity of light (any electromagnetic wave) in vacuum is a constant regardless of the reference frame from which the measurement is made. This is due to the metric of the spacetime manifold.

The torque example shows vectors in two dimensions which generate a vector in a third. We can generalize from this example to vectors in "vector spaces" used to describe actions in our universe.

The nature of spacetime is the essential concept that Evans presents. Just as Minkowski spacetime replaced Newtonian and Riemann replaced Minkowski, so Evans spacetime (the manifold of differential geometry) replaces Riemann. Differential geometry allows curvature and spin of the metric. These two must be part of the spacetime itself. Spacetime cannot have just curvature and extraneously overlay spin.

Chapter 2 General Relativity

> ...the general principle of relativity does not limit possibilities (compared to special relativity); rather it makes us acquainted with the influence of the gravitational field on all processes without our having to introduce any new hypothesis at all.
> Therefore it is not necessary to introduce definite assumptions on the physical nature of matter in the narrow sense.[10] In particular it may remain to be seen, during the working out of the theory, whether electromagnetics and the theory of gravitation are able together to achieve what the former by itself is unable to do.
>
> Albert Einstein in *The Foundation of the General Theory of Relativity*. Annalen der Physik (1916)

Introduction

Einstein developed general relativity in order to add accelerating reference frames to physics.[11] Special relativity requires an inertial reference frame, that is, there can be no increase or decrease in velocity for the equations to work. It applies to laboratories at rest and to particles at velocities near c, but there can be no change in speed.

One of Einstein's brilliant discoveries was that acceleration and gravitation are almost the same thing. If an observer is in a closed, windowless room, he cannot tell the difference between being on the planet earth and being in space increasing speed at 9.8 meters per second per second. The effects are the same except for geodesic deviation. (If acceleration is occurring, all force is exactly parallel to the direction of travel; if in a gravitational field, there is a slight difference between one side of the room and the other since the line of force

---

[10] Einstein means that once the geometry is defined there is no need of any other specifics apart from $R = -kT$.

points to the center of the earth. This is geodesic deviation. Differences are too small in any local region like a room to be measured.

Where special relativity describes processes in flat spacetime, general relativity deals with curvature of spacetime due to the presence of matter, energy, pressure, or its own self-gravitation – energy density collectively. An accelerating reference frame is locally equivalent to one within a gravitational field.

Gravitation is not a force although we frequently refer to it as such.

Imagine that we take a penny and a truck, we take them out into a region with very low gravity, and we attach a rocket to each with 1 kg of propellant. We light the propellant. Which – penny or truck – will be going faster when the rocket stops firing? We are applying the same force to each.

The penny. At the end of say 1 minute when the rockets go out, the penny will have traveled farther and be going faster.

It has less mass and could be accelerated more.

If instead we drop each of them from a height of 100 meters, ignoring effects of air resistance, they will both hit the ground at the same time. Gravity applies the same force to each, as did the rocket. Thus, gravity seems like a force but is not a force like the rocket force. It is a false force. It is actually due to curved spacetime. Any object – penny or truck – will follow the curved spacetime at the same rate. Each will accelerate at the same rate. If gravity were a force, different masses would accelerate at different rates.

This realization was a great step by Einstein in deriving general relativity.

Gravity is not force; it is spatial curvature. In general relativity, all energy is curvature; this applies to velocity, momentum, and electromagnetic fields also.

Observe that a falling person does not feel any force. (Well, you have to be in free fall for a while to realize this. After falling off a ladder, one doesn't typically have enough time to observe feeling "no force" before feeling the impact of the ground.) The falling person or object follows the curvature of the spacetime. This is the geodesic.

---

[11] David Hilbert also developed general relativity almost simultaneously.

However, acceleration and a gravitational field are nearly indistinguishable. By following the separation of geodesics – lines of free fall – over a large distance, they can be distinguished. Nevertheless, locally, they are the same.

If an object is accelerated, a "force" indistinguishable from gravitation is experienced by its components. During the acceleration, the spacetime it carries with it is compressed by the increase in energy density - Lorentz contraction. After the acceleration stops, the increase in energy density remains and the object's spacetime is compressed as it moves with a higher velocity than originally. The compression is in the two dimensions in which the object is accelerated, not in the two perpendicular to movement. This is special relativity.

If the same object is placed in a gravitational field, the object's spacetime is again compressed by the gravitational field. However, when the object is removed from the field, there is no residual compression. The energy is contained in the region around the mass that caused the field, not the object. The compression occurs within all four dimensions.

In any small region, curved spacetime is Lorentzian – that is it nearly flat like the spacetime of special relativity and it obeys the laws of special relativity. Overall it is curved, particularly near any gravitational source. An example is the earth's surface. Any small area appears flat; overall large areas are curved. The earth's surface is extrinsically curved – it is a 2-dimensional surface embedded in a 3-dimensional volume. The surface is intrinsically curved - it cannot be unrolled and laid out flat and maintain continuity of all the parts. The universe is similar. Some curved spaces, like a cylinder, are intrinsically flat. It is a 2-dimensional surface that can be unrolled on a flat surface. The mathematics of the flat surface is easier than that of a curved surface. By imagining flat surface areas or volumes within a curved surface or volume, we can simplify explanations and calculations.

For some readers, if the 4-dimensional vectors and equations in the text are incomprehensible or only barely comprehensible, think in terms of Lorentz

Chapter 2 – General Relativity

contraction. Instead of contraction in two dimensions, compression occurs in all four dimensions.

Einstein used Riemannian geometry to develop general relativity. This is geometry of curved surfaces rather than the flat space Euclidian geometry we learn in basic mathematics. In addition, spacetime had to be expressed in four dimensions. He developed several equations that describe all curved spacetime due to energy density.

The Einstein tensor is:

$$G = 8\pi T \qquad (1)$$

This is one of the most powerful equations in physics. It is shorthand for:

$$\mathbf{G} = 8\pi GT/c^2 \qquad (2)$$

Where the first, bold **G** is the tensor and the second G is Newton's gravitational constant. It is assumed that the second G = c = 1 since these depend on the system of units used, and written **G** = 8πT. We do not normally make **G** bold and we let context tell us it is a tensor distinguishable from the gravitational constant G. In some texts, tensors are in bold.

Einstein showed that four dimensions are necessary and sufficient to describe gravitation. Evans will show that those same four dimensions are all that are necessary to describe electromagnetism and particles in unified field theory.

$8\pi G/c^2$ is a constant that allows the equation to arrive at Newton's results in the weak limit – low energy density gravitational fields like the earth or the sun's. Only when working in real spacetime components does one need to include the actual numbers. Thus G =kT is another way to express the equation.

More often in Evans' work we see the mathematicians' language. Then R = - kT is seen where k = $8\pi G/c^2$ which is Einstein's constant. **G** is equivalent to R, the curvature. Physicists use **G** as often as R.

37

T is the stress energy tensor or energy density. T stands for the tensor formulas that are used in the calculations. In a low energy limit T = m/V. T is the energy density, which is energy-mass per unit of volume. The presence of energy in a spacetime region will cause the spacetime to curve or compress. See Figure 2-1.

By finding curves in spacetime we can see how gravity will affect the region. Some of the curves are simply orbits of planets. Newton's simpler equations predict these just as well as Einstein's in most instances. More spectacular are the results near large bodies – the sun, neutron stars, or black holes. The particle is a region of highly compressed energy; curvature describes its nature also. General relativity is needed to define the affects of such dense mass-energy concentrations.

What is spacetime?

For all we speak in this book of "spacetime" it remains undefined.

Mathematically it is distance between events. That is it is distance between points. It is the metric.

This is vague; however spacetime is not substance or matter as we experience it in everyday life.

Time is space in general relativity. Arguments can be made that there is no such thing as time as we experience it. Rather there is only space. This is beyond the subject of this book; however, it is probably the physical reality. Somehow, beyond present understanding, spatial interaction of matter and waves – the forces – leads to a concept of time, or at least our subjective experience.

While spacetime remains undefined physically, mathematically it is easier. This is part of the insistence of Einstein and others that physics is geometry.

Chapter 2 – General Relativity

Figure 2-1

Energy density reference frames
T = m/V in our low energy density
reference frame.

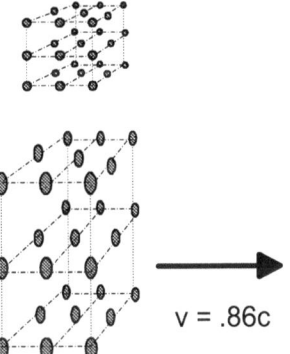

a) Low energy density spacetime can be viewed as a lattice

b) High energy density spacetime is compressed in all dimensions -GR

c) Accelerated energy density reference frame is compressed in only the x and t dimensions - SR

v = .86c

Einstein proposed R = -kT where k = 8πG/c². R is the curvature of spacetime – gravitation. R is mathematics and kT is physics. This is the basic postulate of general relativity. This is the starting point Evans uses to derive general relativity and quantum mechanics from a common origin. Chapter 6 will introduce the Evans equations starting with R = -kT.

In Figure 2-2 there are three volumes depicted, a, b, and c. We have suppressed two dimensions – removed them or lumped them into one of the others. While we have not drawn to scale, we can assume that each of the three volumes is the same as viewed by an observer *within* the frame. If we were to take c and move it to where a is, it would appear to be smaller as viewed from our distant, "at infinity," reference frame.

In Figure 2-2, the regions are compressed in all four dimensions.

Figure 2-2

Energy density reference frame in general relativity with 2 dimensions suppressed

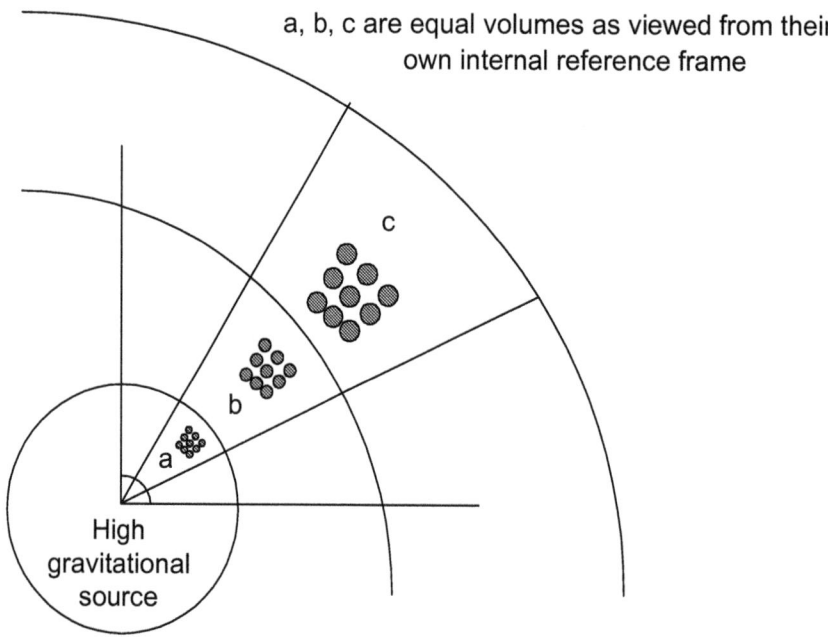

a, b, c are equal volumes as viewed from their own internal reference frame

From within the spacetime, the observer sees no difference in his dimensions or measurements. This is the same in special relativity for the accelerated observer – he sees his body and the objects accelerated with him as staying exactly as they always have. His spacetime is compressed or contracted but so are all his measuring rods and instruments.

When we say spacetime compresses we mean that both space and time compress. The math equations involve formal definitions of curvature. We can describe this mechanically as compression, contraction, shrinking, or being scrunched. There is no difference. Near a large mass, time runs slower than at infinity - at distances where the curvature is no longer so obvious. Physicists typically speak of contraction in special relativity and curvature in general relativity. Engineers think compression. All are the same thing.

Chapter 2 – General Relativity

Figure 2-3   A curved space with a 4-vector point and a tangent plane.

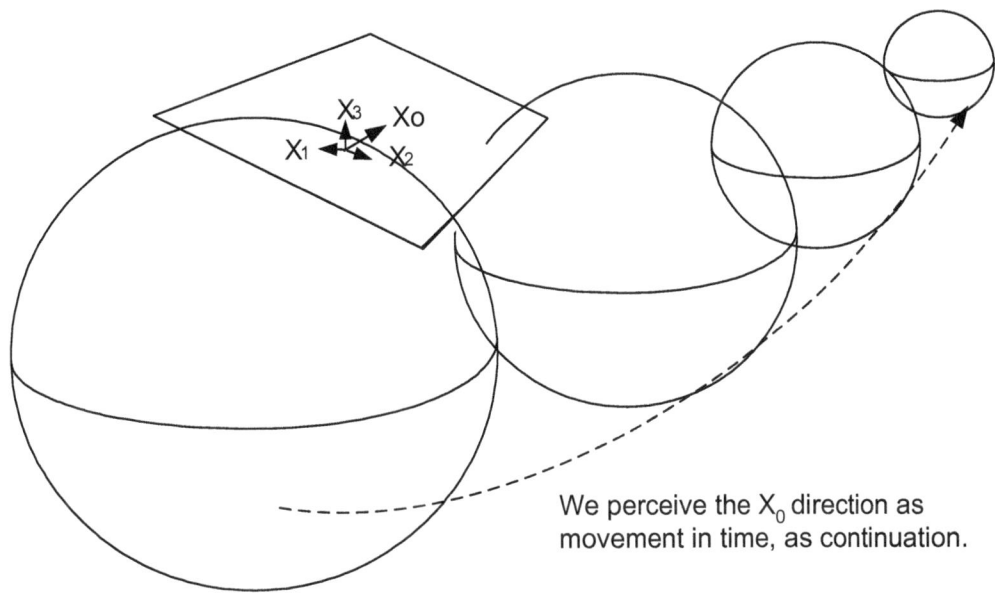

We perceive the X_0 direction as movement in time, as continuation.

In general relativity, there is no distinguishing between space and time. The dimensions are typically defined as x_0, x_1, x_2, and x_3. x_0 can be considered the time dimension. See Figure 2-3.

By calculating the locations of the $x_{0, 1, 2, 3}$ positions of a particle, a region, a photon, a point, or an event, one can see what the spacetime looks like. To do so, one finds two points and uses them to visualize the spacetime. All four are described collectively as x_μ where μ indicates the four dimensions 0, 1, 2, and 3.

Curved Spacetime

Mass energy causes compression of spacetime. In special relativity we see that contraction of the spacetime occurs in the directions of velocity and time movement. When looking at geodesics, orbits around massive objects, paths of shortest distance in space, we see the space is curved.

Figure 2-4   Visualizing Curved Spacetime

Flat space with no energy present. Two dimensions are suppressed.

X
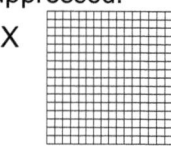
t

A dense mass in in the center of the same space. The geodesics, shortest lines between points, are curved.

X
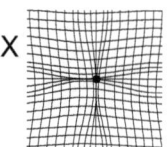
t

Visualization in the mind's eye is necessary to imagine the 3 and 4 dimensional reality of the presence of mass-energy.

Past / Future

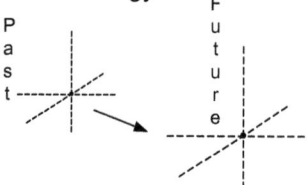

The geodesics are curved; they are the lines a test particle in free fall moves along. From its own viewpoint, a particle will move along a line straight towards the center of mass-energy.

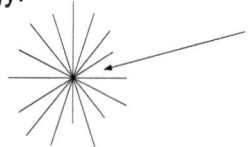

Figure 2-5   Stress Energy =

No mass, no stress energy, T=0.
No curvature, R=0.

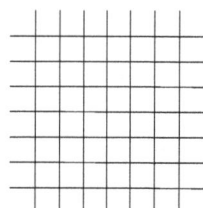

Medium mass per volume, T = moderate
Moderate curvature, R = medium.

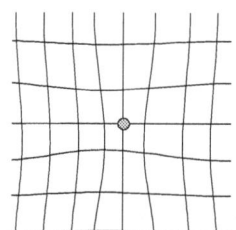

Dense mass, high stress energy, T is large. R is large.

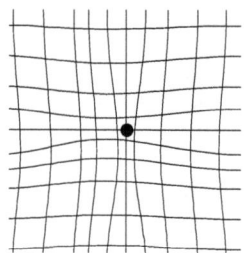

The mass density to amount of curvature is not linear. It takes a lot of mass density to start to curve space. As mass builds, it, compresses itself. Pressure then adds to the curvature.

This can be taken to the region of a particle density also. Mass causes curvature.

In Figures 2-4 and 2-5, spacetime is depicted by lines. There is only curvature and unpowered movement must follow those curves for there is no straight line between them. Those curves are geodesics and are seen as straight lines from within the reference frame.

The amount of R, curvature, is non-linear with respect to the mass density, m/V = T. Figure 2-6 is a graph of the amount of compression with respect to mass density. A black hole occurs at some point.

When we hear the term "symmetry breaking" or "symmetry building" we can consider the case where the density increases to the black hole level. At some density there is an abrupt change in the spacetime and it collapses. The same process occurs in symmetry breaking.

Mass causes compression of the spacetime volume at the individual particle level and at the large scale level of the neutron star or black hole. Spacetime is compressed by the presence of mass-energy.

Pressure also causes curvature. As density increases, the curvature causes its own compression since spacetime has energy. This is self-gravitation.

Figure 2-6   Graph of Mass Density in Relation to Compression

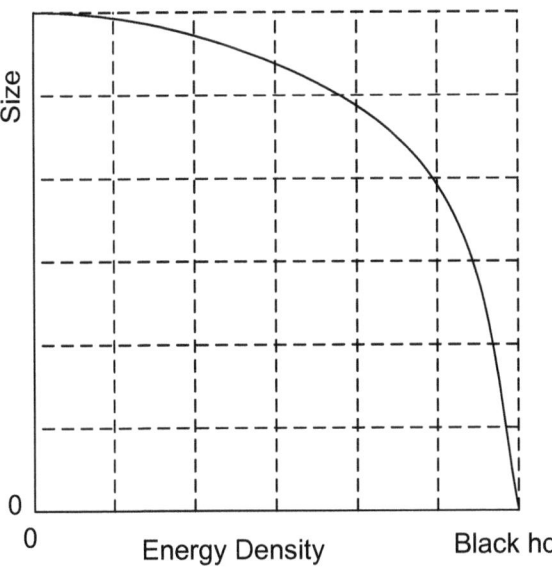

In general, as the energy density increases, the reference frame of the mass shrinks. From its own viewpoint, it is unchanged although at a certain density, some collapse or change of physics as we know it will occur.

Curvature

Curvature is central to the concepts in general relativity and quantum mechanics. As we see the answers the Evans equations indicate we start to see that curvature is existence. Without the presence of curvature, there is no spacetime vacuum.

The Minkowski spacetime is only a mathematical construct and Evans defines it as the vacuum. Without curvature, there is no energy present and no existence. The real spacetime of the universe we inhabit is curved and twisting spacetime. See Figure 2-7.

Figure 2-7   Curvature

The measure of the curvature can be expressed several ways. R is curvature in Riemann geometry. In a low limit it can be a formula as simple as $R = 1/\kappa^2$ or $R = 2/\kappa^2$.

In four dimensions the curvature requires quite a bit of calculation.

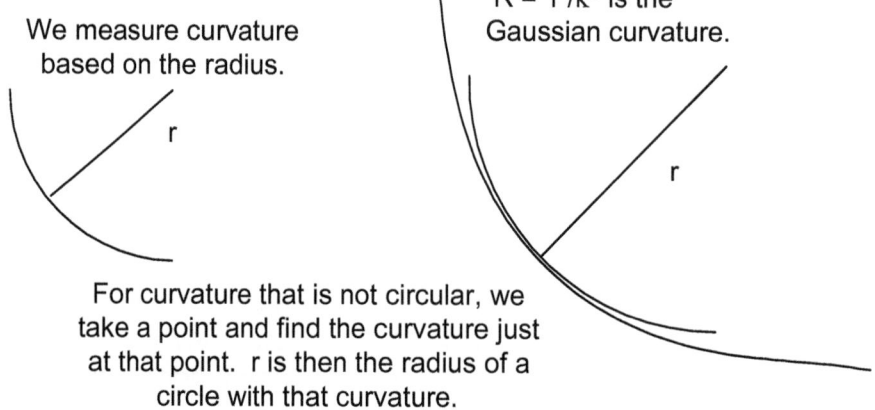

We measure curvature based on the radius.

$R = 1/\kappa^2$ is the Gaussian curvature.

For curvature that is not circular, we take a point and find the curvature just at that point. r is then the radius of a circle with that curvature.

Chapter 2 – General Relativity

Concept of Field

The field is a mathematical device to describe force at a distance from its source in spacetime. Where curvature exists, there is a field that can be used to evaluate attraction or repulsion at various distances from the source. Likewise where an electrical charge exists, a field is imagined to be present around it. The whole idea of "unified *field* theory" is to combine the gravitational and electrical descriptions. Both are very similar with respect to the field description.

We use the term potential since there is an ability to do work on a charge or a mass even if none is present.

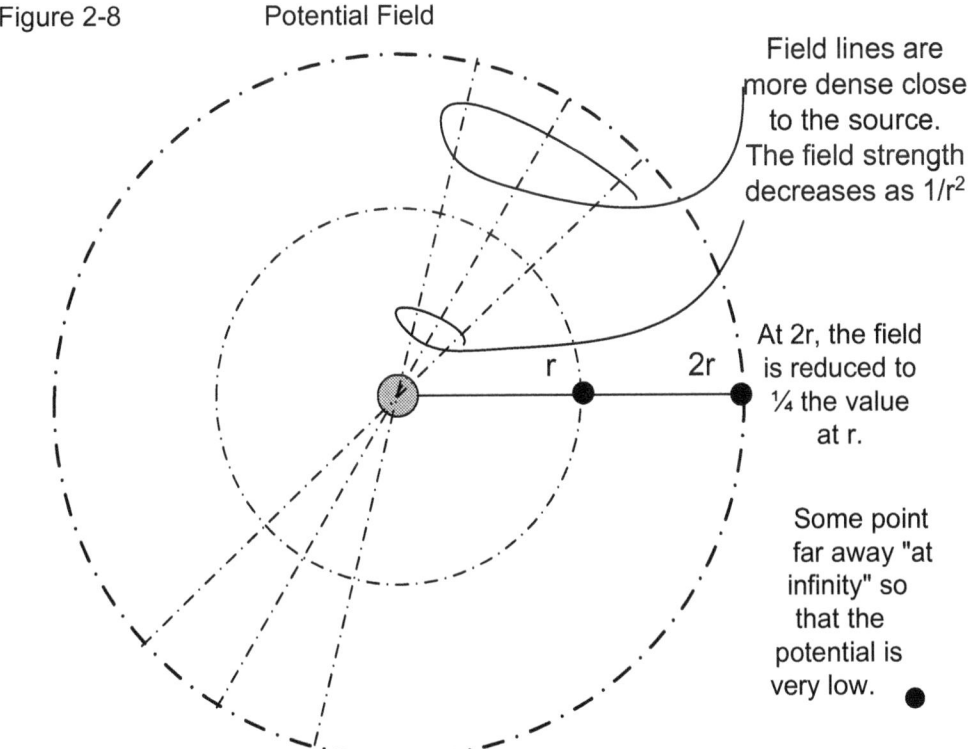

Figure 2-8   Potential Field

Field lines are more dense close to the source. The field strength decreases as $1/r^2$

At 2r, the field is reduced to ¼ the value at r.

Some point far away "at infinity" so that the potential is very low.

The field strength at a distance from a charge or a mass decreases as the inverse square of the distance.   Potential = a constant / r

45

In Figure 2-8 assume there is a source of energy in the center gray circle. Build a circular grid around it and establish a distance r. The *field* varies as the inverse of the distance squared, $1/r^2$. The *potential* itself is a scalar and is proportional to $1/r$. The field is a vector. There is a direction in which the potential "pushes" or "pulls." The force resulting from the field will vary as $1/r^2$.

We showed that gravitation is curvature, not a force, in the introduction at the beginning of this chapter. The same is going to be seen in unified field theory for electromagnetism and charge – electrodynamics collectively.

We showed that gravitation is curvature, not a force, in the introduction at the beginning of this chapter. The same is going to be seen in unified field theory for electromagnetism and charge – electrodynamics collectively.

Vacuum

In the past there has been some question about the nature of the vacuum or spacetime. Maybe vacuum was composed of little granular dots. Was spacetime just a mathematical concept or is a more of a reality? Just how substantial is the vacuum?

There are a number of views about the vacuum versus spacetime. Quantum theory tells us that at the least the vacuum is "rich but empty" – whatever that means. The stochastic school sees it as a granular medium. Traditional general relativity sees it as an empty differentiable manifold. In any event, virtual particles come out of it, vacuum polarization occurs, and it may cause the evaporation of black holes. It is certainly full of photon waves and neutrinos. It could be full of potential waves that could be coming from the other side of the universe.

At the most, it is composed of actual potential dots of compressed spacetime. These are not a gas and not the aether, but are potential existence. Vacuum is not void in this view. Void would be the regions outside the universe – right next to every point that exists but in unreachable nowhere.

The word vacuum and the word spacetime are frequently interchanged. They are the common ground between general relativity and quantum mechanics. In Evans' use, the word vacuum is Minkowski spacetime everywhere with no curvature and no torsion anywhere; i.e., it is everywhere flat spacetime. The key will be seen further into this book in the development of Einstein's basic postulate that R = -kT. Where spacetime curvature is present, there is energy-mass density, and where spacetime torsion is present there is spin density.

"Vacuum" is flat quantum spacetime. "Spacetime" is a more general word and can be curved.

Manifolds and Mathematical Spaces

At every point in our universe there exists a tangent space full of scalar numbers, vectors, and tensors. These define the space and its metric. This is the tangent space, which is not in the base manifold of the universe. The tangent within the same space as a curved line in a three dimensional space can be pictured clearly, but in a four dimensional spacetime, a curved line becomes a somewhat vague picture. Typically we see it as time and can envision slices of three dimensional spaces moving forward. The tangent space allows operations for mathematical pictures. In general relativity this is considered to be a physical space. That is, it is geometrical. Figure 2-3 was an attempt to show this.

In quantum mechanics the vector space is used in a similar manner, but it is considered to be a purely mathematical space.

At any point there are an infinite number of spaces that hold vectors for use in calculations. See Figures 2-9 and 2-10.

The tetrad is introduced here and will be developed further in later chapters. The tetrad is a 4 X 4 matrix of 16 components that are built from vectors in the base manifold and the index as shown in Figure 2-9.

Evans Equations of Unified Field Theory

Figure 2-9   Tetrad and Tensors
See text for description.

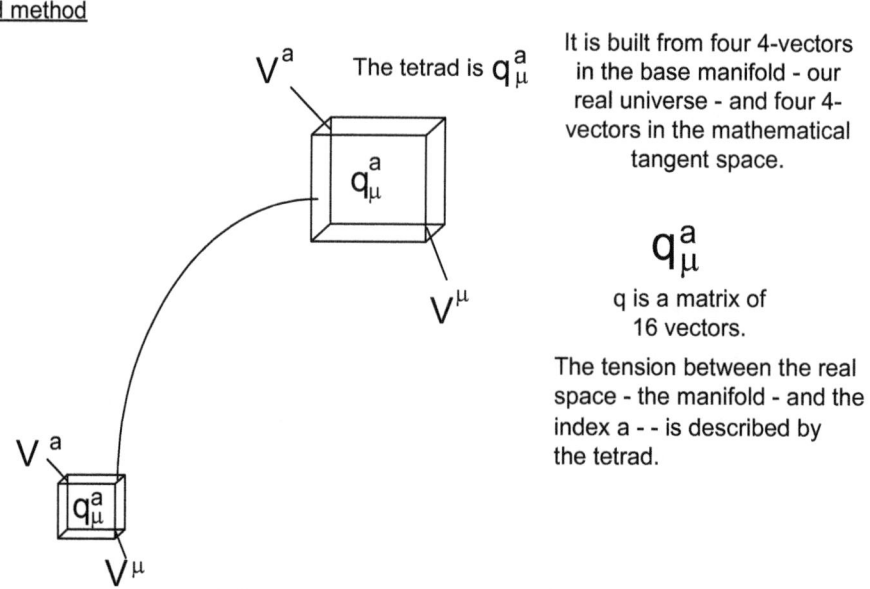

Tensor method

A is a spacetime far from any strong gravitational source.

Our universe is depicted above the curve.

D is a spacetime close to a black hole. It is squished and warped.

Tangent space

Below the curve is the mathematical tangent space.

Index, α

The index represents the vectors transported from one gravitational field to another.

In order to describe electromagnetism in this gravitational scenario, the tensor must be spun on top of the curved spacetime.
Doesn't work.

Tetrad method

The tetrad is q^a_μ

It is built from four 4-vectors in the base manifold - our real universe - and four 4-vectors in the mathematical tangent space.

$$q^a_\mu$$

q is a matrix of 16 vectors.

The tension between the real space - the manifold - and the index a - - is described by the tetrad.

The tetrad can itself be spun. Since it is essentially composed of spacetime itself, the spacetime is spinning in the electromagnetic description.
This works.

Chapter 2 – General Relativity

Cartan's tetrad theory analyzes a four-dimensional spacetime using alternate differential methods called frames. The tetrad is like a Riemann tensor in a different guise.

To the non-mathematician, this is difficult to understand. It is not critically important to understand it, but the tetrad is the key mathematical method Evans uses to develop unified field theory.

For practical purposes, imagine our four dimensional spacetime as the three dimensions we are familiar with and time as a path that the three dimensions move along. In differential geometry, general relativity, and unified field theory, the time path follows the same mathematical constructions that spatial dimensions follow. Time is not well understood; time can become a spatial dimension in Kerr black hole mathematics, Einstein put space and time on an equal footing in special relativity.

In Figure 2-9 at the top is the tensor method of general relativity. Tensors are mathematical constructions that can be moved from one gravitational field to another, say from a space well outside a black hole to almost touching the horizon. Far away the space is nearly flat as shown as A, a box at the top right. However, near the black hole the space compresses and warps. This is shown as D at the lower left. The basis vector is **e**. It is four vectors which define the distances in the box. It is in a tangent space and is rectangular. This is shown as B. Those vectors are squished by gravity as seen at C.

How the tensors are used is that spacetime A can be mapped as a tensor B. B gets moved into a new gravitational field and is compressed and becomes C. C can then be reverse mapped to the spacetime D.

$q^a{}_\mu$ is a tetrad. "a" indicates the mathematical index in the tangent space. "μ" indicates the spacetime of our universe. The components inside the tetrad draw a map from the manifold of our real space to the index space. There are 16 components because each of the four spacetime vectors defining the curvature at a point are individually multiplied by each of four vectors in the index space. Without going into how calculations are made, it is sufficient to know that the tetrad allows one curved spacetime to be visualized in a different gravitational field and define the resulting curved spacetime. The path it takes is inside the tangent space. Figure 10 shows another view of the spacetime mapped to mathematics.

For example, in Evans' theory an electromagnetic field is turning spacetime itself. It can move from one gravitational field to another and the tetrad provides the method of calculation.

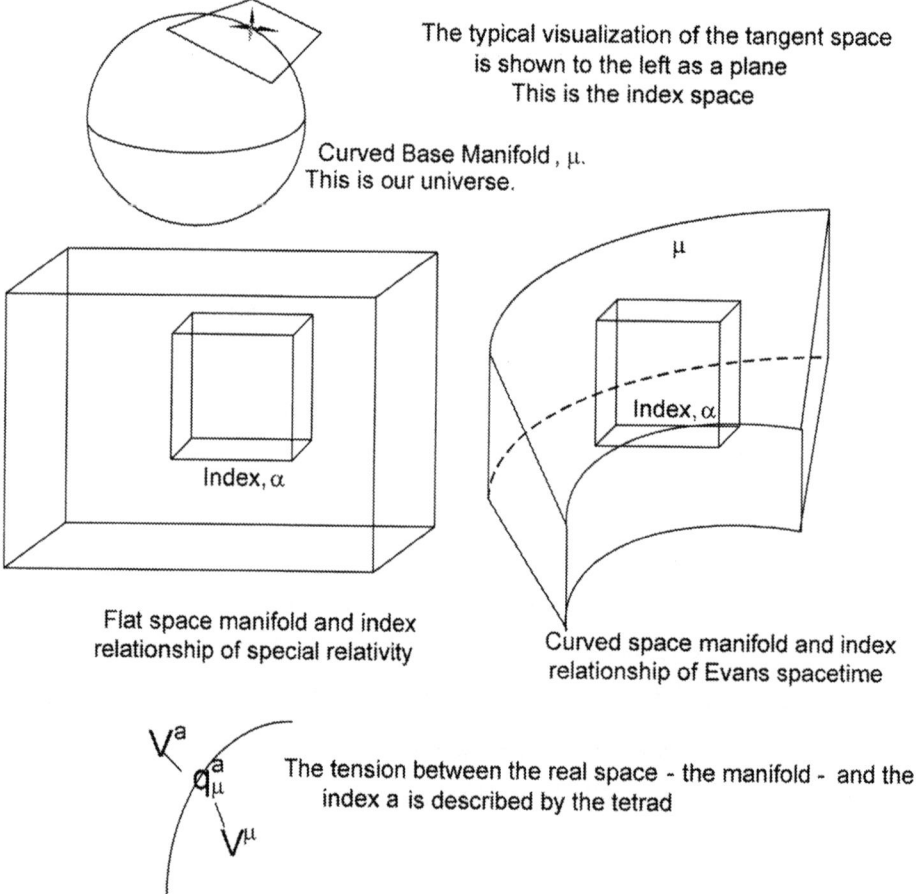

Figure 2-10 Base Manifold with Euclidean Index

Evans used vectors in his first paper of unified field theory, but soon changed to the tetrad formulation which allowed greater freedom to uncover physical properties of spacetime. The tetrad is known in general relativity as the Palatini variation – an alternate to the Riemann geometry that Einstein used.

We will see later that the tetrad can be spun. It describes gravitation along the curve and electromagnetism by simultaneously spinning. Electromagnetism is spinning spacetime. Gravitation is curving spacetime.

The Metric and the Tetrad

The metric is a map between vectors (and a vector and a one- form). The metric of our four dimensional universe requires two 4-vectors to create a definition.

A metric vector space is a vector space which has a scalar product. A scalar is a real number that we can measure in our universe. Essentially, this means that distances can be defined using the four dimensional version of the Pythagorean Theorem. Minkowski spacetime is an example of a metric vector space; it is four dimensional, but it is flat – gravitation does not exist within it. Lorentz transformations allow definition of the distances between events. It cannot describe accelerated reference frames or gravitation. This is the limitation which led Einstein to develop general relativity.

Tensor geometry is used to do the calculations in most of general relativity. Evans uses more differential geometry than commonly found.

The metric vectors are built from vectors inside the manifold (spacetime of our universe) by multiplying by the metric. Below **e** indicates basis vectors and η_n indicates the metric with $\eta_n = (-1, 1, 1, 1)$. This is complicated if one is actually doing the calculations, but it is not conceptually difficult. The basis vectors below are similar to the invariant distance concept in the preceding chapter. Four vectors make up what we call a "4-vector." They can describe physical properties of the spacetime at a point in the universe.

$\mathbf{e}_{0 \text{ metric}} = \mathbf{e}_0 \, \eta_0$

$\mathbf{e}_{1 \text{ metric}} = \mathbf{e}_1 \, \eta_1$

$\mathbf{e}_{2 \text{ metric}} = \mathbf{e}_2 \, \eta_2$

$$e_{3\,metric} = e_3\,\eta_3 \tag{3}$$

At every point in spacetime there exist geometrical objects. The metric tensor is the one necessary to measure invariant distances in curved four dimensional spacetime. In order for us to see what occurs to one object, say a simple cube, when it moves from one gravitational field to another, we need linear equations. Movement in curved spacetime is too complicated.

We put the object in the tangent space.

We use basis vectors to give us the lengths (energy, etc.) in an adjustable form.

Then we move the vectors in the linear tangent space.

Finally, we can bring those vectors out of the tangent space and calculate new components in real spacetime. This is what basis vectors and tensors do for us.

Two metric vectors can establish the metric tensor. The metric produces the squared length of a vector. It is like an arrow linking two events when the distance is calculated. The tangent vector is its generalization.[12]

It takes two 4-vectors to describe the orientation of a curved four dimensional spacetime. This is similar to the need for two lines to describe a plane in three dimensional space.

The metric 4-vector is the tetrad considered as the components in the base manifold of the 4-vector q_a. Thus, the metric of the base manifold in terms of tetrads is:

$$g_{\mu\nu} = q^a{}_\mu\, q^b{}_\nu\, \eta_{ab} \tag{4}$$

where η_{ab} is the metric. This is sometimes expressed as $g_{\mu\nu}x^\mu y^\nu$. A four-vector is multiplied by the factors –1, 1, 1, 1 to find the metric four-vector. All we are doing is using the 4-dimensional Pythagorean theorem to find distances. See Figure 2-11. The distance between two metric 4-vectors needs the negative time to calculate correctly. It can be done with vectors or with tetrads also. When a metric vector is used, the signs are placed on the vector.

---

[12] See http://en.wikipedia.org/wiki/Metric_space

Figure 2-11   Metric Vectors

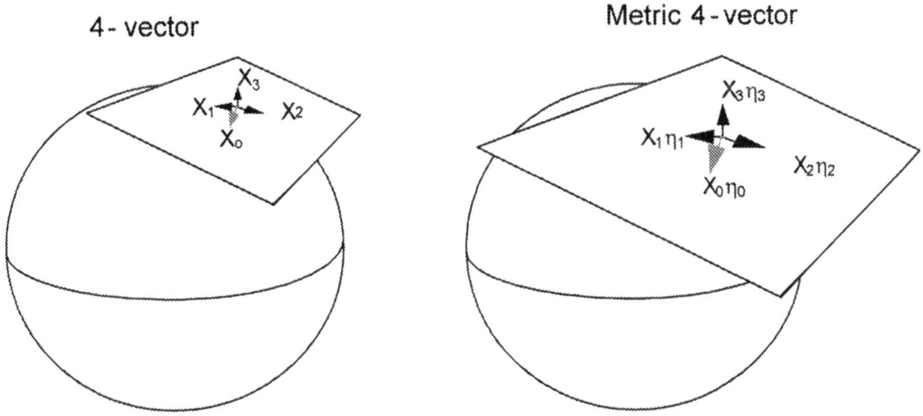

4- vector

Metric 4-vector

The concept of the tetrad is definitely advanced and even the professional physicist or mathematician is not normally familiar with it.

Einstein worked with the metric as his fundamental field although he did use the tetrad when he attempted unified field theory. There is another fundamental field called the Palatini variation which uses the tetrad as fundamental field. Evans uses the Palatini variation.

Tensors and Differential Geometry

Tensor calculus is used extensively in relativity. Tensors are mathematical machines for calculations. Their value is that given a tensor formula, there is no change in it when reference frames change. In special relativity, fairly easy math is possible since there is only change in two dimensions. However in general relativity, all four dimensions are curved or compressed and expanded in different ways as the reference frame changes position in a gravitational field. More sophisticated math is needed.

Tensors are similar to vectors and can use matrices just as vectors do for ease of manipulation.

The *metric tensor* is important in general relativity. It is a formula that takes two vectors and turns them into a distance – a real number, a scalar.

$g = (\mathbf{v}^1, \mathbf{v}^2)$ is how it is written with $\mathbf{v}^1$ and $\mathbf{v}^2$ being vectors. This is the general abstract math variation. All this means is "the distance."

If a calculation with real numbers is being made, then:

$$g = (\mathbf{v}_1, \mathbf{v}_2)\eta_{\alpha\beta} \qquad (5)$$

where $\eta_{\alpha\beta}$ is the metric indicating that the results are to use (-1, 1, 1, 1) as factors multiplied against the values of the vectors.

g is the invariant distance in four dimensions. It is a scalar distance. The calculation is the four dimensional Pythagorean formula with the metric applied.

The mathematics used to compute the correspondences in general relativity is differential geometry. For example dx / dt is a simple differential equation. "dx / dt" can be stated as "the difference in x distance per the difference in time elapsed." If we said dx / dt = 100 km / hr over the entire 3 hours of a trip, then we could calculate an equivalent by multiplying by the 3 hours. dx / dt = 300 km / 3 hrs. Alternately, one could calculate the speed at any individual point. In Chapter 4 on geometry we will go into more detail.

It is mentioned here to show that while working in four dimensions is complicated, the basic idea is one we see frequently and is not all that difficult. Miles per hour or kilometers per hour is a differential equation. The Evans equations use differential geometry. We will explain some as we go along and both the Glossary and appendices go deeper into the subject. However, full understanding of the math is not needed and pictures and verbal descriptions are given.

Equivalence Principles

This will be discussed again since the Evans equations extend the concept even further.

The *weak equivalence principle* is the equality of inertial mass and gravitational mass. There is no obvious reason why this should be true. Inertia refers to the tendency of a body to resist change in velocity. Once moving, it continues without outside influence.

In the equation

$$\mathbf{F} = m\mathbf{a} \qquad (6)$$

m is inertial mass. It is the resistance to acceleration or deceleration. F is the force that results and a is acceleration (change in velocity) in meters/second/second. Bold case indicates vector quantities since a definite direction exists.

In the equation

$$g = mMG/r^2 \qquad (7)$$

m is gravitational mass. It causes gravitation and it responds to gravitation.

The equality of inertial mass and gravitational mass has no proof in contemporary physics, but in every experiment they are shown to be identical.

The *strong or Einstein equivalence principle* is that the laws of physics are the same in every reference frame. This applies to both special relativity - that is to velocity - and general relativity – acceleration and gravitation.

The implication of the Evans formulation is more extensive. We will be able to define a very strong equivalence principle that essentially states that everything is spacetime and the laws of physics are the same in every portion of spacetime – particle, vacuum, electromagnetic field.

Chapter 3 Quantum Theory

> I can safely say than nobody understands quantum mechanics ... Do not keep saying to yourself , "But how can it be like that?" .... Nobody knows how it can be like that.
>
> Richard Feynman

Quantum Theory

Quantum theory[13] includes quantum mechanics which is the foundation, quantum electrodynamics which includes electromagnetic phenomena, and quantum chromodynamics which adds the quark color theory. Quantum mechanics is the study of basic particles, photons, electrons, and the vacuum at the smallest atomic levels. Among the things which up until now relativity has not been able to describe while quantum mechanics has, are the quantum packets of energy, the particle-wave duality of existence, and the angular momentum (spin) of particles.

The standard model maintains that particles are composed of smaller discrete quarks and gluons and that existence is probabilistic at the smallest levels. This is unproven and the Evans development in general relativity indicates otherwise. Evans shows that curvature and torsion are the essential mathematical forms and provides a theoretical framework for experiments that have shown the Heisenberg uncertainty is not accurate.

---

[13] Among those who contributed to the development of quantum theory were Planck and Einstein for the origin, but especially Bohr, Born, Schrodinger, Heisenberg, Hilbert, Dirac, Compton, Pauli, and de Broglie.

Evans does not reject quantum theory, he shows that it emerges from general relativity and with a few paradigm changes, unification occurs.

Just as relativity changed the viewpoint of physics and natural philosophy about the nature of existence, quantum theory had further impact.

Quantum theory has developed mathematical ability to make most precise predictions of the results of experiments concerning the mutual interaction between particles, electrons, photons, forces, masses, molecules, and the polarization of the "vacuum."

When dealing with the very small, it is necessary to visualize its subject matter with abstractions and mathematics. We know the pictures we draw are not true, but we live in the big world with gazillions of particles clumped together tied together by forces making up our environment and bodies and minds. It is impossible to draw a picture of a photon, we can only abstract to explain. Sometimes the photon is traveling, say from this page to your eye; sometimes it is "inside" an atom having struck an electron and excited it where it stays converted to spacetime mass.

In studying the black body problem, Max Planck discovered that energy is found in very small, but discrete, packets, h. When we measure a temperature, we see it as continuous. It could be 69.004°F (20.558°C) or 69.005°F (20.558°C) at a point in space. Our large-scale measurements show that a totally continuous range of temperatures can occur. However, at the small scale, it was discovered that there is a limit. At the level of the individual structures or systems in nature, change can occur only in discrete energy increments or decrements. The amount of energy it takes for a system to change is a discrete amount, which varies from one state of the system to another state and from one system to another system. The smallest discrete amounts of energy are called "quanta" of energy. This was published in 1899, but the new constant he devised, h, was largely ignored until 1905 when Einstein used it to explain the photoelectric effect. Additions to the theory were slow until the 1920's and 1930's when it became well established. While he was one of the founders of the basics of quantum theory, Einstein never accepted the probabilistic interpretations that developed.

There are a number of basic concepts that underlie quantum theory.

1. The quantum. Planck's quantum hypothesis that all energy comes in packets is well received. The phenomenon is seen in all energy transmission. The quantum is h. All changes must be multiples of h.
2. Wave-particle duality of photons and particles is well established. Photons and all matter have properties that in the large-scale world would be either wave like or particle like. Waves spread out, particles are discretely in one place. (Einstein maintained and now Evans and others show that the duality is simultaneous. The standard model assumes they are neither particle nor wave until an interaction occurs and then become either wave or particle, but never both.)
3. It is necessary to simply accept that below the atomic level, the nature of things is a duality. It is an over simplification, but in general photons travel like waves and interact like particles. See Figures 3-1 and 3-2.
4. Energy and momentum are related to their frequency by $E = nhf$. That is energy = any discrete integer number times h times the frequency.
5. Using Einstein's $E = mc^2$ and Planck's $E = hf$ one can set $mc^2 = hf$. (In terms of radians this is often expressed as $E = h\omega$.) Then $m = hf/c^2$. This is the duality of mass and energy. The particle is on the left side of the equation and the right side is expressed as a wave with frequency.
6. A "state" of a particle is all the information needed to describe it. Energy (mass), velocity, spin, and position are among these. The Schrodinger equation is used to predict these things at the atomic level. Quantum theorists accept some vagueness in the state; relativists believe that a more causal explanation is possible.

Figure 3-1   Electron Location Probability

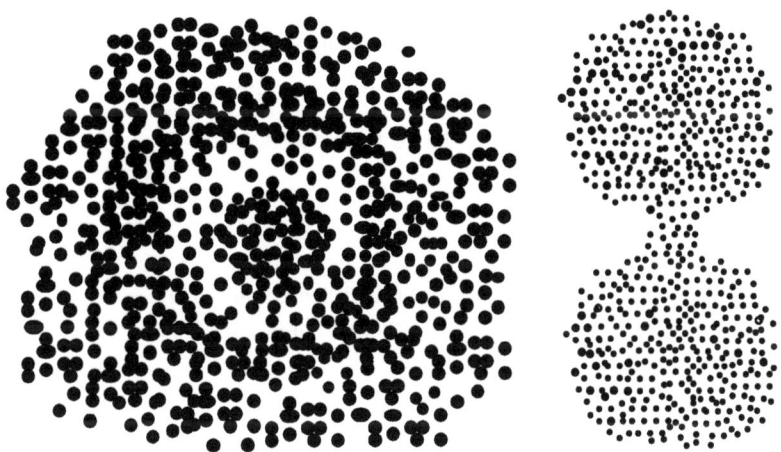

One asks where an electron is in an atom. Quantum mechanics believes that it has no specific location. An electron is a cloud. There are definite probability distributions for an electron in an atom. The electron is not in one place, it is spread out; it is more likely located in some regions than others. The regions where it is most likely to be are quite regular, but the electron is probably not a little ball of electricity; rather it is a diffuse, roughly localized, potential that has various numbers that can be assigned to it to use in predicting its behavior. It is quantum, but it is not a discrete object as one would find in the macro world our immediate senses show us. See Figure 3-1 for just one of many electron probability distributions. ($|\psi_{100}|^2$ and $|\psi_{210}|^2$ are shown.)

Quantum theory's statistical nature has always concerned some physicists. One way of looking at it is that given the very small areas it operates in, it is not surprising that we cannot predict the location precisely. We are looking at vacuum, photons, and electrons that are 10^{-15} cm with physical probes that are thousands of times bigger – that is why we use mathematics that can look at more detail. It is like trying to study a marble's mass and trajectory after dropping it off the Empire State Building, letting it bounce once, and only then using a magnifying glass to watch it. It is amazing that we have such accuracy as we have achieved.

The meaning of quantum mechanics has been argued over the years as physicists and philosophers have tried to decipher it. Einstein never accepted it as complete. The Evans equations indicate he was correct.

Figure 3-2   Double slit experiment

The photon travels like a wave, experiences interference like waves, and interacts with the screen discretely, localized, like a particle.

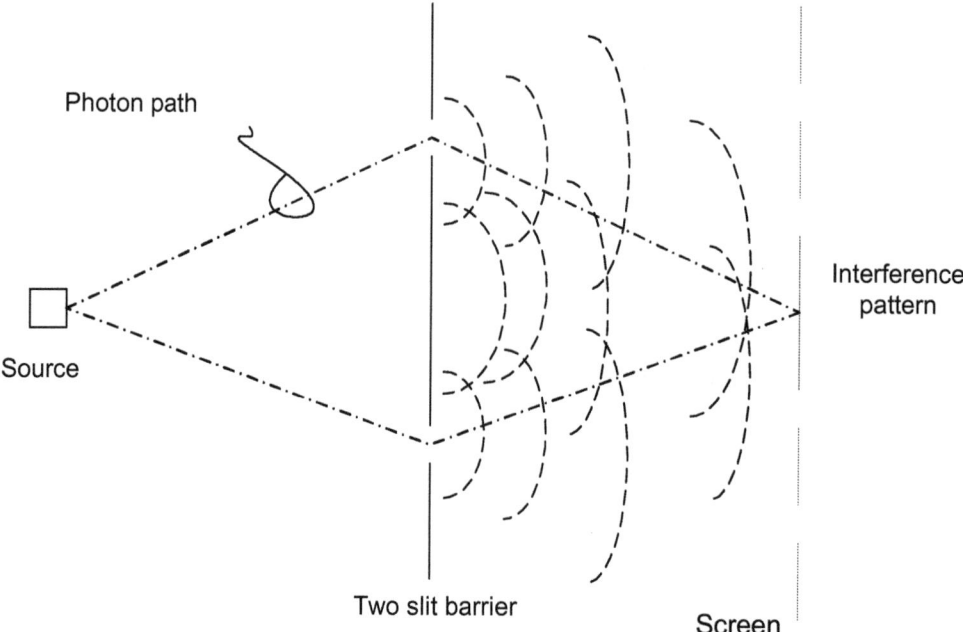

Quantum theory's statistical nature has been a significant area of contention. One way of looking at it is that given the very small areas it operates in, it is not surprising that we cannot predict the location precisely. We are looking at vacuum, photons, and electrons that are 10^{-15} cm with physical probes that are thousands of times bigger – that is why we use mathematics that can look at more detail. It is like trying to study a marble's mass and trajectory after dropping it off the Empire State Building, letting it bounce once, and only then using a magnifying glass to watch it. It is amazing that we have such accuracy as we have achieved.

Another way of looking at it is that at the smallest levels, the particles are jumping around a little and they move position in each 10^{-44} or so seconds (the minimum Planck time). Meanwhile we are up here at the top of thousands of billions of the same actions operating our brains and instruments and only a statistical measurement can be made.

These last two paragraphs are not what the Copenhagen School meant by probabilistic. The Copenhagen school refers to those quantum theorists who believe that the *nature of spacetime itself is inherently probabilistic*. This was the most common interpretation of the experiments into basic physics and reality. Light is considered to be made of particles and waves of *probability*. In contrast, the Einstein-de Broglie School saw light as waves and particles simultaneously. They did not define what waves and particles are; Evans does define them as spacetime. This interpretation is causal as opposed to probabilistic.

Schrodinger's Equation

An equation designed by Erwin Schrodinger gives $\psi(r, t)$ = function of i, ℏ, r, t, and m. $\psi(r, t)$ is the wave function of a particle defined over r space and t time. If squared it becomes a probability. In quantum theory ψ^2 is used to find the probability of an energy, a position, a time, an angular momentum.

Essentially, ψ^2 = probable position = a function of velocity and time. This is shown in Figure 3-3.

Figure 3-3 Probability Distribution

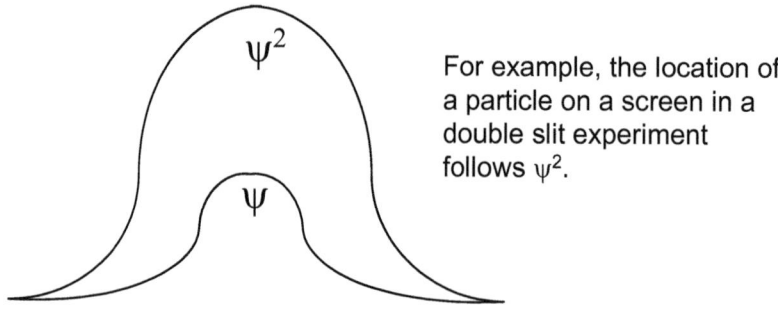

For example, the location of a particle on a screen in a double slit experiment follows ψ^2.

Vector Space

Just as general relativity uses vectors and geometric spaces to envision curvature in four dimensions, so also quantum mechanics uses imaginary vector spaces. The formulation is beyond the scope of this book, but essentially formulas have been found that operate on vectors producing scalars and vectors that exist in vector space. Then those are used to predict position, momentum, etc. with great precision.

In quantum theory, the metric is not used and it is a special relativistic theory. It cannot deal with gravitation and the quantum equations are primarily flat spacetime equations; gravitation has not been connected.

In the calculations that are performed, it is necessary to use imaginary spaces to find the answers. These are math tricks and no physical meaning is given to the spaces. Hilbert space is the vector space of quantum mechanics. Special relativity's spacetime is called Minkowski spacetime. The spacetime of general relativity is Riemannian spacetime. The Evans formulation is in Riemann spacetime, but with the torsion tensor reconnected so we could say he uses Cartan Differential space. In some of his papers he calls it "non-Riemann spacetime."

This author proposed "Evans spacetime" which has been accepted by Professor Evans.

Dirac and Klein-Gordon Equations

Paul Dirac developed a relativistic equation to explain "spin" or angular momentum of the electron. There is some kind of magnetic field and rotation of the electron, a bit like a planet. The electron has spin ½ (magnetic dipole moment) equal to $\hbar/2$. Planck's $h/2\pi = \hbar$ (h-bar). Think of h as a circumference of a circle and $\hbar$ as the radius.

One result of Dirac's equation was the prediction of positive electrons. The positron was eventually discovered. This is an excellent example of a mathematical description of one experimental result leading to increased knowledge and prediction of other experimental results. The *Dirac equation* can be used to obtain information at relativistic speeds. It was not compatible with general relativity, just special relativity.

The Klein-Gordon equation is also a special relativistic equation that up until Evans' work was considered flawed. It was interpreted as a probability and had negative solutions. Since probability must be between zero and positive one, Klein-Gordon was considered incoherent.

The Dirac and Klein-Gordon equations are dealt with in Chapter 9. They can be derived from the Evans Wave Equation of general relativity and Klein-Gordon is no longer interpreted as a probability.

A similar equation is the Proca equation. This replaces the d'Alembert equation when the photon has mass. It states $\Box A_\mu = -(m_0 c/\hbar)^2 A_\mu$ where A_μ is the electromagnetic potential 4-vector. When mass of the photon is zero, the Klein-Gordon, d'Alembert, and Dirac equations can be written $\Box A_\mu = \Box \phi_B = \Box \psi_B = 0$. Here ϕ_B is a scalar field, A_μ is the electromagnetic gauge vector field, and ψ_B is a spinor field. See Chapter 5 for discussion of the d'Alembertian, $\Box$.

We will see new general relativistic formulations of Dirac and Klein-Gordon equations in Chapter 9.

The Quantum Hypothesis

The Planck constant h is usually stated to be the smallest quantum of energy possible. (The reduced Planck constant, also called Dirac's constant, is h/2π and is actually the smallest measurable quantity from within the reference frame where measurement is made.) Planck's constant is called the unit of action. Action is time multiplied by energy; one can think of it as the duration of existence.

A number of physical quantities can be defined in terms of Planck's constant, G the gravitational constant, and c the speed of light. For example, the Planck length = $l_p = (G \hbar / c^3)^{1/2} = 10^{-35}$ m. This is about 10^{-20} times the size of a proton. Its importance is that it may be the smallest discrete distance between points in any reference frame's spacetime.

Figure 3-4   Quantum Postulate

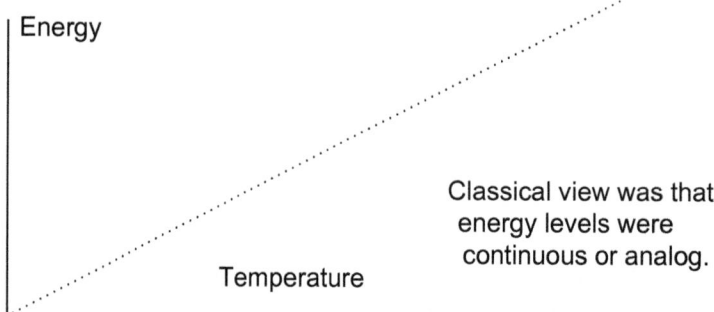

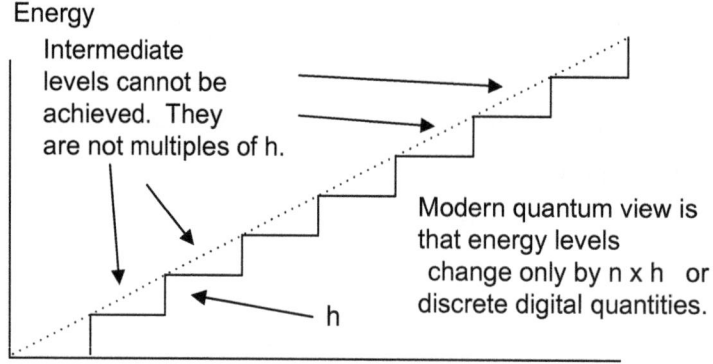

Note the similarity in Figure 3-4 between the concept of discrete multiples of h, one-forms as a topographical map, and eigenvalues. This will become clearer as we progress.

The Planck length has been considered the smallest distance with any meaning. It is the scale at which classical ideas about gravity and spacetime cease to be valid, and even quantum effects may not be clear. The Planck time is the time it would take a photon traveling at c to cross a distance equal to the Planck length. This is a quantum of time, equal to 10^{-43} seconds. No smaller division of time has any meaning in present physics.

Heisenberg Uncertainty

As measurements are made it is found that accuracy in measurement of one quantity can interfere with accuracy of another. These come in pairs of "complementary" relationships. Time of a process and its energy cannot be measured accurately: $\Delta t \times \Delta E$ is about equal to $\hbar$. Nor can the position and the momentum be arbitrarily measured simultaneously: $\Delta x \, \Delta p_x \leq \hbar$.

This expresses the Heisenberg uncertainty relationship. It has recently been called into question by J.R. Croca[14] who gives experimental evidence that Heisenberg uncertainty is not valid in all cases. Evans derives a relation that indicates the uncertainty principle is incorrect. See Chapter 10.

The Heisenberg uncertainty concept has permeated all of quantum mechanics and with the statistical interpretation of the Schrodinger wave equation has dominated thinking since the 1930's. Einstein-Cartan-Evans physics rejects this.

---

[14] Towards a Nonlinear Quantum Physics, J. R. Croca, World Scientific Series in Contemporary Chemical Physics - Vol. 20. Also "Experimental Violation of Heisenberg's Uncertainty Relations by the Scanning Near-Field Optical Microscope," J.R. Croca, A. Rica de Silva and J.S.Ramos, 1996.

Quantum Numbers

The equations below are meant to teach the concepts only and give a feeling for quantum mechanics. Certainly the use of such equations is the realm of the professional. It gives the reader a "feel" for the abstraction of quantum theory.

The location of the electron in a hydrogen atom is described by:

$$\psi(r) = 1/(\pi r^3)^{1/2} \, e^{-r/r_0} \tag{1}$$

where $r_0 = \hbar^2 \varepsilon / \pi m e^2$); r is the radius of location of electron with nucleus at center. $\psi(r)$ is the wave function of position and $|\psi|^2$ is the electron cloud density. Note that the electron can appear anywhere in the universe given that $\psi(r) \to 0$ only as $r \to \infty$. This indicates the vacuum may be composed partially of "potential" electrons from the entire universe. (And potential everything from everywhere.) This concept is questionable; however the equation above works very well in predicting the results of experiments.

n is the principal quantum number and is 1, 2, 3…the integers to infinity. n is simply a number without particular physical meaning.

n is used to find physical descriptions, for example:

Energy level of an electron = $13.6V / n^2$ \hfill (2)

The other quantum numbers are described below in the Quantum Mechanics section.

All of the quantum numbers are represented by vectors that occupy the state space in quantum calculations. This is a mathematical space where calculations are performed, and when the new vectors are found, accurate predictions of new experiments are possible. No one knows why yet.

Chapter 3 – Quantum Theory

Quantum Electrodynamics and Chromodynamics

Quantum electrodynamics evolved from earlier quantum mechanics. It dealt with field theory, electromagnetics and introduced renormalization. Renormalization has been criticized, but it was the best method we had for making corrections to certain calculations. The problem involved considering the particle size to be zero. If zero, then energy density is infinite. Renormalization arbitrarily sets a minimum volume.

The Evans principle of least curvature indicates that there is a calculable minimum volume for every particle. The method of renormalization was necessary since minimum volume was unknown; now a real volume can be applied instead of an arbitrary one.

This is a wonderful example of how general relativity and quantum electrodynamics can work together. The mathematical methods come from quantum theory and the minimum volume is defined in general relativity. Together a greater accuracy and understanding are achieved.

With the discovery of SU(3) symmetry and the hypothesis of the existence of quarks as the basic building blocks of protons and neutrons, *quantum chromodynamics* developed. It added the quality called color to the scheme. Color is some unknown quality of quarks and gluons that causes them to appear in our mathematics in 3 colors. Quarks always come in triplets or doublets of a quark and antiquark. Any quark that is isolated in the vacuum "drags" another out giving up some of its energy to cause another to form.

SU(3) symmetry was observed in the relationships among certain particle energies and the quark model was developed. It would appear that this is only a mathematical model and that quarks may not exist as particles, rather as energy levels.

Quantum Gravity and other theories

The *electroweak theory* unifies the weak nuclear force and the electromagnetic force. The standard model indicates that the weak force and electromagnetism are, at very high energies, the same force. At some even higher energy, it is expected that gravitation will be shown to be part of the same primal force. This is quantum gravity. Given that in a black hole all mass and energy become homogeneous, we might expect that everything that takes various forms in the universe will, at sufficiently high energies – high compression – be manifestations of the same primordial cause. This would have been the situation just before the Big Bang took place, assuming it did.

Since the 1920's physicists have searched for a combination of quantum theory and general relativity that would explain more at the very smallest levels at the highest energy densities. So far this has eluded them. The term Grand Unified Field Theory (GUFT) describes the combined theories.

The Evans equations are equations of GUFT.

Quantum gravity is any of a variety of research areas that have attempted to combine gravitational and quantum phenomena. This usually involves the assumption that gravity is itself a quantum phenomenon.

Up until the Evans equations were developed, it was unknown how the quantum and gravitational theories would be combined. There have been four general concepts: quantize general relativity, general relativitise quantum theory, show that general relativity comes from quantum theory, or find a totally new theory that gives both quantum theory and general relativity in the appropriate limits.

This last method has been the most extensive. String theory, super symmetry, super gravity, superstring theory, loop quantum gravity, and M theory have been attempts to find basic mathematical formulations leading to gravity and quantum mechanics. While they have all found some interesting mathematical formulations, it does not seem that they are physical enough to lead to either general relativity or quantum mechanics. So far they have produced no results and appear to be superfluous to a unified theory.

The Evans equations show that the 3-dimensional quantum description emanates from Einstein's general relativity. More than that, the four basic forces in nature are combined and shown to come from Einstein's basic postulate of general relativity, R = -kT. Only four dimensions are needed. String theory's nine, 10, 11, or 26 dimensional approaches are mathematical, but not physics.

The Evans Wave Equation mathematically combines the two theories rigorously. How much new physics will come out that completely gives a GUFT remains to be seen. Already Evans has found a number of significant explanations for processes.

In the next chapter we discuss some of the mathematical language that is needed to understand Evans in the context of quantum theory and general relativity.

While the reader is not expected to work the problems or do any math, it is necessary to be familiar enough to read about them.

Planck's Constant

Planck's quantum hypothesis, the foundation of quantum theory is

$$E = nh\nu \text{ or } E = nhf \qquad (3)$$

$$E = n\hbar\omega \text{ since } \omega = 2\pi\nu \qquad (4)$$

where E is energy, n is the quantum number, h is Planck's constant, ν and f are frequency, $\hbar$ is the reduced Planck's constant $h/2\pi$, and ω is angular frequency. The Planck constant is the minimum amount of action or angular momentum in the universe. Einstein used h to explain the photoelectric effect. h is the quantum of action which is energy x time. $h = 6.625 \times 10^{-34}$ joule-seconds. A joule is measured in electron volts or N-m. A joule is a watt-second making h = watt x second x second. Action can have units of torque times seconds, energy x time, or momentum x distance, or angular momentum x angle. It is measured in joule seconds or N-m-sec.

Evans Equations of Unified Field Theory

The significance of Planck's constant is that radiation is emitted, transmitted, or absorbed in discrete energy packets – quanta. The energy E of each quantum equals Planck's constant h times the radiation frequency. $E = hf$ or $h\nu$ (Greek nu). The Dirac constant, $\hbar$ (h bar) is frequently used; $\hbar$ equals $h/2\pi$. ω, lower case omega, the frequency of orbital motion is used frequently and $1/2\pi$ provides a conversion. $\hbar$ is more fundamental than h itself.

E is energy, sometimes given as En. n (any integer) is the principle quantum number. f is frequency in cycles per second. λ is wave length in meters, p is momentum in kilogram-meters/second, γ is gamma $= 1/\sqrt{[1-(v/c)^2]}$, the Lorentz-Fitzgerald contraction factor based on the ratio of v to c, v is velocity, c the speed of light, , and ω is angular velocity. See Figure 3-5.

Figure 3-5   Pythagorean Relationships

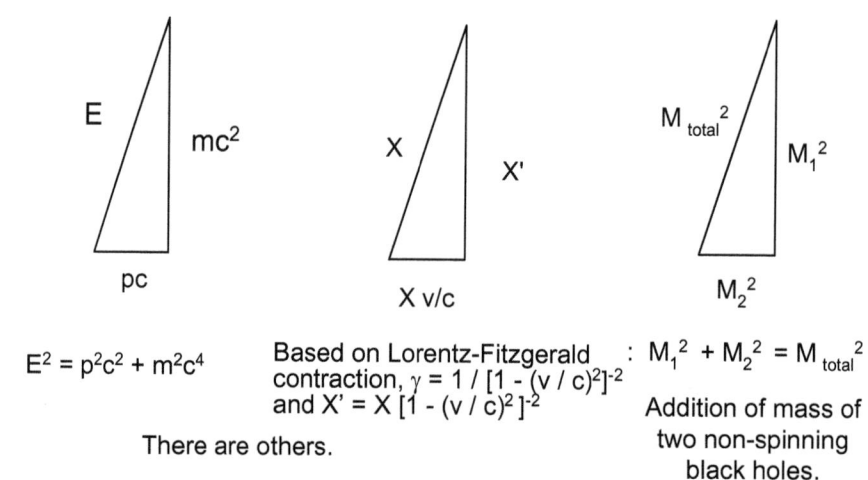

Generalizing, we see that the square of the total energy, mass, or distance in spacetime is the sum of the components squared. We can see an origin of distance in spacetime relating to velocity in pc and X v/c.

Some of the most common uses of h and equations that relate to it follow.

$$E = nhf \qquad (5)$$

Sometimes stated as $E = hf$ with n understood. The energy of a structure (atom, electron, photon) is equal to a number n (n = 1, 2, an integer) times h times the frequency of the wave or particle. *This is the basic quantum hypothesis.*

$$f\lambda = v \tag{6}$$

f is the frequency and with λ the wavelength, we define the frequency times the length of the wave to equal the velocity. This is true for any wave from one on a string or in the ocean to an electron or a probability wave. For the photon, $f\lambda = c$, where c is the speed of light. That is, the frequency (number of cycles) times the length of the cycle = the speed, which for electromagnetic waves is c.

$$E = pc \tag{7}$$

Energy equals the momentum times c. This is the energy in a photon. It may have no mass during transit, but it has momentum. It has mass energy when captured by an atom. Given $E = pc$ and $E = hf$, then $p = h/\lambda$.

$$\mathbf{p} = mv \tag{8}$$

Newton's definition of momentum for an object or particle.

$$\mathbf{p} = \gamma mv \tag{9}$$

This is Einstein's relativistic definition of momentum. At low velocities, it reduces to Newton's equation, $\mathbf{p} = mv$.

$$E = mc^2 \tag{10}$$

Einstein's energy mass relationship. Mass and energy are, under the right circumstances, interchangeable. $E = m$ can be stated with c^2 a conversion factor into joules. See equation (14).

A number of relationships that exist for h can be derived. Depending on the need, a different arrangement can be substituted into any equation.

$$\lambda = h/p = h/\gamma mv = h/\gamma mc = hc/\gamma mc^2 \tag{11}$$

Here the wavelength of a photon is $h/\gamma mc$ but the wavelength of a particle is $h/\gamma mv$. The only difference is that the velocity of a particle can be slower than c.

$$\mathbf{p} = mv = E/c = hf/c = h/\lambda = mc = (E/c^2)c \tag{12}$$

$E = pc$ and $E = \hbar\omega$ and therefore $pc = \hbar\omega$ and

$$p = \hbar\omega/c = \hbar(2\pi/\lambda) = \hbar k \tag{13}$$

where k is the wave number. The momentum of a particle increases with velocity or mass. The momentum of a photon increases only with increase in frequency since its velocity is fixed. We can then equate mass with frequency.

A very significant relationship, the de Broglie guiding theorem is:

$$E = h\nu = m_0 c^2 \qquad (14)$$

Here equations (5) Planck and (10) Einstein are set equal to each other. We see that the energy of a particle is mass; and that mass is frequency. The particle and the wave are equivalent. In double slit experiments we see that these convert seemingly instantaneously (or maybe in one Planck time) from one to the other. More likely, although not absolutely clear, is that the particle is a standing wave of spacetime. The Evans equations indicate that the electromagnetic wave is spacetime spinning; the next step is to realize that the particle is spacetime also.

In equation (13) we stated $p = \hbar\omega / c$. Thus the photon has momentum. In the standard theory the photon has zero mass, but seemingly contradictory, we know it has momentum. Evans' $\mathbf{B}^{(3)}$ equations imply it indeed has mass and the equations above are to be taken literally.

In special relativity the definition of 4-momentum is $p_\mu = m_0 c v_\mu$ where v_μ is 4-velocity. This is meaningless if the photon has no mass.

Quantum states can have momentum defined when they are in motion. The energy of the state of a moving particle is: $E^2 = p^2 c^2 + m^2 c^4$, again a Pythagorean relationship pointing at the geometric nature of the physics. The momentum, p, can be extracted and the wave frequency of the particle found by f = pc/h. With p = 0 the particle is at rest and the equation becomes $E = mc^2$ or simply E = m, energy equals mass.

Note that energy is not mass, rather they can be converted into one another. They are different aspects of the same thing.

Chapter 3 – Quantum Theory

Geometricized and Planck units

Geometric units convert physical quantities into lengths, areas, curvature, and dimensionless ratios. This simplifies formulas and gives another way to look at abstract physics. Geometric units set c and G = 1. Since these are scaling factors, it simplifies the equations. In addition the Coulomb force and Boltzman constant are set to one; we will not deal with these. Planck units set $\hbar$ = 1 also. Real calculations use SI units.

Time is expressed as the distance light travels in the chosen interval. They are on an equal basis which is consistent with relativity.

Geometricized units set values in meters of light travel time. $c = 2.998 \times 10^8$ meters per second. A second can then be expressed as 2.998×10^8 meters. Given that time is x_0 and is a spatial dimension in general relativity, this makes perfect sense.

For mass we would have $G/c^2 = 7.425 \times 10^{-28}$ meters / kilogram. The sun would be 1.5 kilometers when expressed as a distance. Now just what that means is a matter for thought.

In geometric units:

Time:	$c = 3 \times 10^8$ meters / second
Mass:	$G/c^2 = 7.4 \times 10^{-28}$ meters / kilogram
Temperature	$G k/c^4 = 1.1 \times 10^{-67}$ meters / degree Kelvin
Energy	$G/c^4 = 8.3 \times 10^{-45}$ meters / Newton

In terms of the Planck length, the commonly used terms are:

Length:	$L_p = (\hbar G / c^3)^{1/2} = 1.6 \times 10^{-35}$ meters
Time:	$T_p = (\hbar G / c^5)^{1/2} = 5.4 \times 10^{-44}$ seconds
Mass:	$M_p = (\hbar c / G)^{1/2} = 2.18 \times 10^{-8}$ kg
Temperature:	$T_p = (c^5 h / G)^{1/2} = 1.4 \times 10^{32}$ Kelvin
Energy:	$E_p = M_p c^2 = 10^{18}$ GeV

When the Planck units are used, it is assumed that the Planck length is the smallest distance at which general relativity is meaningful and that quantum mechanics takes over at smaller distances. Since this is roughly 10^{-20} smaller than the size of the proton, we believe spacetime is very finely structured. By combining relativity and quantum theories, Evans' work indicates that we can start to look at these distances.

For more detail see http://en.wikipedia.org/wiki/Planck_units.

Notice that in discussing this subject and others like reference frames, we cannot directly look at them. Even when we get some test black holes and spaceships that travel at near light speeds, we will need some way to look at both high and low energy density reference frames simultaneously from outside them both. To get this Godlike viewpoint, we need mathematics.

Quantum Mechanics

Quantum mechanics, electrodynamics, and chromodynamics have developed very precise mathematical methods for particle and energy physics. Interestingly, the reasons behind the methods are not all known. A great deal of discussion over the years has not clarified the "why" beneath the mathematics. The "how" is well developed. Those "why" reasons are the classical descriptions that we humans wish to express.

There are five quantum numbers that describe a particle's state.

"n" is the principle quantum number. $E = nhf$ is the basic quantum formula. Energy always comes packaged in quanta that are the product of n, Planck's number, and the frequency of a wave or particle. n is any positive integer – the steps which the quanta must take.

Figure 3-6   Principle Quantum Number

n = 2   n = 4   n = 2.5 not allowed

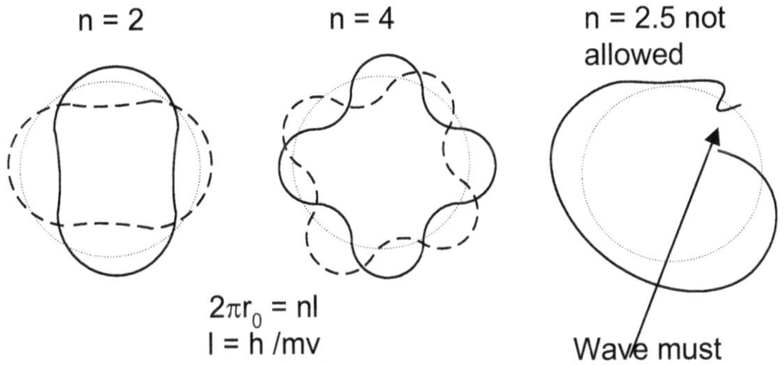

$2\pi r_0 = nl$
$l = h/mv$
$mvr_0 = nh/2\pi$
The number of wavelengths is also the principle quantum number.

Wave must reconnect with itself.

The allowable "orbits" of electrons in an atom are functions of n. (The most likely location of an electron is in the region indicated by n; the electron is probably a standing wave spread out over a large region around the atom. When it moves through space it is like a particle in as much as it has a somewhat discrete region in which it exists. The formula

$$E_n = 13.6 \text{ eV} / n^2 \tag{15}$$

predicts the energy level of an electron. As a wave, the electron can only take those paths where the wave is complete and reconnects to itself. A spherical standing wave must close on itself with a whole number contained in the wave. See Figure 3-6.

"l" is the orbital quantum number. This is related to the angular momentum. It can have values from 0 to n-1.

M_i is the magnetic quantum number. It can have values from –l to +l. It indicates the direction of the angular momentum.

L is the magnitude of the orbital angular momentum.

$$L = \sqrt{(l(l+1)}\,\hbar) \tag{16}$$

where l is the orbital quantum number. L is space quantization and is typically related to the z-axis. It is the orientation of the angular momentum.

$$L_z = m_l\,\hbar \tag{17}$$

m_s is spin quantum number. It is a type of angular momentum that is not clearly defined in any way that we can describe classically. It is not a measure of spinning motion; more it is a turning or torque. $m_s = +1/2$ or $-1/2$. This is sometimes said to be spin up or spin down. Spin angular momentum S is given by:

$$S = m_s (m_s + 1)^{\hbar/2} \tag{18}$$

The various vectors that describe the state of a particle are placed in Hilbert space. This infinite capacity mathematical space can be manipulated to find new results to explain experiments or predict the outcome.

Summary

Quantum theory is a mathematical description of physics which has had great success in predicting the results of reactions. It operates in poorly understood vector spaces and is open to physical interpretations. A lot of arguing has gone on for 70 years.

Quantum theory is a theory of special relativity. It can deal with the high velocities of particle interactions, but cannot deal with gravitation. This drawback is serious for two reasons:

1. One of our goals is to understand the origin of the universe when gravity was extreme.
2. Particles have high density and therefore gravitation must be high in the local region. While still up in the air as of the time of this writing, it seems that the particle will prove to be a region of spacetime with both high gravitational curvature and torsion producing frequency.

Unified field theory will combine quantum and gravitational theories and should give us equations explaining more of these phenomena.

Chapter 4 Geometry

> "One can simplify physics to a certain level, but after that it loses precision. Similarly one can express a poem in words, but you lose all metre, rhyme and metaphor.
>
> Similarly one can keep asking questions like what is charge etc., but at some stage a kind of blotting paper process takes place where one begins to understand in more depth and realizes what questions to ask. Without mathematics, this understanding will always be empirical. Great instinctive empiricists like Faraday could get away with it, but no ordinary mortal."
>
> Myron Evans, 2004

Introduction

Physics is geometry. Geometry does not simply describe physics; rather one cannot separate them. In the Evans equations and as Einstein believed, all physics is based on geometry.

The mathematics in this and the next chapter will be too difficult for many readers. If so, read it anyway. Skip the impossible parts for now. Then come back and use it and the Glossary when necessary.

Mathematics is a language and to a certain extent, learning to *read* it is possible without knowing *how to do* it. There are three ways to look at geometry – mathematically, verbally, and visually. Concentrate on the latter two as necessary. Information here is meant to try to give a physical explanation to the math. Nevertheless, physics needs mathematics to model the results of experiments and to describe events. One cannot avoid mathematics entirely. Learning to read it and associate a picture with it will give it meaning.[15]

We take a formula, J = r x p. Does this have any physical meaning? No. It is just a formula. But what if we give the letters meaning? One ties a stick to a string and swirls it around. Let **J** be the angular momentum, a vector describing the tendency of the stick to move in a circular direction. (Bold letters indicate that the value is a vector, that it has direction.) Let r be the radius – the length of the string from the center of rotation. And let **p** be the stick's mass times its velocity – the momentum, **p** = mv. Now we have meaning:

$$\mathbf{J} = \mathbf{r} \times \mathbf{p} \text{ or } \mathbf{J} = \mathbf{r} \, m \, \mathbf{v} \tag{1}$$

Now the equation has physical meaning. We have given a definition to each letter and – more importantly – the equation predicts real results. The equation is correct. After all, some equations are wrong.

Eigenvalues

While beyond the study of this book, we will see that eigen equations, eigen values, "eigen...." are frequently mentioned in Evans' work. Eigen in this context indicates that the result has real physical meaning. "Eigen" is German for "proper."

For example, the orbits of the electron in an atom must take on only integer values according to E = nhf, Planck's quantum hypothesis. While an oversimplification, we can say that values of E obeying this formula are eigenvalues. Any energy that cannot result from the equation is not real and cannot have a physical reality.

---

[15] For a wealth of material on mathematics and physics there are any number of web sites. Among them is www.mathworld.wolfram.com. (Accessed July, 2007). It has valuable cross links within it from mathematics to physics.

Chapter 4 – Geometry

Riemann Curved Spacetime

Riemann geometry was developed to allow geometric calculations in curved spaces of any number of dimensions. The drawing in Figure 4-1 represents a 2-dimensional surface similar to a sphere inside a 3-dimensional space. When dealing with 4-dimensional spacetime, Riemann geometry has the methods necessary to find distances, geodesic curves, directions, areas and volumes. Einstein used Riemann geometry in his theory of relativity as the foundation for the geometrical explanation of physics.

One assumption of general relativity is that the spacetime of our universe is everywhere fully differentiable. There are everywhere points in the spacetime that allow complete continuous identification of a curved line, area, or volume. The quantization of spacetime would imply that spacetime is everywhere differentiable down to a very small distance, at least to the level of the Planck distance.

The Riemann space is a 4-dimensional non-linear spacetime. In any coordinate system, *vectors* can be defined at every point. The vectors exist in orthogonal tangent space, which is a mathematical space that holds the vectors. Those vectors define the curvature and allow us to evaluate gravitational and electromagnetic fields as the space changes. The vector space is orthogonal to real space, which is to say it would be perpendicular if there were no curvature. See Figure 4-1.

There are vectors at every point giving the compressive stress due to mass or energy, direction of forces and fields, gradient of change of these forces, and rotation of spacetime in the region.

There are numerical scalars at every point. Mass is a scalar. It does not have a direction like a vector.

The collection of all these vectors and scalar numbers is called the *tangent bundle* in relativity. In quantum theory the collection of vectors describing events is envisioned to be in *gauge space*. Evans indicates that these are the same space; this is a mathematical unification of relativity and quantum theory.

Figure 4-1   Curved Space

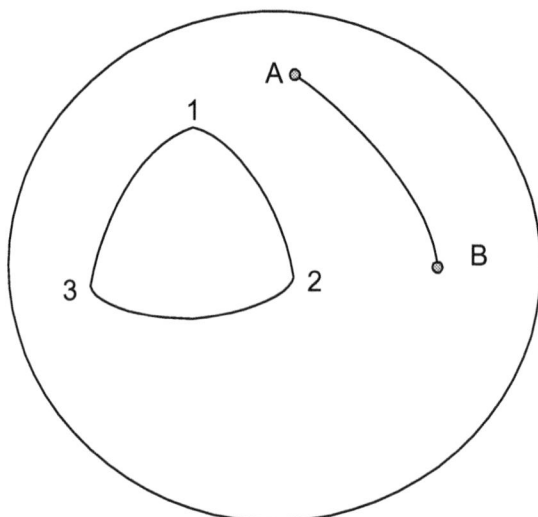

Space is curved either by mass or energy. In a curved space the shortest distance between two points is a curve because there are no straight lines.

The shortest distance between points A and B is shown.

It is a geodesic. An airplane traveling a long distance on the Earth follows such a line.

The triangle 1-2-3 cannot be drawn with straight lines. The line must follow the curve of the space in which it is placed. The triangle will not have angles totalling to 180 degrees in a curved space.

Orthogonal vectors

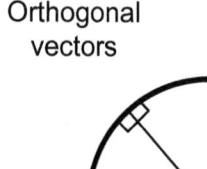

Perpendiculars to a curve are referred to as "orthogonal" in multiple dimensions.

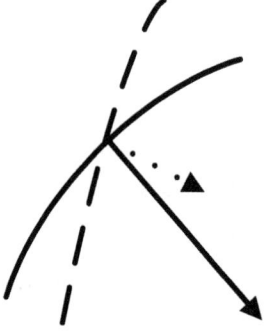

Where a curved surface is in three dimensons, two orthogonal vectors are defined at each point.

Chapter 4 – Geometry

Torsion

Torsion is the twisting of a manifold – any spacetime like our universe. It is sometimes called the second type of curvature. Riemann is the first curvature and is along the distance of a manifold. Torsion is curvature of the manifold as it turns upon itself.

Derivatives

Differential geometry is the study of derivatives. See Figure 4-2. The derivative of an equation for a curve gives the slope or the rate of change at a point. The whole curve may be wandering around, but at any one point the rate of change is specific. The basic concept for change is given by dx/dt.

That is, the change in x, say the number of kilometers driven, with respect to the change in time, say one hour. 100 kilometers an hour = dx/dt = 100 km/hr. While the formulas can look complicated, the basic idea is always simple.

Figure 4-2 Slope as a derivative

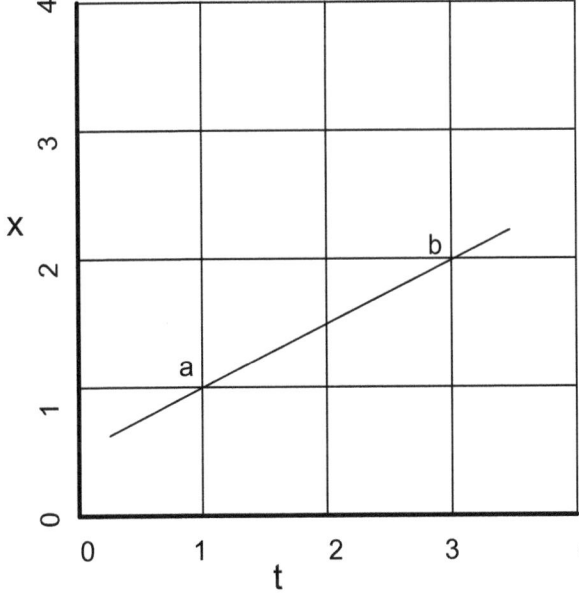

$dx/dt = ½$

a and b are used to get real numbers.
x = 2-1 = 1
t = 3-1 = 2
dx/dt = 1/2

81

If a curve is described by $y = a + bx + cx^2$, then the first derivative gives the slope. Acceleration, or the rate of change of the change is the second derivative. If one is accelerating, then the speed is increasing at a certain rate. This would be indicated by dx/dt^2. That is the change per second per second. The slope of the acceleration curve at any point is the speed.

A car may accelerate from 0 to 100 in 10 seconds. However the rate of increase is not constant. Differential calculus allows a more precise evaluation of the change at any moment. We may detail it knowing, for example, that the rate of change is lower during the first 4 seconds and faster the last 6 seconds.

> The existence of a metric implies a certain connection between points on the space. The curvature may be thought of as that of the metric. When going from one space to another, there are mathematical connections between them. The metric is a map. The map will warp, twist, shrink in one dimension, and expand in another. The mathematics allows us to calculate the changes properly.

∂ The Partial Differential

"∂" is "partial," the partial differential symbol. When there are two or more variables that are unknown, one can perform operations on them by changing one variable at a time. Partial differentials are used where there are several unknowns that change together.

This is one of a zillion possible partial derivatives:

$$\partial^\mu = \frac{1}{c}\frac{\partial}{\partial t} - \frac{\partial}{\partial X} - \frac{\partial}{\partial Y} - \frac{\partial}{\partial Z} \tag{2}$$

It results in one number, ∂^μ, after some calculation. It could be the slope of one of the vectors in Figure 4-3.

Chapter 4 – Geometry

Figure 4-3   Spacetime Manifold

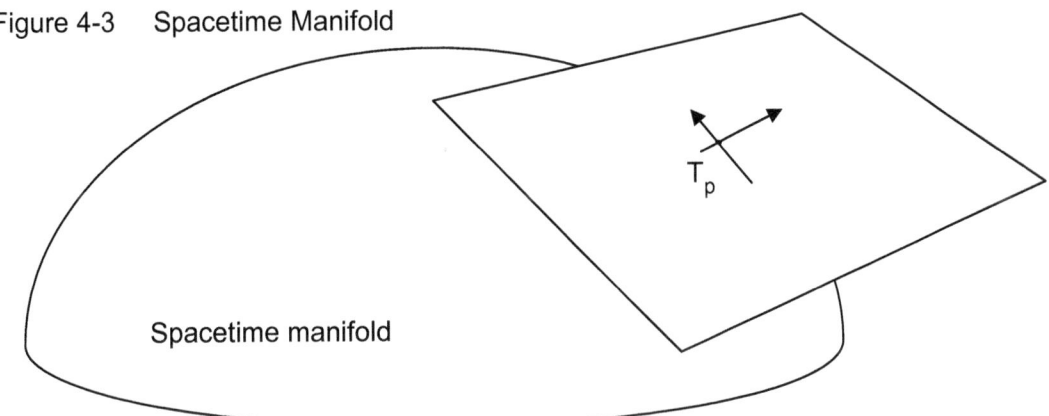

At every point in space, there is a vector space, T_p

Figure 4-4   Dot Product

The dot or scalar product is a scalar, just a distance without direction. It can be used to get work or other values also.

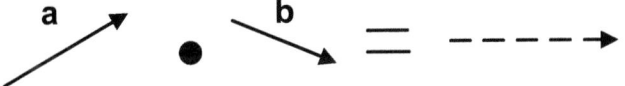

Vectors

Most vectors are arrows. They have direction and a numerical value. A force acts like a vector and is represented by a vector. If one pushes against an object, the force has a magnitude and a direction. **a** • **b** is the dot product. Given the vectors in Figure 4-4, this is defined as the scalar number $|a||b|\cos\theta$. $|a|$ indicates the absolute value of **a**. An absolute value is always positive.

83

Scalar, or Inner Product

The *dot product* is only defined in three dimensions. In four dimensions it is called the *scalar product or inner product*. The vectors in general relativity are 4-dimensional. This is indicated by say, q^μ, where the Greek letter mu indicates four dimensions and can adopt letters or numbers such as 0, 1, 2, 3. If a Latin letter is used for the index, then 3 dimensions or parameters are indicated or the four orthonormal[16] indices are implied – ct, x, y, z. In four dimensions the term *inner product* is used. (Two four-vectors define a four dimensional space and the product of two is *in* the space.)

If the vectors are perpendicular or orthogonal to one another, the dot or inner product is zero. The projection does not extend along the second vector because between perpendicular lines their mutual projection on each other is zero.

The dot product is commutative: **x** • **y** = **y** • **x**. It is associative: a (**x** • **y**) = a **y** • **x**. It is distributive: a • (**x** + **y**) = a • **y** + a • **x**. Here x and y are vectors and "a" is a constant. The dot product is not defined for three or more vectors and the inner product is used. The dot product is invariant under rotations – turning of the reference frame. The dot is not typically shown; it is understood. If an inner product is defined for every point of a space on a tangent space, all the inner products are called the Riemann metric.

In components, one could define the four-vectors as **a** = a_1, a_2, a_3, a_4 and **b** = b_1, b_2, b_3, b_4 and the dot product = $a_1 b_1 + a_2 b_2 + a_3 b_3 + a_4 b_4$. The component calculations are usually not shown. This is a four-dimensional Pythagorean distance in a space where the metric is not applied since it does not refer to our spacetime.

---

16 Orthonormal means orthogonal (perpendicular as far as the curved space is concerned) and normalized with respect to the basis vectors (the vectors are expressed as multiples of a unit vector e).

Chapter 4 – Geometry

> **a • b** means $|a||b|\cos\theta$ and is the two-dimensional version dot product. $a^\mu b_\mu$ is the dot product of two 4-vectors; this is a scalar.
>
> The inner product of two tetrads is: $q_{\mu\nu} = q^a{}_\mu q^b{}_\nu \, \eta_{ab}$. This is a tensor, the symmetric metric, $g_{\mu\nu}$, which is the distance between two points or events in 4-dimensional Riemann space – the space of our universe.

Cross or Vector Product

The cross product is also called the vector product. This is a different kind of multiplication. The result of **a x b** is a vector. It can determine an area in two dimensions. Torque calculation is another example. The torque delivered by a ratchet wrench is perpendicular to the handle and the force pushing on it. If three vectors are multiplied, the result is a volume in three dimensions.[17] See Figure 4-5.

When one sees cross products in 3-vectors, assume that a vector perpendicular to the three dimensions is established if the unit vector "e" is used. Without the unit vector considered, the cross product of two-dimensional vectors gives a parallelogram and if 3-vectors are used a parallelopiped is defined. See Figure 4-6.

Magnetic forces and the torque of spinning and rotating objects can be described by cross products.

---

[17] www.physics.syr.edu/courses/java-suite/crosspro.html has a nice applet that allows one to get a real feel for the cross product. (Accessed July, 2007)

Figure 4-5 Cross Product

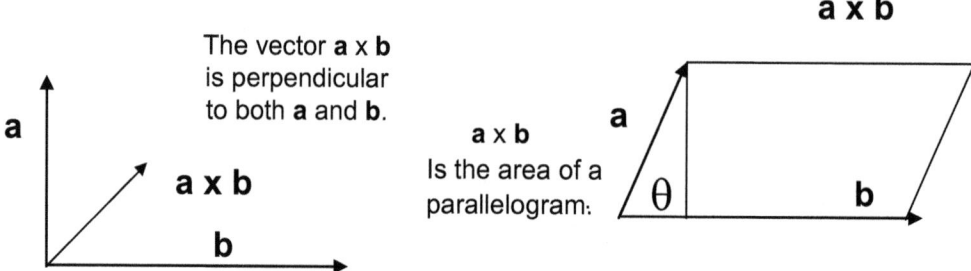

Figure 4-6 Product of three vectors

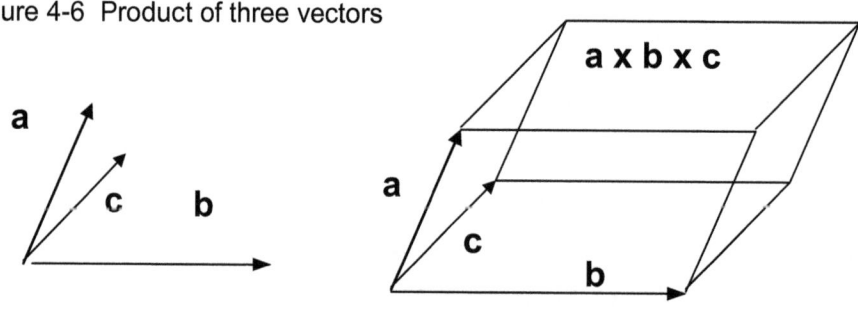

Curl

The curl of a vector tells if it is rotating. See Figure 4-7. See also The Gradient Vector and Directional Derivative, ∇ further in this chapter

Figure 4-7   Curl

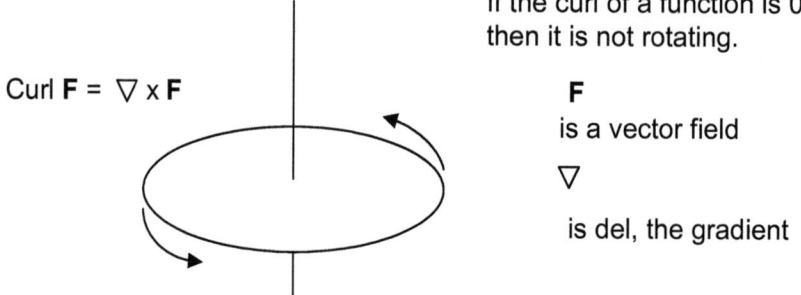

If the curl of a function is 0, then it is not rotating.

Curl **F** = $\nabla \times$ **F**

F is a vector field

∇ is del, the gradient

Chapter 4 – Geometry

Divergence

The divergence of a vector tells the amount of flow through a point. See Figure 4-8.

Figure 4-8   Divergence

div **F** = ○───◄

Divergence measures the rate of change of flow through a point.

div **F** is a scalar field

Vectors are used to find curvature, directions, and magnitude properties of the physical space itself and of particles in general relativity.

Wedge Product

The point of the preceding figures is to make the wedge product a bit more real to the reader. We have seen some types of vector products and the physical meaning that can be applied to them.

The wedge product is also a vector product. It is the four dimensional version of the cross product and can be used for vectors, tensors, or tetrads.

The wedge product can be used to compute volumes, determinants, and areas for the Riemann metric. If two tensors, vectors or matrices are labeled A and B, then the wedge product is defined as:

$$A \wedge B = [A, B] \tag{3}$$

Where $A^a{}_\mu \wedge B^b{}_\nu = A^a{}_\mu B^b{}_\nu - A^a{}_\nu B^b{}_\mu$. When one sees the $\wedge$ (wedge symbol), think of the egg crate shapes of spacetime in Figure 4-9. The wedge product produces surfaces that cut vectors or waves. The result is a egg-crate type of shape also shown in Figure 4-10.

Figure 4-9   Wedge Product

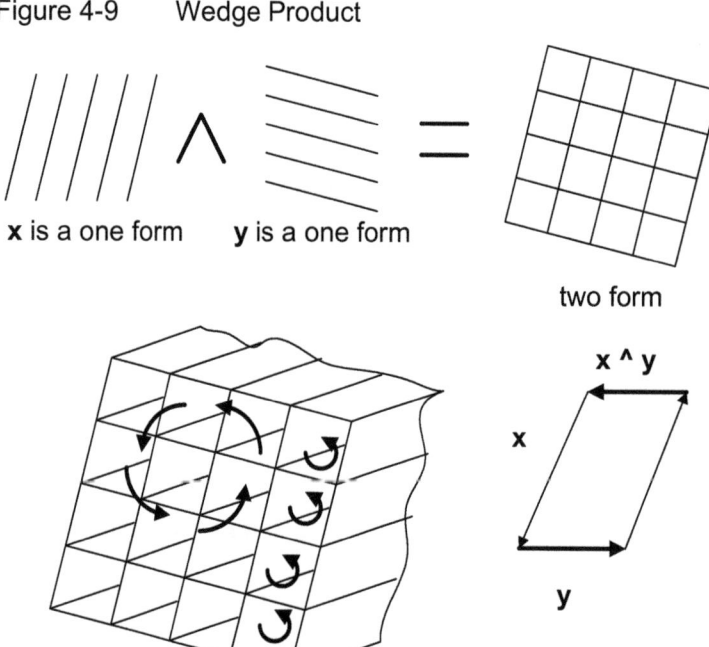

The wedge or exterior product can be used in several ways. It gives the circulation or the perimeter of regions.

The wedge product is a two form. The vectors **x** and **y** are one forms.

Figure 4-10   Wedge Product extended

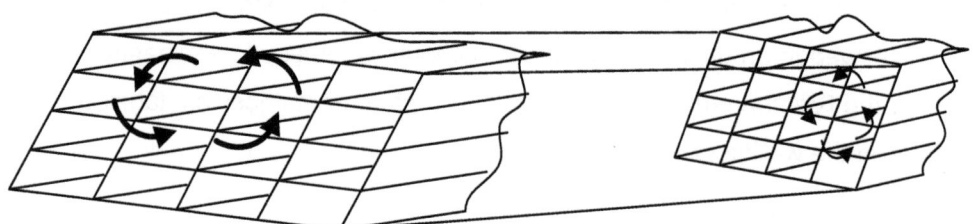

The wedge product is an antisymmetric operation called the exterior derivative performed on differential forms: $dx_i \wedge dx_j = -dx_j \wedge dx_i$. The lines in the figures also define electromagnetic lines of force in physics.

Outer, Tensor, or Exterior Product

The outer product is defined for any number of dimensions. It produces a matrix from row and column vectors. One of the confusions is that mathematicians and physicists have slightly different terminology for the same things. The exterior algebra was invented by Cartan.

The exterior wedge product allows higher dimensional operations.

For example, for four dimensions, the 1-dimensional subspace of functions called 0-forms, and the four dimensional spacetime of 1-forms, can be multiplied to construct other dimensional spaces. The algebra here is closed – the results are all in a subspace of their own. This produces a topological space that is a manifold. It is a subspace to the next higher dimensional subspace.

The gradient operator is a similar function that can construct a gradient vector field.

Scalars, vectors, and tensors exist in these exterior algebra subspaces. The exterior derivative is a rule of differentiation that transports an element of a lower dimensional exterior algebra subspace to the next higher dimensional subspace. The primitive example is the use of the gradient operator acting on a function to construct a gradient vector field.

$$A \wedge B = C \quad (4)$$

where C is also called a bivector.

Below, the column vector subscripts are μ, the base manifold. The row vector subscripts are a, the Euclidean spacetime.

Outer Product

$$\begin{bmatrix} q_0 \\ q_1 \\ q_2 \\ q_3 \end{bmatrix} \begin{bmatrix} q_0 & q_1 & q_2 & q_3 \end{bmatrix} = \begin{bmatrix} q_0 q_0 & q_0 q_1 & q_0 q_2 & q_0 q_3 \\ q_1 q_0 & q_1 q_1 & q_1 q_2 & q_1 q_3 \\ q_2 q_0 & q_2 q_1 & q_2 q_2 & q_2 q_3 \\ q_3 q_0 & q_3 q_1 & q_3 q_2 & q_3 q_3 \end{bmatrix}$$

(with μ labeling rows and a labeling columns on the left factor)

4-Vectors and the Scalar Product

Any two 4-vectors can form a Lorentz invariant quantity. This is a scalar product.

For example, $p^\mu = (En/c, \mathbf{p})$ indicates the 4-momentum of a particle where **p** is a vector. The four components would be multiplied as the outer product above and then the values added to find the momentum. In three + 1 dimensions, the momentum is defined as $\mathbf{p} = m\mathbf{v}$ or in special relativity as $\mathbf{p} = \gamma m\mathbf{v}$. In general relativity it is a bit more detailed.

$v^\mu = dx^\mu / d\tau$ where $d\tau$ indicates the proper time of a particle. This is the 4-velocity. The proper time is the time as measured in the particle's reference frame, not that at rest or at some other velocity.

In general, coordinates and momentum have their indices up and derivatives have their indices down. This takes a while to relate to and confuses most of us at sometime.

Tensors

> Do not let the details make you loose sight of the purpose of these equations. Keep in mind that they all calculate curvature, density, or directions of spacetime. One can read the equations without knowing how to operate them.

Tensors are equations or "mathematical machines" for calculations in Riemann metric geometry. For the most part they take the form "Tensor of x and y = xy – yx" or $ds^2 = dx^2 - dt^2$. An invariant distance, mass, scalar is found using a Pythagorean type of relationship.

All the vector operations above can be performed for tensors.

Tensors manipulate vectors, 1-forms, and scalars.

A simple example is g, which is the metric tensor:

$$g_{\mu\nu} = (\mathbf{u},\mathbf{v}) \qquad (5)$$

That is, g = a function of (4-vector **u** and 4-vector **v**). It is understood that the product has the metric, $\eta_{\mu\nu} = (-1, 1, 1, 1)$ applied. It gives the distance between two events in components. An event is a point with time and distance considered together; that is, it is a 4-vector. $\eta_{\mu\nu}$ converts the vectors to metric vectors in real spacetime. This gives us the components. This is the 4-dimensional Pythagorean theorem with a negative time dimension which defines distances between events in spacetime.

In a triangle $c^2 = a^2 + b^2$ where c is the hypotenuse. This Pythagorean relationship is true for many basic physical processes. Combine the masses of two non-spinning stationary black holes: $M_1^2 + M_2^2 = M_{total}^2$. The total is the Pythagorean sum. Mass can be added like a distance. $E^2 = m^2 + p^2$ is another simple but powerful relationship. Energy equals the Pythagorean sum of mass plus momentum. Rather than distance, we can think in abstract geometrical terms in general relativity.

This indicates the basic geometric relationships of our physical world and the equivalence between distance and mass-energy relationships.

We can generalize the Pythagorean theorem to any number of dimensions; in physics we need four. We refer to the space we are in and the distances between points – or events – as "the metric." The 4-dimensional distance is found from $ds^2 = dx^2 + dy^2 + dz^2 - dt^2$. The answer must be positive definite – it is a real distance. In some cases we make the dt positive and the distance negative to ensure that the answer is positive.

$g_{\mu\nu}$ in equation (5) is the metric tensor. g is a type of 4-dimensional Pythagorean relationship, but it is invariant when the reference frame changes dimensions – time, velocity, compression in size or change in shape due to gravitational effects or the presence of energy. The squared length between two events – points where time is included – is called the separation vector.

$\eta_{\mu\nu}$ is defined as a matrix that is valid in any Lorentz reference frame:

$$\begin{vmatrix} -1 & 0 & 0 & 0 \\ 0 & 1 & 0 & 0 \\ 0 & 0 & 1 & 0 \\ 0 & 0 & 0 & 1 \end{vmatrix}$$

$$ds^2 = \eta_{\mu\nu} dx^\mu dx^\nu = -dt^2 + dx^2 + dy^2 + dz^2 \qquad 6)$$

Sometimes the – and + are reversed. This is the inner product. Ignoring the mathematic rigor, it is a version of the 4-dimensional Pythagorean theorem and gives us a distance in the 4-dimensional spacetime. Time is typically a negative in this metric and could be written $(ct)^2$.

See particle collisions in glossary for another example of Pythagorean invariance.

Basis vectors are like common denominators. Assume two vectors have respective lengths of 2.5 and 4.5. The basis vector **e** = .5. and the 2.5 vector can be expressed as 5**e** and the 4.5 vector as 9**e**. When the vectors are moved to new highly scrunched reference frames, **e** may change to say .03. but the ratio of 2.5 and 4.5 is maintained. See Figure 4-11.

Chapter 4 – Geometry

Figure 4-11   Tensors and basis vectors

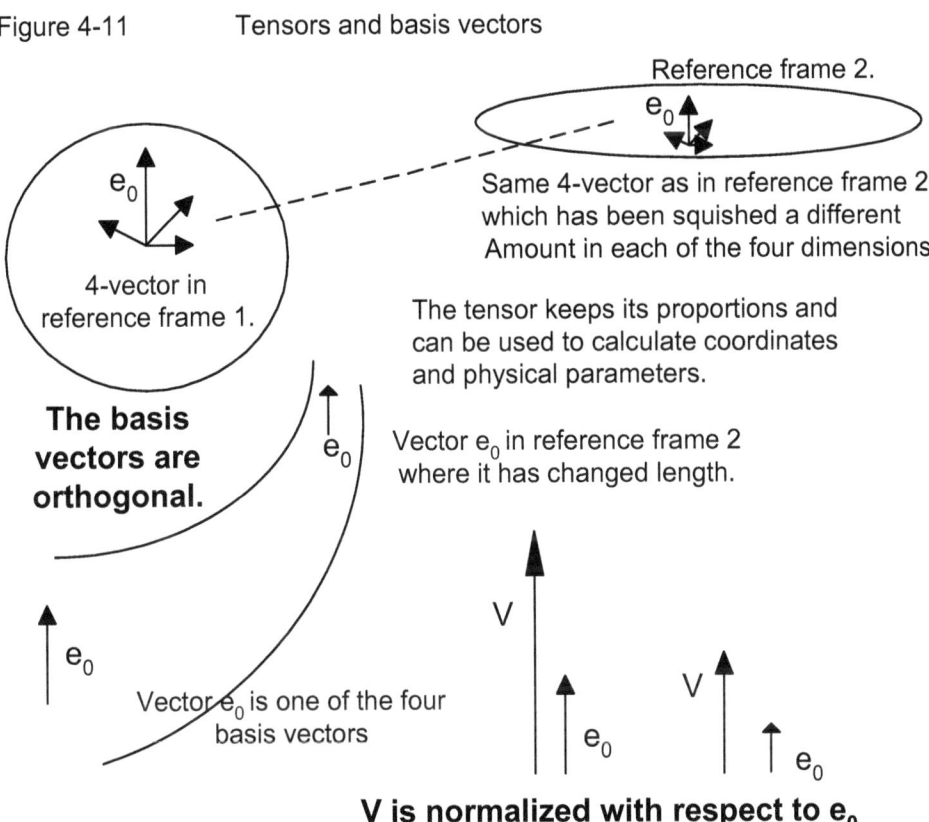

V is normalized with respect to e_0

Each basis vector influences the length and direction of the others. When moving from one reference frame to another, all other vectors, V here, are calculated from the basis vectors.

Some Other Tensors

g is the metric tensor described above; it gives distance.

Riemann is the foundation for multidimensional curved geometry. It can take three vectors and produce new vectors that give geodesic deviation or relative acceleration. This amounts to the rate of separation of world lines or geodesics in a gravitational field. $R^{\rho}{}_{\sigma\mu\nu}$ is the Riemann tensor.

Einstein is the tensor defined as G = 8πT. It gives the average curvature over all directions. R = 8πT is an alternate description. R is geometry. The left side is mathematics, the right side is physics.

T is the *stress energy tensor*. It is produced from the energy density per volume occupied. It is T = (,) where one or two vectors can be inserted. In the weak field limit, T = m/V, that is mass per volume which is simple density.

If the 4-velocity vector is put in one slot and the other is left empty, then it produces the 4-momentum density.

If the 4-velocity is put in one slot and a vector **v** in the other, then it produces the 4-momentum density in the direction of the vector, **v**.

It can perform other functions. One that we will see is T in the weak field limit – low energy density. This simplifies to T = m/V = mass / volume = simple Newtonian mass density.

The **Ricci** tensor is $R_{\mu\nu}$ is used to find the scalar curvature; it appears in the Einstein tensor. (Scalar curvature is the trace, the sum of the diagonal elements, of the Ricci curvature.)

Faraday defines lines of electrical force using the wedge product of vectors as described above. The results can be pictured the same as shown for vectors alone.

There are other tensors used for various purposes.

All tensors are frame independent. They adjust to changes in the spatial dimensions. Thus we say they are generally invariant.

The total number of indices of a tensor gives its rank. A 0-rank tensor is a scalar, a rank-one tensor is a vector, and a tensor has two or more indices and can be rank-two, -three, etc.

In this book, knowledge of tensors is unnecessary, however if one reads Evans or Einstein, it is necessary to at least know the generalities of the subject, although not how to calculate results.

Chapter 4 – Geometry

Matrix Algebra

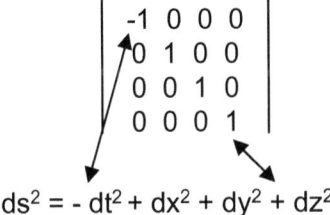

$$ds^2 = -dt^2 + dx^2 + dy^2 + dz^2$$

The matrix here simply gives the sign to be placed in
front of the distances between two events.

The matrix above shows the metric tensor, $\eta_{\mu\nu}$.

A matrix is a group of functions, vectors, or numbers that work together with operators to achieve a calculation. Matrices simplify actual calculations. Tensors can use matrices and be expressed in terms of a matrix.

There are many rules that need to be followed, but the tensors in relativity can be depicted as 4 X 4 matrices.

Any square asymmetric matrix can be decomposed into two matrices – symmetric and antisymmetric ("skew-symmetric" in mathematics and Einstein's terminology). The Symmetric can be decomposed again producing two matrices. There are then a total of three parts – the traceless symmetric matrix, the trace, and traceless antisymmetric matrix.

Evans uses this to decompose the tetrad into parts. Those parts define gravitation and electromagnetism. He uses a purely mathematical function that is well known and generalizes to the physical process.

$$q_\mu^a = q_\mu^{a\,(S)} + q_\mu^{a\,(A)}$$
$$= \text{(symmetric)} + \text{(antisymmetric)}$$
$$= \text{gravitation} + \text{electromagnetism}$$
$$= \text{curvature} + \text{torsion}$$
$$= \text{distance} + \text{turning}$$

With some operations one must add the individual elements of the matrix to find a new value. Some elements in the matrix will cancel each other out because they have the same absolute value as, but are negatives of, other elements. Some will be zero as in the metric tensor above where only the diagonal has non-zero values.

Matrices are used to manipulate linear transformations – functions that obey normal addition and multiplication rules. One can multiply each element by some constant. Or one can add each element of one matrix to each element of another.

The turning of spacetime can be described by a matrix that is multiplied by a "rotation generator." The matrix changes its elements according to a formula that describes rotation.

The Tetrad, $q^a{}_\mu$

If we have a base manifold, there are two ways to choose coordinates. We could pick a four-vector in the manifold and it would be related to that metric. Or we could choose an orthonormal four-vector in the mathematical index space. These are the same vector, but described in different ways. The tetrad is a way of relating these two choices. Tetrads can connect and relate the different expressions of the vector in the base manifold and in the tangent index space.

In other words, the tetrad relates a point in the real universe with the same point in the flat Minkowski mathematical spacetime.

In the tetrad matrix each element is the product of two vectors. One vector is in the base manifold and one in the index. $q^0{}_0$ indicates q^0 times q_0.

$$q^a{}_\mu = \begin{bmatrix} q^0{}_0 & q^0{}_1 & q^0{}_2 & q^0{}_3 \\ q^1{}_0 & q^1{}_1 & q^1{}_2 & q^1{}_3 \\ q^2{}_0 & q^2{}_1 & q^2{}_2 & q^2{}_3 \\ q^3{}_0 & q^3{}_1 & q^3{}_2 & q^3{}_3 \end{bmatrix}$$

Chapter 4 – Geometry

Let V^a be the four-vector in the tangent orthonormal flat space and V^μ be the corresponding base four-vector, then:

$$V^a = q^a{}_\mu V^\mu \tag{7}$$

$q^a{}_\mu$ is then the tetrad matrix. Here a and m stand for 16 vectors as shown in the matrix above. Each element of $q^a{}_\mu$, the tetrad, is $V^a V^\mu$.

> The tetrad is the gravitational *potential*. The Riemann form is the gravitational *field*. With an electromagnetic factor of $A^{(0)}$ the tetrad is the *electromagnetic potential*. The torsion form of Cartan differential geometry is then the *electromagnetic field*.

Formally, the tetrad is a set of 16 connectors defining an orthonormal basis.

Any vector can be expressed as a linear combination of basis vectors. One can describe old basis vectors in terms of the new ones. The basis vectors are not derived from any coordinate system. At each point in a base manifold a set of basis vectors $\hat{e}_{(a)}$ is introduced. They have a Latin index to show that they are unrelated to any particular coordinate system. The whole set of these orthonormal vectors is the tetrad when the base manifold is four-dimensional spacetime.

$$V^a = q^a{}_\mu V^\mu$$

$$q^a{}_\mu = \begin{bmatrix} q^0_0 & q^0_1 & q^0_2 & q^0_3 \\ q^1_0 & q^1_1 & q^1_2 & q^1_3 \\ q^2_0 & q^2_1 & q^2_2 & q^2_3 \\ q^3_0 & q^3_1 & q^3_2 & q^3_3 \end{bmatrix}$$

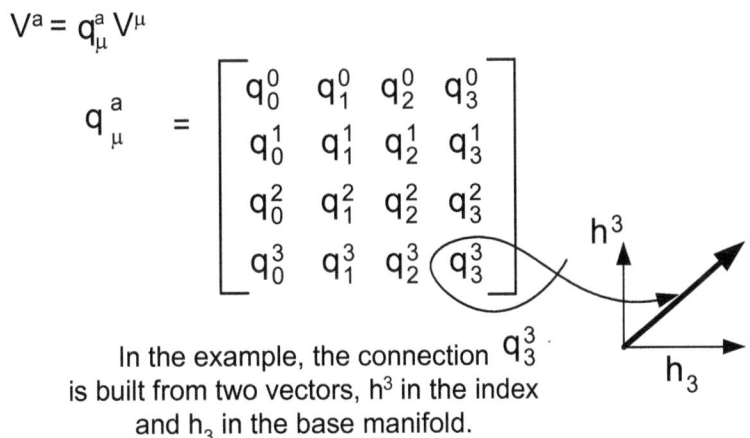

In the example, the connection q^3_3 is built from two vectors, h^3 in the index and h_3 in the base manifold.

Alternately, a set of basis matrices such as the Pauli matrices can be used in place of vectors. This is done in gauge theory.

The metric can be expressed in terms of tetrads:

$$g_{\mu\nu} = e^a{}_\mu \, e^b{}_\nu \, \eta_{ab} \tag{8}$$

There is a different tetrad for every point so the mathematical space is quite large. The set of all the tetrads is a tetrad space.

The point here is that differential geometry is valid in all spacetimes – most importantly, our physical universe.

Cartan's concept that the values in the curved spacetime can be connected and considered in the flat index space. It is also called "moving frames" or the Palatini variation.

$$q^a{}_\mu$$

q can be defined in terms of a scalar, vector, Pauli two-spinors, or Pauli or Dirac matrices. It can also be a generalization between Lorentz transformation to general relativity or a generally covariant transformation between gauge fields.

In differential geometry, the tetrad is a connection between spaces or manifolds. It was developed by Elie Cartan and is an alternate method of differential geometry. It has also been called the orthonormal frame. It is the Riemann tensor in a different form of mathematics. It is a connection matrix mathematically, but Evans shows that in physics it can represent the gravitational field. The tetrad obeys tensor calculus rules.

Take a person casting a shadow on the ground. The person is one manifold, the shadow is the other manifold. The connections are angles and lines describing the path where no photons from the sun hit the ground directly. The tetrad would describe those connections.

It describes the angles that connect the spacetime to various other processes describing the four forces. It provides the connections. Those connections can be quite complicated.

> The tetrad mixes two vector fields, straightens or absorbs the non-linearities, and properly relates them to each other.

In $q^a{}_\mu$ the q can symbolize spinors, matrices, gravitation, quark strong force, electromagnetism, or the weak force. The a indicates the indices of the tangent *index manifold*. The μ is the index of the *base manifold*, which is the spacetime of the universe that can be thought of as the vacuum. The tangent space to a manifold is connected by a "bundle." That bundle is a group of equations that define the relationships. The fiber bundle of gauge theory in quantum mechanics is shown by the tetrad to be the same as the tangent space of general relativity.

Some problems exist which are solved by the tetrad:

1) The gravitational field in the standard model is curved spacetime, while the other three fields (electromagnetic, weak and strong) are entities on flat spacetime. The tetrad can represent both.

2) The electromagnetic field in the standard model is Abelian. Rotating fields are non-Abelian and the standard model's Abelian electromagnetic field cannot be generally covariant. The tetrad can be spun and then represent the electromagnetic field.

These are barriers to uniting the four forces. We must quantize gravitation and find a way to simultaneously describe the electromagnetic, weak, and strong fields in conjunction with gravitation. The Evans Equations cure these two problems by expressing all four fields as entities in curved spacetime within a non-Abelian structure.

The tetrad mathematics shows that well known differential geometric methods lead to the emergence of quantum theory from general relativity.

Figure 4-12

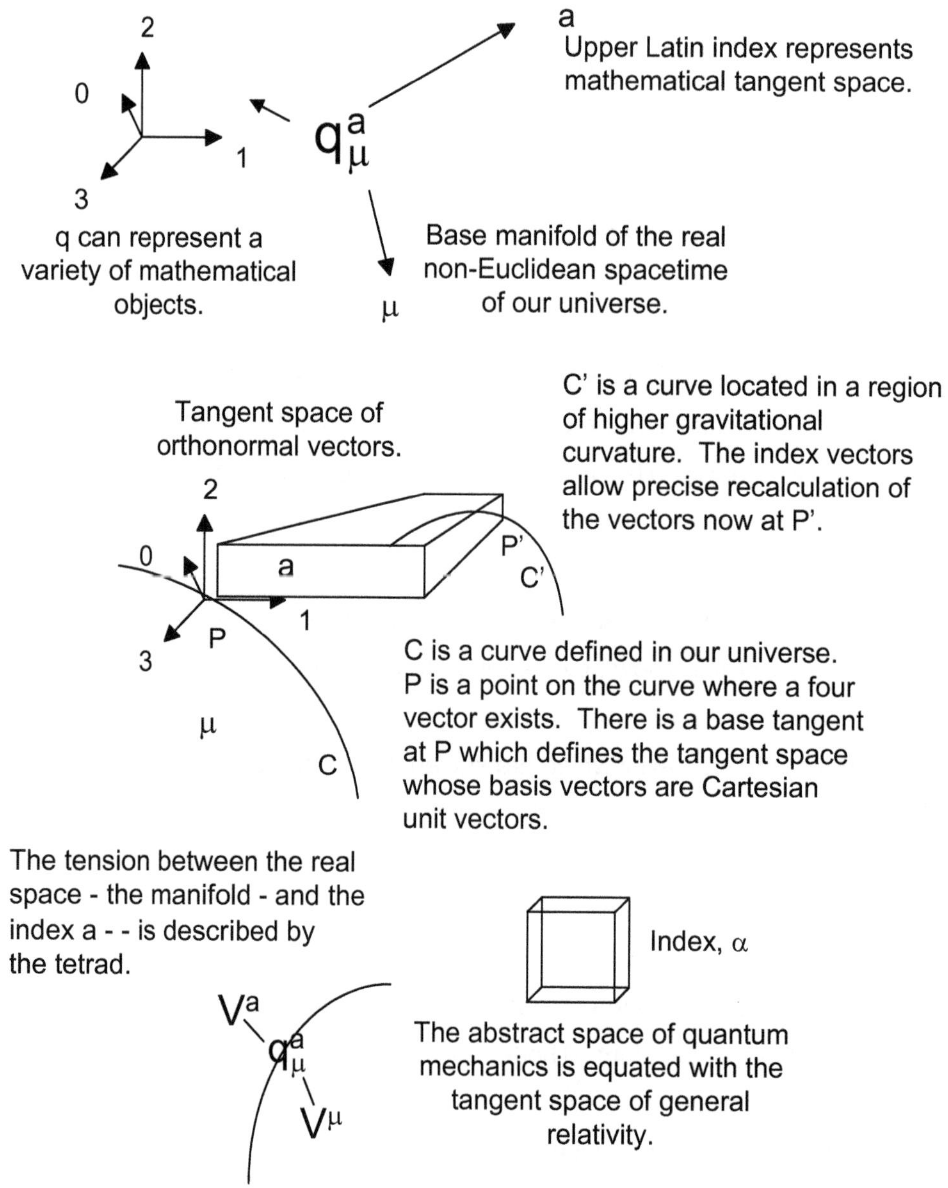

q can represent a variety of mathematical objects.

q^a_μ

a — Upper Latin index represents mathematical tangent space.

μ — Base manifold of the real non-Euclidean spacetime of our universe.

Tangent space of orthonormal vectors.

C' is a curve located in a region of higher gravitational curvature. The index vectors allow precise recalculation of the vectors now at P'.

C is a curve defined in our universe. P is a point on the curve where a four vector exists. There is a base tangent at P which defines the tangent space whose basis vectors are Cartesian unit vectors.

The tension between the real space - the manifold - and the index a - - is described by the tetrad.

V^a
q^a_μ
V^μ

Index, α

The abstract space of quantum mechanics is equated with the tangent space of general relativity.

In differential geometry, the tetrad is valid regardless of the manifold with or without torsion – the electromagnetic connections. This is the distinguishing feature from Einstein's work. He used tensors to get the invariance in spaces. Evans uses vectors and differential geometry equivalent to tensors. Einstein and others used tensors that represented spin and put them on top of the curvature.

This did not work. Evans, following Cartan and Einstein in a new way, spins the spacetime itself. The tetrad with O(3) electrodynamics, which we will cover in later chapters, allows spinning spacetime.

Nothing Einstein developed is lost, but a great deal is gained.

$\omega^a{}_b$ is called the spin connection. This is absent in Riemann geometry but Cartan differential geometry allows it. This gives the ability to use spinors and to represent the turning of spacetime. The spin connection appears if the Riemann base manifold is supplemented at any given point by a tangent spacetime. This spin connection shows that spacetime can itself spin. More later.

Given two topological spaces, A and B, a fiber bundle is a continuous map from one to the other. B is like a projection. In the example above, if A is a human body and B is a shadow, the invisible, imaginary lines from A to B would be the fiber bundle. The space B could be a vector space if the shadow is a vector bundle. When moved to a new reference frame, B could reproduce A.

Gauge theory uses fiber bundles. Spinor bundles are more easily described by the tetrad than by the more traditional metric tensors.

> The tetrad is the eigenfunction – the real physical function - in general relativity. The Riemann form is then the outer product of two tetrads and defines curvature of the spacetime; the torsion form is the wedge product of two tetrads and defines twisting of the spacetime.

Note that basis vectors in tangent spacetime are not derived from any coordinate system. They will travel from one base manifold to another without change. The basis vectors are orthonormal to the real spacetime base manifold. This is the same concept used in gauge theory. The tetrad is shown in Figure 4-12.

Contravariant and covariant vectors and one-forms

The material here is for the student who wishes a verbal description of the mathematics used in unified field theory. This is covered in more detail in the Glossary for those reading Evans' work at www.aias.us. His books Generally Covariant Unified Field Theory (Volumes 1-3, arima publishing) contain detailed introductory chapters covering the mathematics.

Contravariant tensors (or vectors are the tangent vectors that define distances.

Covariant tensors (or vectors) give all the same information. These are dual vectors. One forms are also dual vectors which are like perpendicular vectors.

Any tangent vector space has a dual space or cotangent space. The dual space is the space with all the linear maps from the original vector space to the real numbers. The dual space can have a set of basis dual vectors. It is also called the dual vector space.

One forms are a type of vector that establishes lines in two dimensions, planes in three dimensions , and volumes in four dimensions. They are akin to the gradient. See Figure 4-13.

In general, derivatives have lower indices while coordinates and physical quantities have upper indices.

A distinction between covariant and contravariant indices must be made in 4-dimensions but the two are equivalent in three-dimensional Euclidean space; they are known as Cartesian tensors.

Geometrical objects that behave like zero rank tensors are scalars.

Ones that behave like first rank tensors are vectors.

Matrices behave like second rank tensors.

Contraction is use of the dot product with tensors. Summation of the products of multiplication is the result. Tensor derivatives can be taken and if the derivative is zero in any coordinate system, then it is zero in all.

Each index of a tensor indicates a dimension of spacetime.

Chapter 4 – Geometry

Differential geometry is difficult to learn, but it has great power in calculations. The preceding material is definitely advanced and the explanation here is incomplete.

The goal here for the non-physicist is to give some idea of the process. The result is that in spite of non-linear twisting and curvature of the spacetime, the geometry can be described when moving from one reference frame to another.

The mathematics is difficult, it is the general idea that we wish to convey here. Gravitation compresses and warps spacetime. We can calculate the results using geometry.

Figure 4-13   The one-form

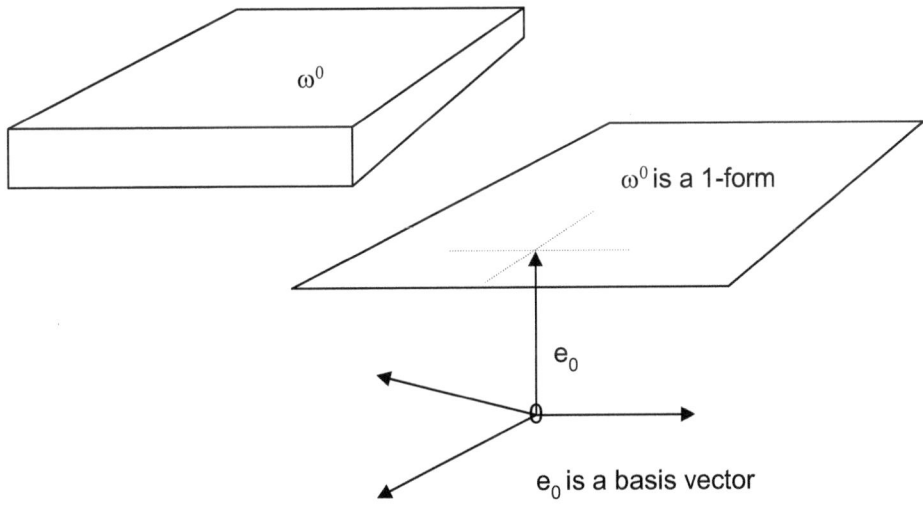

The 1-form is one basis vector away from the origin, 0.
The 1-form and the basis vector are "duals" of each other. They contain the same information in different ways.

> **The scalars, whether real or imaginary numbers, are more basic than the vectors and tensors used to arrive at them.**
>
> The invariant quantity of the metric in four dimensions is the distance, a scalar. We see so much math, but we should not lose sight of the goal. For example, the distance in spacetime's four dimensions is the invariant. Regardless of what reference frame one is in – high or low density energy, gravitational or velocity – that distance is invariant. So too are the mass, energy, and momentum.
>
> **The invariants represent real existence as opposed to derived quantities which are not invariant.**

Wave Equations

Wave equations can be based on simple geometric calculations. They look daunting, but underneath they are simple.

A basic wave equation is y equals the sine of x or "y = sin x" in math terminology. The one-dimensional wave equation is a partial differential equation:

$$\nabla^2 \psi = \frac{1}{v^2} \frac{\partial^2 \psi}{\partial t^2} \qquad (9)$$

It would allow calculation of the position of variables in Figure 4-14.

∇^2 is the *Laplacian*. Another version is indicated as , $\Box^2$ and is the *d'Alembertian*, the 4 dimensional version.

The Gradient Vector and Directional Derivative, ∇

"Del" is the gradient operator or gradient vector also called "grad." It is used to get the slope of a curved surface or the rate of change of a variable in three dimensions.

∇ in an orthonormal basis e is:

$$\nabla = e^i \partial_i \qquad (10)$$

Figure 4-14   ψ = sin θ

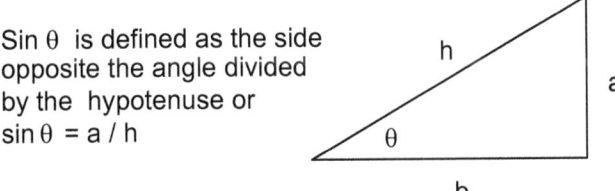

Sin θ is defined as the side
opposite the angle divided
by the hypotenuse or
sin θ = a / h

ψ is the Greek symbol for y
So while it may look difficult, it is just an x and y axis
with the curved graphed on it.

sin θ = a / h
As θ varies from 0 to 90 degrees, the sine varies as a/h goes
from 0 to 1. This is graphed below.

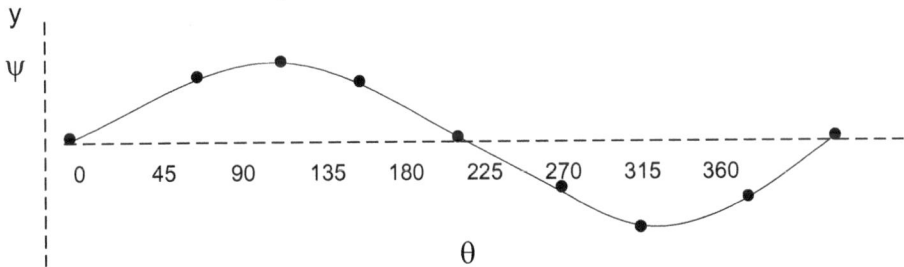

The *directional derivative* is the rate of change of unit vectors. It is simply the slope of a line in a specific direction.

To evaluate the directional derivative it is necessary to use the gradient vector. It is used so much that it has its own name. It is a slope of a function in multiple dimensions with multiple unknowns. The 4-dimensional version is the d'Alembertian.

The d'Alembertian is □ which is:

$$\nabla^2 - \frac{1}{c^2}\frac{\partial^2}{\partial t^2}$$

(11)

Figure 4-15   Directional Derivative

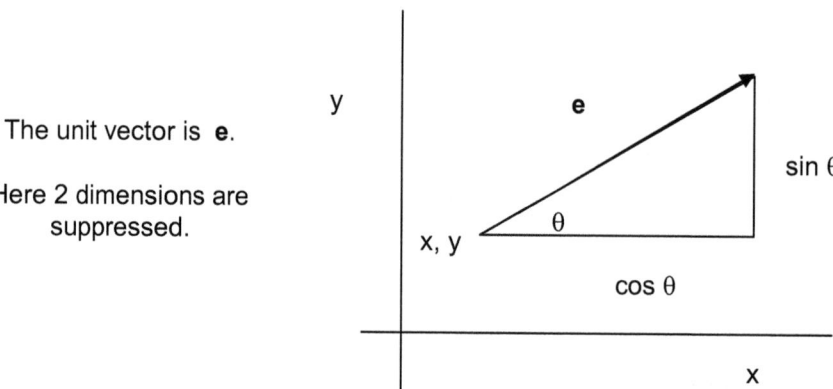

The unit vector is **e**.

Here 2 dimensions are suppressed.

$$\nabla^2 - \frac{1}{c^2}\frac{\partial^2}{\partial t^2}$$

can be written $\square$ and that is the convention we use in this book.[18] It can also be written as $\partial^\mu \partial_\mu$ or as $\nabla^\mu \nabla_\mu$. It is Lorentz invariant.

Here $\nabla = \mathbf{i}\frac{\partial}{\partial x} + \mathbf{j}\frac{\partial}{\partial y} + \mathbf{k}\frac{\partial}{\partial z}$.

It could be written as $\square = (\nabla_\mu)^2$ where

$$\nabla_\mu = \mathbf{i}\frac{\partial}{\partial x} + \mathbf{j}\frac{\partial}{\partial y} + \mathbf{k}\frac{\partial}{\partial z} + \mathbf{l}\frac{1}{c}\frac{\partial}{\partial t} \qquad (12)$$

and the square is as in equation (11). Here **i, j, k**, and **l** are the basis vectors of a Cartesian four-dimensional coordinate system.

i is the vector from (0,0,0,0) to (1,0,0,0)

j is the vector from (0,0,0,0) to (0,1,0,0)

k is the vector from (0,0,0,0) to (0,0,1,0

l is the vector from (0,0,0,0) to (0,0,0,1)

These vectors are the Cartesian vectors, which form a basis of a manifold that can be called R^4. Any 4-dimensional vector **v** from (0,0,0,0) to (x, y, z, w)

---

[18] Physicists tend to use $\square$ while mathematicians use $\square^2$. For example Evans, Carroll, and Misner et. al. use $\square$.

Chapter 4 – Geometry

can be written as **i**, **j**, **k,** and **l**. Or **v** = x**i** + y**j** + z**k** +w**l**. This is a linear combination.

Keep in mind that is simply the four-dimensional gradient between one point and another. See Figure 4-16.

Figure 4-16 Tangent basis vectors

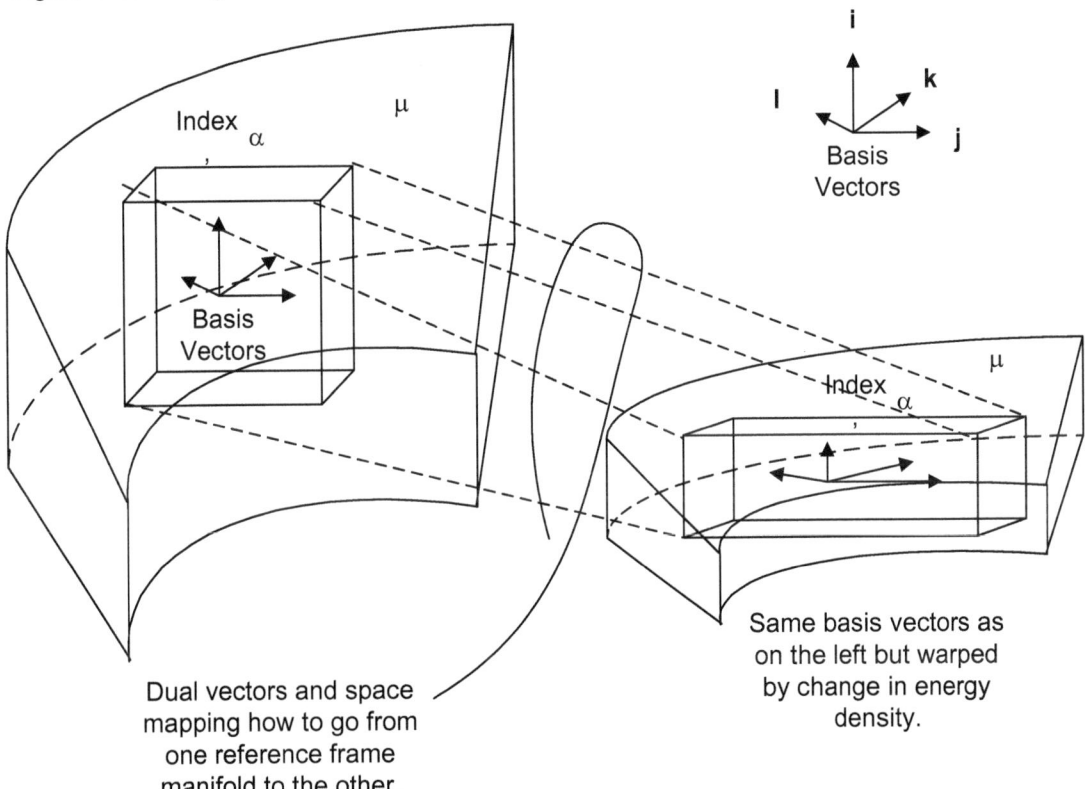

Dual vectors and space mapping how to go from one reference frame manifold to the other.

Same basis vectors as on the left but warped by change in energy density.

Exterior Derivative

The exterior derivative is the same as the gradient of a function. It is noted as "**df**" or as "d ∧". This is a stricter meaning of the idea of differential. "**df**" is a one form and it gives the direction of change. The exterior derivative is a wedge product but there is no spin connection.

Gradient, curl, and divergence are special cases of the exterior derivative. See Figure 4-17,

Figure 4-17   Exterior derivative

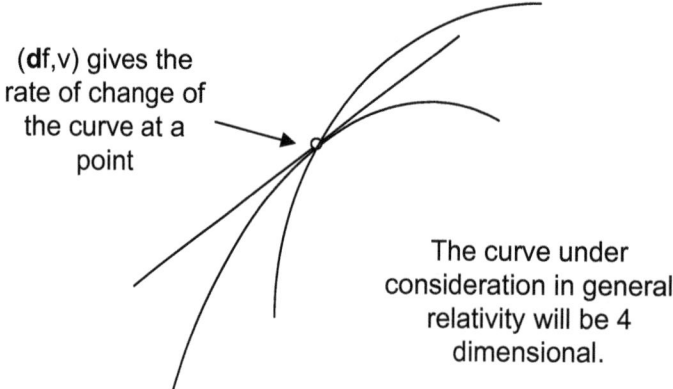

(**df**,v) gives the rate of change of the curve at a point

The curve under consideration in general relativity will be 4 dimensional.

Covariant Exterior Derivative D ∧

The covariant exterior derivative acts on a tensor. It takes the ordinary exterior derivative and adds one term for each index with the spin connection.

The exterior derivative does not involve the connection, the torsion never enters the formula for the exterior derivative.

The *covariant* exterior derivative acts on a form by taking the ordinary exterior derivative and then adding appropriate terms with the spin connection. It does express torsion. This is critical to the development of the Evans equations.

Vector Multiplication

Figure 4-18 summarizes some multiplication of vectors and forms.

Figure 4-18 Vector Multiplication

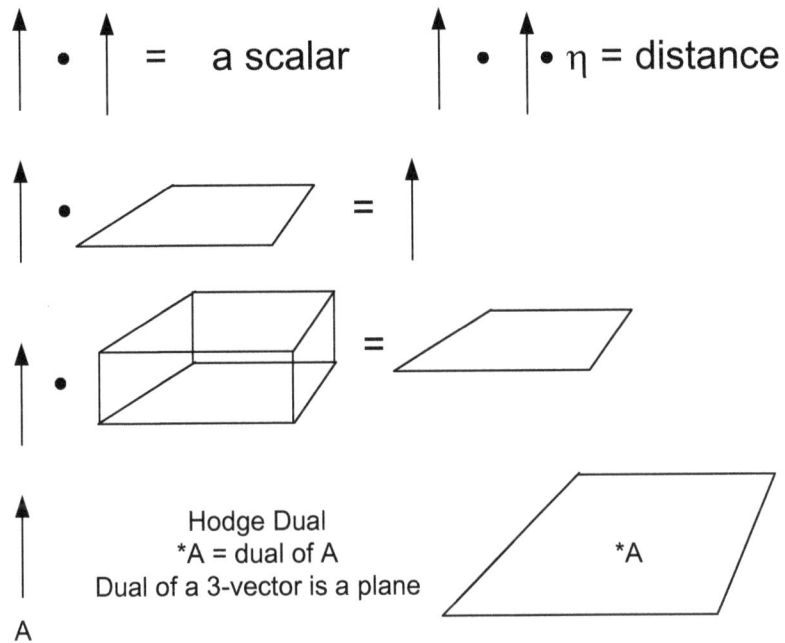

Summary

As said in the beginning, this is a difficult chapter. It is advised that the reader concentrate on the parts that are intelligible for him or her. Not many of us have studied differential equations much less the more erudite differential geometry.

The pictures and verbal descriptions should help in establishing the vocabulary necessary for the understanding of the balance of this book and much of the material at www.aias.us or www.atomicprecision.com.

Chapter 5 Well Known Equations

> These necessarily retain their meaning for all times and for all civilizations, even extraterrestrial and non-human ones, and can therefore be designated as natural units.
>
> Max Planck, 1899

Introduction

Just as the Planck units have been found to be basic in physics and have special meaning, there are equations in physics that are needed for explanations. This chapter presents and explains some of the essential equations. Again, it is not necessary to be able to mathematically manipulate each with facility, but it is necessary to have some conception of what they mean.

The explanations here are meant to help the non-physicist to comprehend the results of the Evans equations. They are not strictly proper in every respect and should be considered a review or a vocabulary lesson.

Newton's laws of motion

1. Newton's First Law of Motion - Inertia

Every object in a state of rest or uniform motion tends to remain at rest or in that state of motion unless an external force is applied to it.

While Newton dealt only with mass, Einstein showed that both mass and energy have inertia. m_i is inertial mass as opposed to m_g, which is gravitational mass. The equality of m_i and m_g is the Weak Equivalence Principle.

Chapter 5 – Well Known Equations

2. Newton's Second Law of Motion - Force

The relationship between an object's mass m, its acceleration a, and the applied net external force F is **F** = m**a**.

This is one of the most widely used formulas in physics. **F** and **a** are vectors. Speed is a scalar quantity or scalar for short, which is a quantity having magnitude, but no direction; velocity is a vector quantity which has both magnitude and direction; acceleration is change in velocity, so it is a vector also.

The most general form of Newton's second law is **F** = d**p**/dt, which states that force equals change in momentum per unit change in time. The net force applied to an object is equal to the rate of change of the momentum over time. The 4-vector that expresses this is $F^\mu = dp^\mu/d\tau$. Tau, τ, here is the proper time – the time measured in the vector's reference frame. By using the 4-vector, the equation keeps the same form under a reference frame transformation. It is covariant. Newton's second law applies to velocities small compared to the speed of light (v<<c) and for regions of spacetime devoid of intense gravitational fields. In regions of high energy, the need for accuracy requires use of covariant general relativity. Special relativity applies for high velocities and general relativity for high gravitation.

3. Newton's Third Law of Motion - Conservation

For every action there is an equal and opposite reaction.

This is the law of conservation of mass and energy. Nothing in physics is created or destroyed.

Electrical Equations

There are a number of basic definitions that are needed if one is to go into the electromagnetic sectors of the Evans equations. It takes some time and hard study to understand the various terms and interrelations. We cover them briefly here.

$V = IR$. A circuit is a completed circle of wire or the equivalent. In a circuit, the voltage V equals the [current that flows in amperes] times [the resistance in ohms]. See Figure 5-1.

Voltage V is force pushing the electrons. A Volt or voltage is a measure of potential difference. A battery has a positive and negative terminal. The negative terminal has an excess of electrons (negative charge) compared to the positive terminal (electrons have been removed). When they are connected, charge flows from negative to positive.

Potential difference can exist without the interconnected wires across capacitors, clouds to earth, or from any spacetime region to another.

Current (I) is flow of electrons or other charge.

The resistance (R) is a property of the material – a copper wire, the vacuum, etc. An Ohm is resistance to the flow of current.

Figure 5-1   Basic Current Flow
$V = IR$

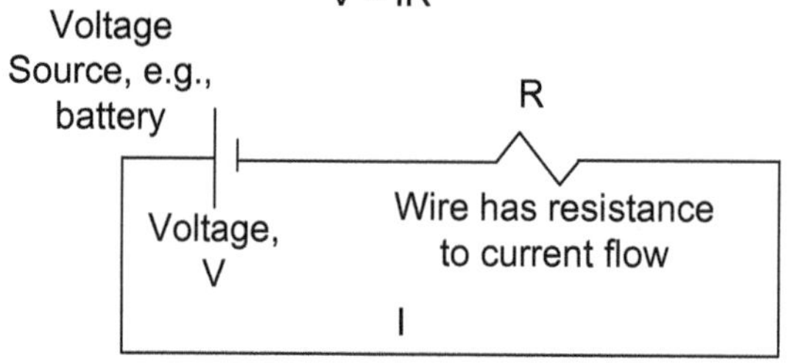

Voltage Source, e.g., battery

Voltage, V

R

Wire has resistance to current flow

I

The amount of current that flows equals V/R.

In spacetime, A is the term we will see used for voltage. It is the essential quantity that allows us to move from mathematics to physics with the electromagnetic field.

$I = dQ/dt$. Current is the flow of charge. Q is measured in Coulombs, Q, per unit of time; a Coulomb is about 10^{19} electrons. So current is the flow of electrons (or positrons).

J is current density, that is I current flow per unit area or ampere per square meter, A/m^2. An analogy would be the number of cars passing through a wide intersection. At times 4 cars go thru, at other times only one car passes.

Electric field strength is measured in volts per square meter or V/m^2.

Electric charge density is the number of electrons (or positrons or protons) in a volume. It is measured in Coulombs per cubic meter or C/m^3.

Flux is lines of force. We imagine that electrical force is proportional to the number of force lines passing through an area. A strong field could have 1000 lines per square centimeter while a weak field could have one line per square centimeter.

Electric flux density is measured in Coulombs per square meter or C/m^2.

In general, energy density is measured as Joules per cubic meter J/m^3. A Joule is a Nm.

The electromotive force (EMF) is often denoted epsilon, ε. Engineers and electricians tend to use E or V, so there is some confusion in terms as one changes disciplines. EMF is commonly called "electric potential" or voltage.

Alternating current circuits have more complicated formulas, but the essentials are the same.

$A^{(0)}$

The letter "A" is used to denote the electromagnetic potential field. $A^{(0)}$ is referred to as a C negative coefficient. This indicates that it has charge symmetry and e⁻ is the negative electron. $A^{(0)}$ is the fundamental potential equaling volt-s / m; that is, $A^{(0)}$ is potential force x time per distance.

A coulomb is approximately equal to 6×10^{18} electron charges, and one ampere is equivalent to 6×10^{18} elementary charges flowing in one second.

The volt is the push behind the flow of current. When current is not flowing and there is a difference in strength, the volt measures the potential difference of the electric field. A force of 1 Nm per Coulomb of electrons equals 1 volt.

The electric field is written E. E is electric field strength in volts / meter. This is sometimes seen as cB.

"A" is a meter times B = V-s/m.

The standard definition of magnetic field is a force that reaches "through" spacetime and affects charges and other magnetic fields.

Magnetic field strength is a measure of ampere per meter or A/m. It is written B.

The SI (Standard International) unit for magnetic field is the tesla.

B is the magnetic flux density in tesla. One tesla = 1 N/A-m = kg/(A-s^2) = Wb/m^2 = J-s/C/m^2. (This is force per amount of current-time squared.)

ϕ is magnetic flux itself in webers. A weber (Wb) measures lines of magnetic flux. One weber = one volt-second = 1 T–m^2 = J/C/m.

A is a meter times B = V-s/m.

The exact electrical and force definitions take time to learn. Suffice it to say, $A^{(0)}$ is the electromagnetic potential field. $A^{(0)}$ is volts-seconds/meter. We will find it in the conversion from geometry to electromagnetic physics in the same way Einstein's R = kT is used to convert from curvature to physics.

See SI units in Glossary for more information.

Maxwell's Equations

Four equations summarize all of classic electromagnetics. They are given in basic form here and more formally in the Glossary under Maxwell's Equations.

1. Gauss's Law for Electric Charge
   The net electric field lines or flux passing through an enclosed region is proportional to the net electric charge Q contained within the region.

   $$\nabla \cdot \mathbf{D} = \rho \qquad \text{Equivalent to Coulomb Law } (E = kQq/r^2)$$

   This may be most easily understood as Coulomb's law, $E = kQq/r^2$. The amount of flux or lines of force at a point is proportional to a constant times the amount of charge and inversely proportional to the distance from the center. This is the inverse square rule for electrical charge. k is a constant of proportionality. Coulomb's law is derivable from Gauss' law. See Figure 5-2.

Chapter 5 – Well Known Equations

Electric charge is a property associated with electrons, protons and their antiparticles.

Two charges exert force on one another. If both are the same polarity (positive or negative), they repel each other. If of opposite polarity, they attract.

Figure 5-2   Gauss's Law for Electric Charge

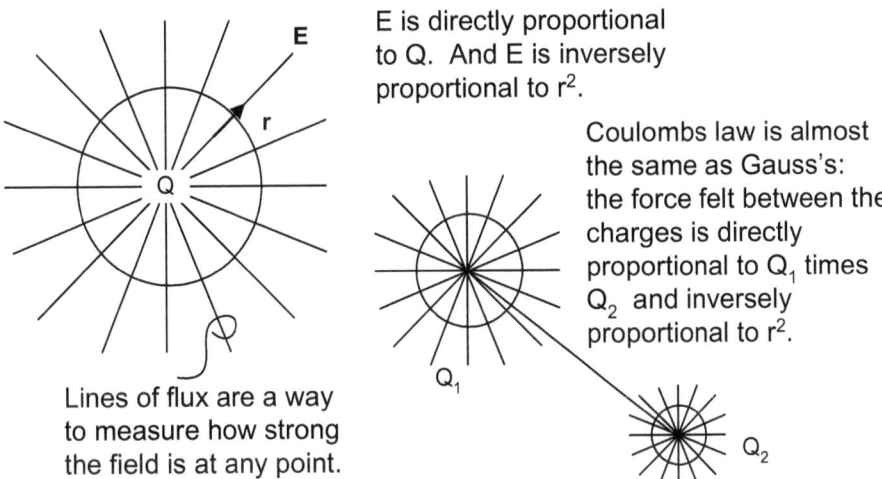

E is directly proportional to Q. And E is inversely proportional to r^2.

Coulombs law is almost the same as Gauss's: the force felt between the charges is directly proportional to Q_1 times Q_2 and inversely proportional to r^2.

Lines of flux are a way to measure how strong the field is at any point.

Field lines of electrical charge in an electrostatic field always end up on other charge. (Or they go to infinity.)

Gauss' electrostatics law states that lines of electric flux start from a positive charge and terminate at a negative charge. The spacetime within which the charges exert force is the electrostatic field.

2.   Gauss's Law for Magnetism

The net magnetic flux passing through an enclosed region is zero. The total positive and total negative lines of force are equal.

$$\nabla \cdot \mathbf{B} := 0 \qquad \text{Gauss Law}$$

Another way of stating this is that magnetic field lines or magnetic flux either form closed loops or terminate at infinity, but have no beginnings or endings. There are equal "positive" and "negative" magnetic field lines going in and out of any enclosed region. The words "positive" and "negative" magnetic field lines, refer only to the direction of the lines through an enclosed surface.

Lines of magnetic flux never end.

Figure 5-3   Gauss's Law for Magnetism

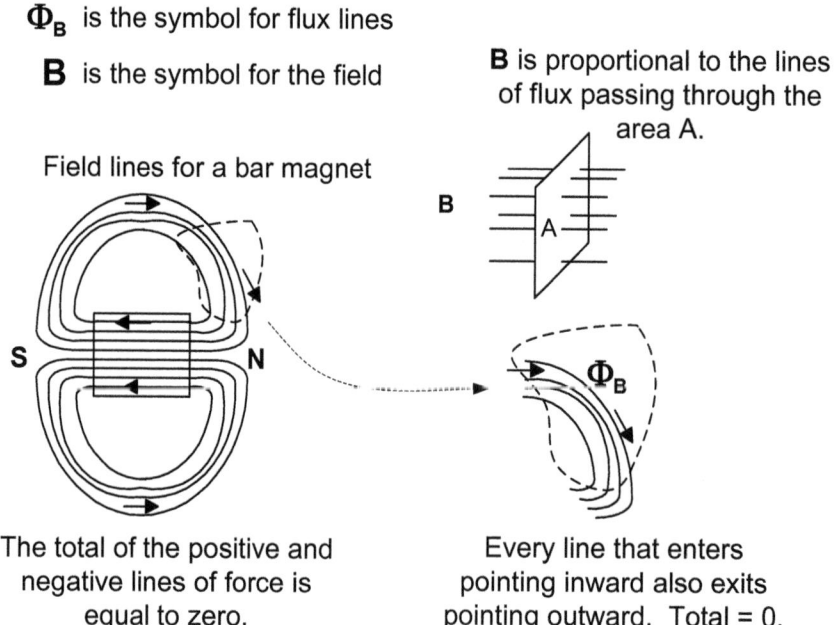

This means there are no magnetic monopoles - magnets always have two poles. Nor do magnetic charges exist analogous to electric charges. The magnetic field passing through a cross sectional area A gives the magnetic flux.

B is measured by putting a wire in a magnetic field and finding the force exerted on the wire. F = I l x **B** where **F** is a force vector due to the magnetic field strength **B,** I is current in amperes, l is the length of the wire, and **B** is the magnetic field.

Magnetic flux is proportional to the number of lines of flux that pass through the area being measured. See Figure 5-3.

Chapter 5 – Well Known Equations

3. Ampere's Law

A changing electric field generates a magnetic field.

$$\nabla \times \mathbf{H} = \mathbf{J} + \partial \mathbf{D}/\partial t \quad \text{Ampère Maxwell Law}$$

where D is electric displacement, J is current density, and H is magnetic field strength.

A voltage source pushes current down a wire. It could be a direct current increasing or decreasing in velocity or an alternating current changing polarity. In either case, the *change* in the electric field causes a magnetic field to appear.

4. Faraday's Law of Induction

A changing magnetic field generates an electric field. The induced emf in a circuit is proportional to the rate of change of circuit magnetic flux.

$$\nabla \times \mathbf{E} + \partial \mathbf{B}/\partial t := 0 \quad \text{Faraday Law}$$

In Figure 5-4, Ampere's Law, the roles of the magnetic field and electric field could be reversed. If a magnetic field changes and the lines of flux touch electrons in the wire, current flows. Faraday's law gives the strength of the results however it is the implication that change in magnetic field causes an electric field that we wish to establish.

Calculation of actual values can be performed using several equations. $\oint \mathbf{E} \cdot dl = -d\Phi_B/dt$ or stated in terms of **B**, $\oint \mathbf{B} \cdot dl = \mu_0 I_{inclosed} + \mu_0 \varepsilon_0 \, d\Phi_E/dt$. We will not be using any laws for calculations.

Charge produces the electric field, E. The potential difference between two regions will be denoted A - not to be confused with A for area. The magnetic field is B.

Figure 5-4   Ampere's and Faraday's Laws

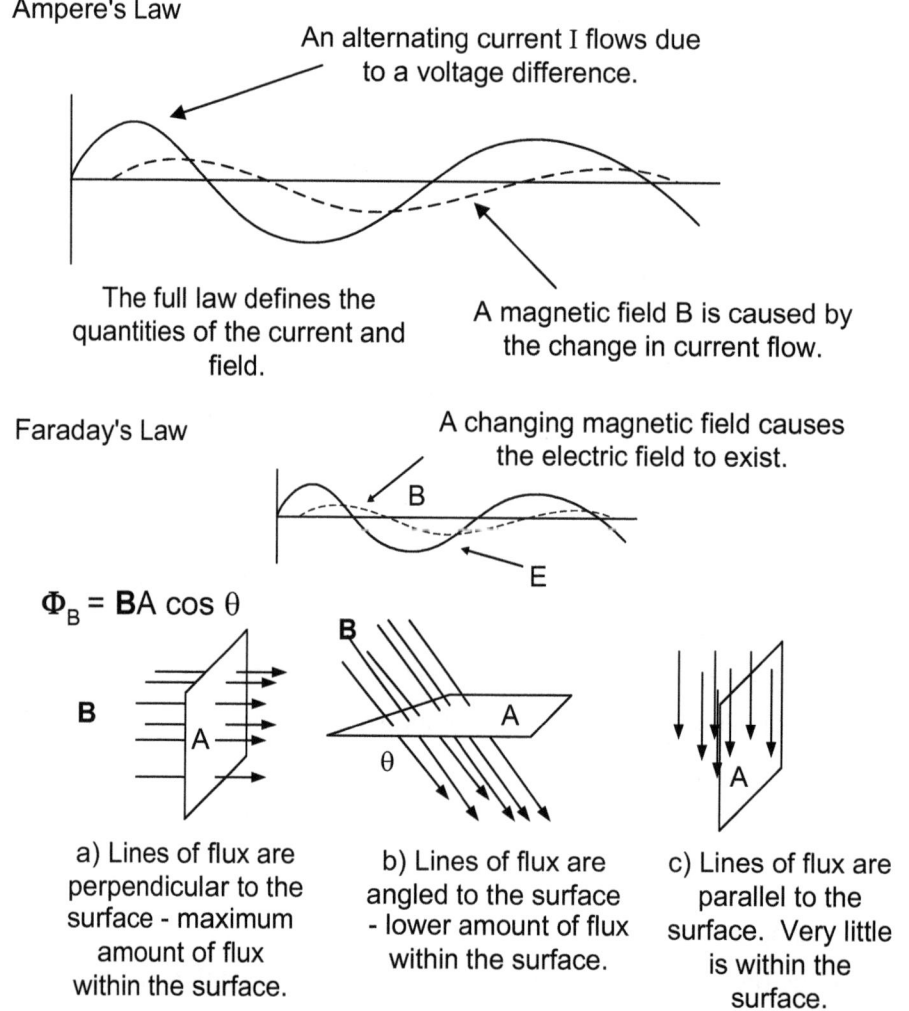

A traveling electromagnetic wave in free space occurs when the B and E fields maintain themselves. The magnetic field causes an electric field which in turn causes the magnetic field. Movement is perpendicular to both the fields. See Figure 5-5.

Chapter 5 – Well Known Equations

Figure 5-5   Translating Photon or Magnetic (B) and Electric (E) Fields

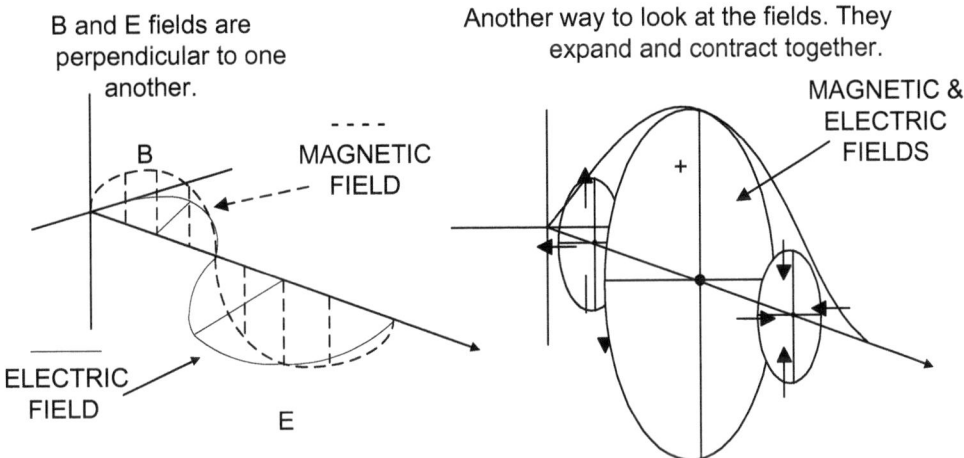

The fields first expand in directions perpendicular to one another. Upon reaching some maximum amplitude, they start to contract. Upon reaching the midline, they reverse polarity and expand in the opposite direction.

The prevailing concept is that the magnetic and electric fields are entities superimposed on spacetime.

The laws are:

$\nabla \cdot \mathbf{D} = \rho$ Gauss Electric, Coulomb Law ($E = kQ/r^2$)

$\nabla \cdot \mathbf{B} := 0$ Gauss Magnetic Law

$\nabla \times \mathbf{E} + \partial \mathbf{B}/\partial t := 0$ Faraday Law

$\nabla \times \mathbf{H} = \mathbf{J} + \partial \mathbf{D}/\partial t$ Ampère Maxwell Law

where ∇ indicates the gradual potential field decreases as one moves from a center, **D** is electric displacement, ρ is charge density, **B** is magnetic flux density, **E** is electric field strength, t is time, **H** is magnetic field strength, and J is current density. See fields in Figure 5-6.

119

Newton's law of gravitation

A mass causes an attraction that is defined by:

$$g = F / m \qquad (1)$$

g is the gravitational acceleration in meters per second per second or m/s². **g** and **F** are vectors since they have a directional component as well as a magnitude. m is the mass in kg.

Gravitational acceleration, **g**, is defined as the force per unit mass experienced by a mass, when in a gravitational field.

This can be expressed as a gradient of the gravitational potential:

$$\mathbf{g} = -\nabla \Phi_g \qquad (2)$$

where Φ is the gravitational potential. It contains all the information about the gravitational field. The field equation is:

$$\nabla^2 \Phi = 4\pi G \rho \qquad (3)$$

which is the Poisson equation for gravitational fields. Φ is gravitational potential, G is Newton's gravitational constant, and ρ is the mass density. ∇ tells us that the force decreases with distance.

The force can be expressed as:

$$F = \frac{GMm}{r^2} \qquad (4)$$

Where $G = 6.67 \times 10^{-11}$ Nm² kg⁻², M and m are masses in kg, and r is the distance in meters between the centers of gravity also known as the center of mass of the two masses. G is just a conversion factor for the units used.

A gravitational field is a force field around a mass that attracts other mass and is itself attracted to other mass. For practical purposes at low mass and energy, this is valid. Einstein showed that gravitation is actually curvature; at high densities of mass and energy, the rules change. The field concept allows us to analyze attractions.

This can be expressed as $\mathbf{g} = -\nabla \Phi_g$ as a gradient potential. A depiction of a gravitational field is shown in Figure 5-6. Near the mass in the center, the field is dense. As one goes farther away, it becomes more tenuous. The amount of attraction depends on the distance, r, squared. The field is always negative meaning it attracts or is directed radially inward toward the center of mass.

The electrical and gravitational fields use a similar formula. This will be explained in the Evans equations as coming from the same process – symmetric spacetime. Evans shows that the Newton gravitational and Coulomb electrical inverse square laws are combined into a unified inverse square law originating in the Bianchi identity of differential geometry. The two laws are respectively:

$$F = \frac{GMm}{r^2} \quad \text{or} \quad F = \frac{kQq}{r^2} \quad \text{(Coulomb's Law)} \tag{5}$$

where G and k are the proportionality constants, designated phi, Φ, in Poisson's equations.

Equation (5) states the attraction between two *masses (charges)* is proportional to the product of their *masses (charges)* and inversely proportional to the square of the distance, r, between them. This identity between gravitation and electromagnetism is explained by the Evans equations as symmetric curvature and symmetric torsion. Figure 5-6 compares the electric and gravitational fields and the gradient.

The Laplacian

∇^2 is a measure of the difference in the change in the electrical or mass field gradient compared with the change in a small region around it. This is a 3-dimensional measurement. The shape ∇ gives an idea of the gradual change in potential from large to small as one moves away from the top.

$$\nabla^2 = \frac{\partial^2}{\partial x^2} + \frac{\partial^2}{\partial y^2} + \frac{\partial^2}{\partial z^2} \quad \text{See Equation (14).}$$

This is a three dimensional equation and is incomplete as used in special relativity since it is not generally covariant.

Figure 5-6   Fields

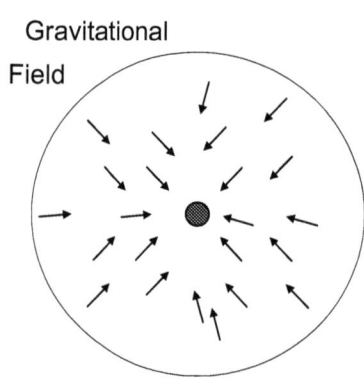

Gravitational Field

The gravitational field is always attractive to all objects. The electrical field can attract or repulse. The similarities indicate they are related.

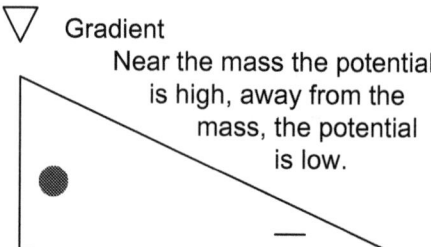

∇ Gradient
Near the mass the potential is high, away from the mass, the potential is low.

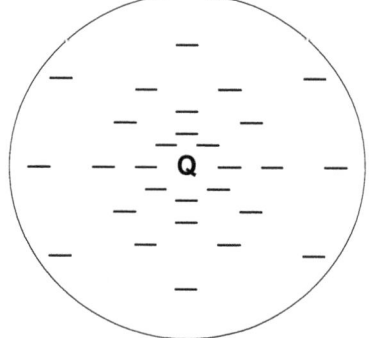

Negative Electrical Field

Attracts positive and repulses negative objects.

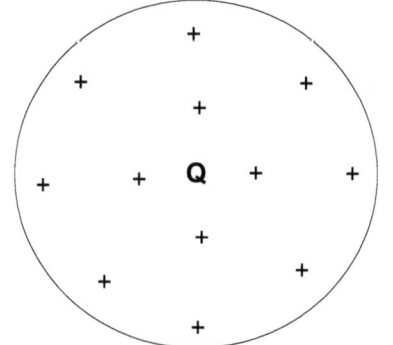

Positive Electrical Field

Repulses positive and attracts negative objects.

Poisson's Equations

Poisson's equation for electrical fields is

$$\nabla^2 \Phi = -4\pi\rho \qquad (6)$$

For electric fields, Φ is the electric potential, ρ is the charge density and the equation is expressed as: $\Phi = -\int \mathbf{E}\, ds$ or as $\nabla^2 \Phi = -\rho_e / \varepsilon_0$ where ε_0 is the permittivity of vacuum equal to 8.854×10^{-12} C^2/J-m. In the cgs system, Φ is the electrical potential in volts (or joules per coulomb) and ρ is the charge density in coulombs. Volts are measurement of potential difference.

For gravitational fields, Poisson's equation is:

$$\nabla^2 \phi = 4\pi G \rho \qquad (7)$$

where ϕ is the gravitational potential, G is the gravitational constant and ρ is the mass density.

∇^2 indicates that the gradient as one goes away from the central mass or charge decreases as the square of the distance. This can be stated as the field force strength = Q / distance squared for electrical fields and m/r^2 for gravitational fields. If at one radius away the field is 4 volts, at two radii away it will be 1 volt.

The mass density generates the potential. This is the view in Newtonian physics. The Evans equations must result in this when the weak limit - low energy, non-relativistic condition - is applied.

Note that $\nabla^2 \phi$ is the notation for a potential field. Both gravitation and electromagnetism are described with the same notation.

The electrogravitic equation results from the Evans wave equation. He shows that $\mathbf{E} = \phi^{(0)} g/c^2$ where $\phi^{(0)}$ is the Evans potential in volts.

The problem with Poisson and Newton's equations is that they propagate instantaneously. They are only three dimensional. When the four dimensional d'Alembertian is used, the fields propagate at the speed of light.

The d'Alembertian $\square$

$$\partial^\mu \partial_\mu = \frac{1}{c^2} \frac{\partial^2}{\partial^2 t} - \nabla^2$$

The d'Alembertian $\square$ or $\square^2$ is the Laplacian, ∇^2, in four dimensions. The signs are reversible. (There is a tendency for $\square$ to be used by physicists and $\square^2$ by mathematicians.)

This is a differential operator used typically in electromagnetism – it operates on other equations without destroying them. See equations (11) and (12) in previous chapter.

It is Lorentz invariant and can be written as $\partial_\mu \partial_\nu \eta^{\mu\nu}$. It is a four dimensional version of the gradient.

Components exist in the real universe, the base manifold or actual reference frame. The tensors and vectors we use to transfer components from one reference frame to another are covariant geometrical objects.

For example, we have a cube with equal sides that weighs 10,000 kg and it rests on the earth. If we move it close to the event horizon of a black hole, it will change shape and weight and real estate value. How can weight and shape and cost be basic if they change depending on the spacetime? They cannot be.

Using mathematics, we can find the inherently real qualities of the "cube." It is better described as a geometrical object. Using basis vectors, we set the dimensions in objects. Then using tensor calculus, we can find the new dimensions when it is moved. They are multiples of the basis vectors, but the components we measure from outside the high gravitation of the event horizon are different from those measured from within.

Depending on the energy density of the reference frame, it is likely that our cube house will become elongated in the dimension pointing into the black hole. There will also be some compression laterally. As a result, we end up with an object shaped more like a tall somewhat pie shaped building with smaller dimensions near the horizon than on the end away from the horizon. This is the view from outside – "at infinity." However from within the cube, the dimensions as measured by a "co-moving" observer are still those of a cube. His measuring instruments have also changed.

Weight is arbitrary depending on gravitational field. Therefore, we do not use it. We use mass instead. As it happens, the mass does not change. It is an invariant, basic, real, physical geometric object. Both the observers will see the same amount of mass – energy – inside the cube.

> We are looking for frame independent descriptions of physical events. Those are "irreducible." They are the reality in physics. The components are the description we experience and they can vary.

Einstein's Equations

Einstein gave us this basic field equation:

$$G_{\mu\nu} = R_{\mu\nu} - \tfrac{1}{2} g_{\mu\nu} R = 8\pi T_{\mu\nu} \qquad (8)$$

where $G_{\mu\nu}$ is the Einstein tensor, $R_{\mu\nu}$ is the Ricci tensor derived from the Riemann tensor; R is the scalar curvature also derived from Riemann; $g_{\mu\nu}$ is the metric tensor and $T_{\mu\nu}$ is the stress energy momentum tensor. $g_{\mu\nu}$ takes the place of Φ in general relativity. The mass density is p and is part of the equation that results in T. The energy density is $T_{00} = mc^2$ / volume in the weak field limit – very low gravitational potential or velocity.

$G_{\mu\nu}$ gives the average Riemann curvature at a point. It is difficult to solve with 10 independent simultaneous equations and 6 unknowns for gravitation.

Without the presence of mass-energy (or equivalently with velocity near zero), the weak limit, the equations reduce to the Newton equations. Except near stars or close to or within particles, Newton's laws work well. General relativity reduces to the equations of special relativity when only velocity – momentum energy - is considered.

The Evans equations show that R = -kT applies to all radiated and matter fields, not just gravitation. This was Einstein's unfinished goal.

> The R is the spacetime shape. The T is matter or the energy density.
>
> R = -kT
>
> *R is geometry*   *T is physics*
>
> *k is a conversion factor*

$G = 8\pi T$ is the basic equation. There are several ways to express this. $R = -kT$ is the basic postulate from which $G = 8\pi T$ is derived. π tells us that we are dealing with a circular volume or curvature.

The point is that spacetime curvature results from the presence of mass, energy, pressure, or gravitation itself. Curved spacetime is energy. In more mechanical terms, curvature is compression in all four dimensions. Energy is compression; force is the expansion of the compression.

All this is a bit vague since we cannot define spacetime in mechanical terms we are more used to in our everyday observations. It is curvature pure and simple. In general relativity time is treated as a 4th spatial dimension. In some calculations, in particular inside a black hole horizon, time and space can reverse roles. Time, X_0, in our equations, becomes a definite spatial dimension; and X_1, a distance, becomes time-like.

We do not yet have a test black hole to check our calculations, but general relativity has proven to correctly predict a number of observations, so we trust our formulas. Something strange happens to spacetime inside real black holes.

All the multiplication in Einstein's equation are dot-outer products. Dot and outer products produce symmetric relationships. These products describe central forces, distances, and gravitation as described in Figure 5-6.

Cross products in three dimensions, and the wedge product in four dimensions produce antisymmetric relationships. These products describe turning, twisting and torque of the quantities used.

Wave equations

The letter y is the Greek letter psi, Ψ. Ψ is typically used to indicate a wave function of the variables x and t – position and time. Ψ (x, t) = a traveling wave. It will have a cosine or sine function along with the velocity and time and

Chapter 5 – Well Known Equations

wavelength and position. That is v, t and λ, x. The wave looks like that in Figure 5-7.

This is because the position (ψ) is a function of the velocity and acceleration, $\partial^2\psi / \partial t^2$.

It allows calculation of the position of x and t. Once the equation producing the curve is known, we plug in the numbers, do some calculations, and out pop the answers.

Figure 5-7 Wave function

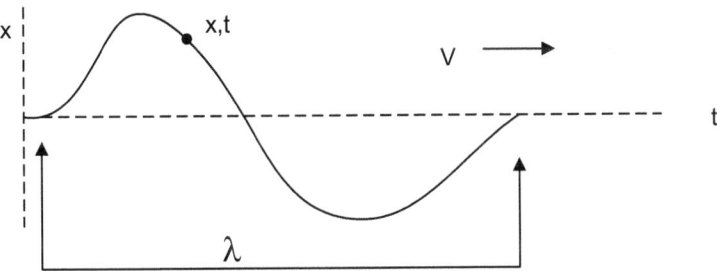

A sine wave showing x, v, λ and t.

The one-dimensional wave equation is a partial differential equation:

$$\nabla^2\psi = \frac{1}{v^2}\frac{\partial^2\psi}{\partial t^2} \qquad (9)$$

Compton and de Broglie wavelengths

The De Broglie wavelength is defined as that of a particle and the corresponding Compton wavelength is that of a photon.

The *Compton wavelength* is found by bouncing a photon off a particle and calculation using the scattering angle gives the wavelength. This wavelength is fundamental to the mass of the particle and can define its value. See Figure 5-8.

The Compton wavelength is:

$$\lambda_c = h / m_e c \qquad (10)$$

where h is Planck's constant, mass m_e is that of the electron (it could be a different particle), and c is the velocity of light.

Where the Compton wavelength is that of a photon, the *de Broglie wavelength* is that of a particle. Instead of $\lambda_c = h / m_e c$, the de Broglie wavelength is:

$$\lambda_{de\,B} = \lambda = h/p = h/mv \qquad (11)$$

where h is Planck's constant, m is mass of any particle, and v is the velocity. The wavelength $\lambda_{de\,B}$ relates to its momentum $p = mv$ in exactly the same way as for a photon. It states the essential wave-particle duality and relates them.

Figure 5-8   Compton Wavelength

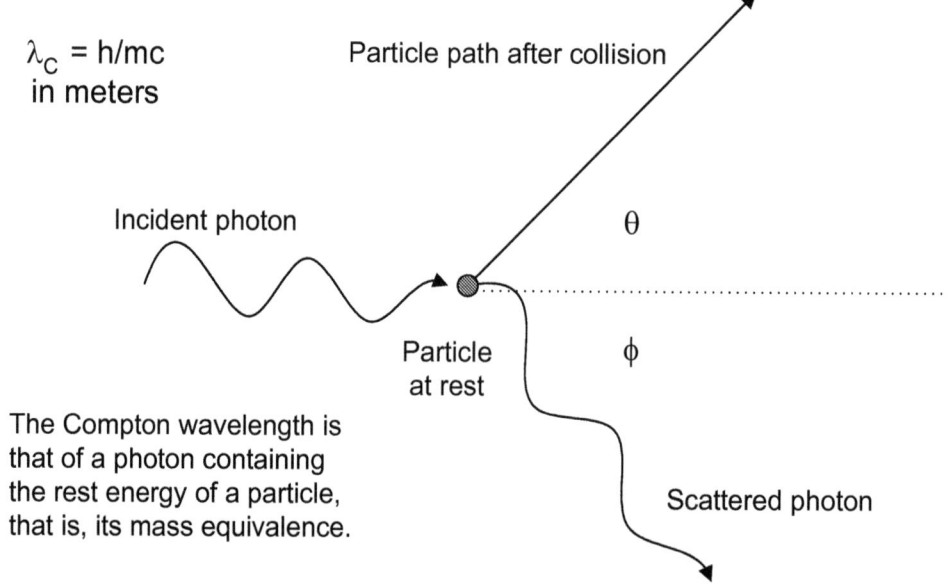

$\lambda_c = h/mc$ in meters

Particle path after collision

Incident photon

Particle at rest

The Compton wavelength is that of a photon containing the rest energy of a particle, that is, its mass equivalence.

Scattered photon

Note that the particle has a very high frequency and consequently a very short wavelength. A particle is theorized to be a standing wave of highly compressed frequency energy, not a little solid ball. The frequency is invariant. It is the dominant concept and the wavelength results from the frequency.

The Compton wavelength is directly related to the curvature of the particle using:

$$\lambda_c = 1/\kappa_0 \qquad (12)$$

We define κ as the wave number also; see Glossary. κ = 1/r and the curvature is then R = κ² = 1/r². Curvature R is then measured in 1/m² or m⁻². This is used frequently in Evans' work. See Figure 5-9.

Figure 5-9

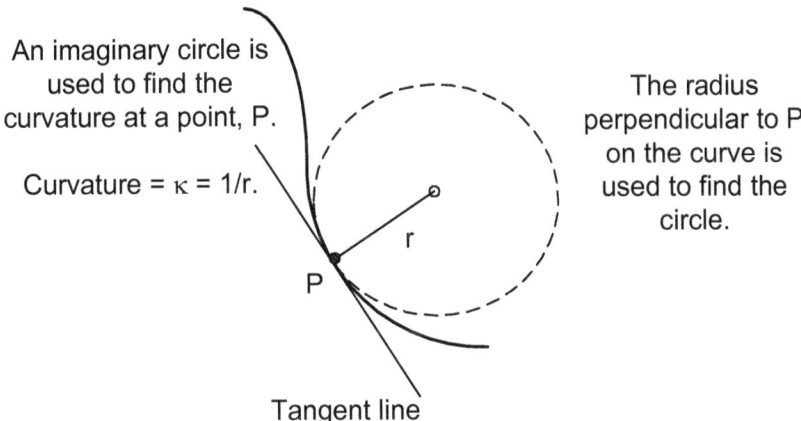

An imaginary circle is used to find the curvature at a point, P.

Curvature = κ = 1/r.

The radius perpendicular to P on the curve is used to find the circle.

Tangent line

Schrodinger's Equation

There are a number of ways this can be expressed.

$$\frac{\hbar^2}{2m} \nabla^2 \phi = -i\hbar \frac{\partial \phi}{\partial t} \qquad (13)$$

Here $\hbar = h/2\pi$ = Planck's constant / 2π = 1.05×10^{34} joule-second; $\hbar$ is also known as the Dirac constant. m is mass of a particle, $i = \sqrt{-1}$, ∂ is the symbol for a partial differential, used here it is essentially the rate of change of ϕ with respect to the rate of change of the time.

∇^2 is the Laplacian operator in 3 dimensions:

$$\frac{\partial^2}{\partial x^2} + \frac{\partial^2}{\partial y^2} + \frac{\partial^2}{\partial z^2} \qquad (14)$$

ϕ is as often seen as ψ. It is the wave function. In quantum theory ψ^2 is used to find the probability of an energy, a position, a time, an angular momentum. While there is much mathematics beyond the level at which we are working, the important concept is that quantum mechanics is statistical in nature. We cannot find the exact value – just an average. However we can say that if we perform an experiment and look at, say 100 results, that 40 of them will be yes and 60 no. Or we can say that the location of a particle is spread over an area and is not ever in one location. 1% of the time it is at position X1, 1% of the time it is at position X2, 1% of the time it is at position X3, ... and 1% of the time it is at position X100.

In general relativity this statistical approach is not accepted. Evans shows that quantum mechanics emerges from general relativity and that the statistical nature is inconsistent.

Einstein never accepted quantum probability interpretations, even though he helped established the quantum itself. Its probabilistic nature did not fit with relativity.

Not explored as of the time of this writing are the implications of the Evans wave equation and the interpretation of the probabilities. These probabilities have been very accurate. The interpretation of quantum mechanics has been deeply argued about over the years and is an interesting subject in itself which will no doubt be reinterpreted in light of Evans' work.

∇^2 measures the difference between the value of a scalar point and the average in the region of the point. In Schrodinger's equation the value is proportional to the rate of change of the energy with respect to time.

Dirac Equation

This is a three spatial and one time ("3+1") dimensional relativistic version of the Schrodinger equation. It predicts antiparticles. It can be correctly written a number of ways.

The Dirac equation in its original form is:

$$(i\gamma^\mu \partial_\mu - mc/\hbar)\psi = 0 \tag{15}$$

$i = \sqrt{-1}$, γ^μ is the Dirac spinor matrix, m is mass, c is the velocity of light, and $\hbar = h/2\pi$, ψ is the wave function. A spinor can be thought of as sort of a square root of a vector. The mathematics comes from the concepts of rotations.[19] The Dirac equation can be derived from general relativity using the Evans equations.[20]

> It has been assumed in the past that while the Dirac equation is correct, it cannot be proven. This is no longer the case. The Dirac equation can be derived from General Relativity using the Evans equations. That means the information the Dirac equation reveals is in fact a subset of what the Evans plus General Relativity equations contain. Quantum theory emerges from relativity.

Mathematics and Physics

To a certain degree, physics is mathematics. We see in Einstein and now in Evans that they both claim that differential geometry is physics. However some equations are more physical than others. It is necessary to find out which equations are physics.

For example, we can say $0 = 0$ (16)

and then $0 = +1 - 1$.   (17)

These are both true mathematically. If we let the mathematics stay on the left side of equation (16) and let physics stay on the right, we could say that the

---

[19] See *Chapter 41 Spinors* in Gravitation by Misner, Thorne, and Wheeler as well as web sources.

[20] From an email from Professor Evans: "Derivation of the Dirac equation from the Evans Wave Equation" (Found. Phys. Lett., submitted, and on www.aias.us) the Dirac equation is given in position representation as eqn. (91), and the Dirac equation in the Klein Gordon form, eqn. (90), is derived from eqn. (91). I will write out all details, and post them on www.aias.us, because these are difficult at first, even for a professional physicist. After a bit of practice though the notation becomes easier to use. The Dirac equation in Klein Gordon form appears as the flat spacetime limit of my wave equation when the metric vector is represented in spinorial form in SU(2)."

sum of energy in the universe was 0. Then we postulate some event or condition that allows the right side to be +1 − 1 giving us equation (17).

The universe does seem to follow 0 = +1 − 1. So far we see almost all creation being a balance between positive and negative, left and right, up and down, etc.

However an ansatz (postulate or conjecture) is necessary to get us from zero to one minus one – that is from nothing to two sums adding to nothing. This conversion from mathematics to physics is a mystery. Some definition, event, or other unknown occurrence is necessary to move from mathematics to physics. Einstein's constant k and we will see Evans constant $A^{(0)}$ serve this purpose.

As much as this author is afraid to argue with Einstein or Evans, it seems more logical that mathematics proceeds from physics. Given some unexplained reason for existence, mathematics was buried inside it. We are finding the relationship between them. True mathematics gives us tools for discovery, but it is hard to see how mathematics preceded physics - a minor point at this juncture.

Summary

This is another difficult chapter, particularly for the layman. However, for anyone looking to understand general relativity and the Evans equations, it is necessary to be at least familiar with the vocabulary.

A surface familiarity is sufficient for basic understanding. If one looks at equation (13) it looks quite daunting at first glance. In addition, for the professional, it is a lot of work to solve. Yet one can look at the parts and see that there are a lot of constants and two numbers, t and x, give ψ, the probability.

It is not necessary to be able to solve it in order to see how it is used.

Figures 8 and 9 are the most important to understand. The Compton wavelength and curvature are concepts will be used in some important unification equations.

If one chooses to study Evans' papers on the electromagnetic sector of physics, then the web site references or a good college physics text is necessary.

Chapter 6 The Evans Field Equation

> In my terminology spaces with a Euclidean connection allow of a curvature and a torsion; in the spaces where parallelism is defined in the Levi-Civita way, the torsion is zero; in the spaces where parallelism is absolute the curvature is zero; thus there are spaces without curvature and with torsion....I have systematically studied the tensors which arise from either the curvature or the torsion; one of those given by the torsion has precisely all the mathematical characteristics of the electromagnetic potential.
>
> Elie Cartan, 1929[21]

Introduction

There are four recognized fields or forces in existence:

Gravitation. This is the curving of spacetime. Einstein presented this in 1916.

Electromagnetism. This is the torsion or spinning of spacetime. Einstein and Cartan believed this to be the case, but neither was able to develop it and keep gravitation in the equations.

Strong force. This is the force that holds the proton together. This is gravitational in nature according to the Evans equations.

Weak field. This holds the neutron together and is associated with the W boson. It is electromagnetic in nature. When existing outside a nucleus after an average of 10.3 minutes a neutron will emit an electron and antineutrino and become a proton.

---

[21] P7, Letters on Absolute Parallelism, 1929-1932, Elie Cartan and Albert Einstein, ed. by Robert Debever, Princeton, 1979.

Chapter 6 – The Evans Field Equation

Force = field	Symbol	Description
Gravitation	G	Described by basis vectors **e** or in the tetrad by $q^a{}_\mu$. Curved spacetime.
Electromagnetism	A	A is the magnitude of electromagnetic potential.
Strong force	S	Presently described by gluons and quarks.
Weak force	W	Weak field of the neutron.

Einstein Field Equation

The complete derivation and explanation of the Evans equations requires a firm knowledge of differential geometry and both general relativity and quantum physics.[22] We do not attempt that here, but rather go over the essential results and give a brief explanation.

We are given $R + kT = 0$ by Einstein.[23] This is the basic postulate of general relativity. "k" is the Einstein constant = $8\pi G/c^2$.

R is a measure of curvature (κ^2) and T is a measure of energy density (mass, pressure, self-gravitation and/or 4-velocity). In the weak field limit, T = m/V, that is mass per volume which is density.

$R + kT = 0$ states that spacetime experiences curvature in the presence of energy. That is:

$$R = -kT \qquad (1)$$

---

[22] For a more complete discussion see www.aias.us, and references for a list of published and preprint papers by Myron Evans and his book GENERALLY COVARIANT UNIFIED FIELD THEORY, The Geometrization of Physics, M. W. Evans, (Springer, van der Merwe Series).

[23] For a good history of the development of the postulate see Misner, Thorne, and Wheeler p 432 ff. It was not always so obvious as it is today.

The negative sign in $-kT$ is a convention, but it can serve to remind us that energy curves spacetime inwards. See Figure 6-1.

Figure 6-1

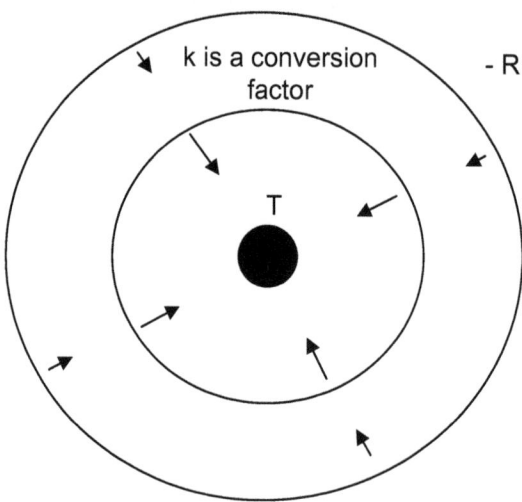

Spacetime is "pulled" towards mass-energy
Two dimensions are suppressed here.

k is a conversion factor

$-R$

T

This is expected, as we know gravity "pulls" things towards its source. A small change in paradigm here is that masses do not pull each other, they pull on spacetime. They *curve* spacetime.

Curvature and Torsion

In Riemann or non-Euclidean spacetime[24] the symmetric metric tensor is $q^{\mu\nu(S)}$.

---

[24] Minkowski space refers to the mathematical tangent spaces and to that of special relativity. The manifold spacetime of our universe in Riemann general relativity is non-Minkowski space. We refer to the unified space with both asymmetric and antisymmetric metric to be Evans space.

Chapter 6 – The Evans Field Equation

The most general asymmetric metric tensor is defined by the outer or tensor product of two tetrads[25]:

$$q^{ab}{}_{\mu\nu} = q^a{}_\mu q^b{}_\nu$$
$$= q^{ab}{}_{\mu\nu}{}^{(S)} + q^{ab}{}_{\mu\nu}{}^{(A)} \qquad (2)$$

The antisymmetric metric tensor is $q^{\mu\nu(A)}$. This defines an area, dA. The antisymmetric metric tensor is defined by the wedge product of two tetrads:

$$q^{ab}{}_{\mu\nu}{}^{(A)} = q^a{}_\mu \wedge q^b{}_\nu \qquad (3)$$

Symmetry indicates centralized potentials – spherical shapes.

Antisymmetry always involves rotational potentials – the helix.

Asymmetry indicates both are contained in the same shape.

In differential geometry, the zero-form ds^2 implies the zero-form dA. This means it is in a way perpendicular. The symmetric vectors give us distances in 4-dimensional spacetime and the antisymmetric vectors give us turning out of the spacetime. See Figure 6-2.

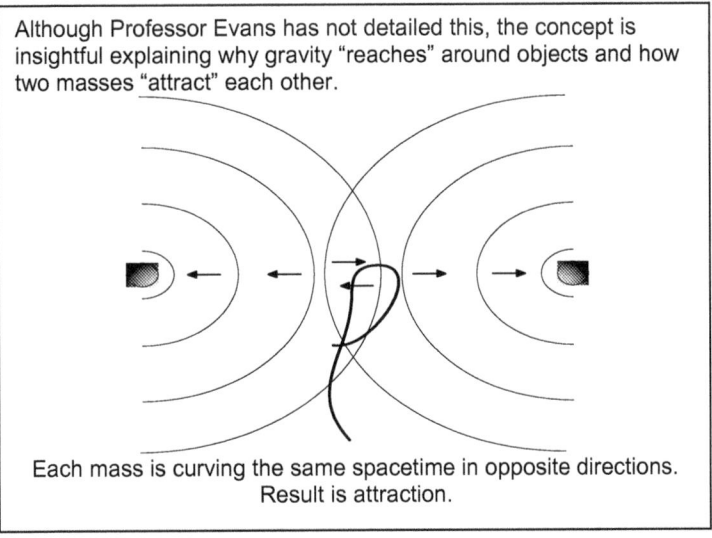

Although Professor Evans has not detailed this, the concept is insightful explaining why gravity "reaches" around objects and how two masses "attract" each other.

Each mass is curving the same spacetime in opposite directions. Result is attraction.

---

[25] Here the tetrad is equivalent to metric four-vectors

Figure 6-2   Curvature and Torsion

$$G^{\mu\nu} = G^{\mu\nu(S)} + G^{\mu\nu(A)}$$

Asymmetric spacetime has both curvature and torsion

Gravitation is curved symmetric spacetime

Electromagnetism = spinning spacetime It is antisymmetric.

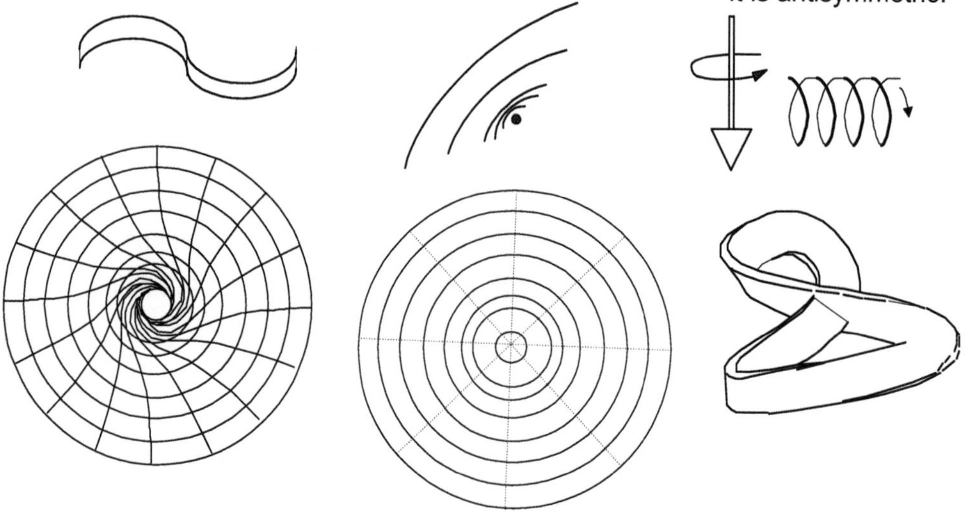

The Evans Field Equation[26]

We have the established Einstein equation $G_{\mu\nu} = R_{\mu\nu} - \tfrac{1}{2} g_{\mu\nu} R = -kT_{\mu\nu}$ which is the tensor version in component form. $R_{\mu\nu}$ is the Ricci tensor and $T_{\mu\nu}$ is the stress energy tensor. The units on each side are $1/m^2$ with curvature on the left side and on the right side is mass-energy density.[27] Essentially, it says that the gravitation equals the stress energy.

The initial Evans equation is in tetrad form instead of Einstein's tensor form:

$$R^a{}_\mu - \tfrac{1}{2} R q^a{}_\mu = k T^a{}_\mu \qquad (4)$$

---

[26] Evans introduced his unified field theory in *A Generally Covariant Field Equation for Gravitation and Electromagnetism*, M.W. Evans, Foundations Physics Letters, Vol. 16, p. 369ff (2003). His wave equation and other papers followed.

[27] Calculation of values of the Ricci tensor takes some 25 lines and of the curvature another 15. We do not want to get involved in components in real spaces due to the complexity.

Equation (4) is the well known Einstein field equation in terms of the metric and tetrads. It gives distances in spacetime that define the curvature.

R^a_μ is the curvature tetrad. R is the scalar curvature. q^a_μ is the tetrad in non-Euclidean spacetime. k is Einstein's constant. T^a_μ is the stress energy tetrad – the energy momentum tetrad which is directly proportional to the tetrad of non-Euclidean spacetime. This was the first equation of the unified field theory. It was published in an email to the aias group in 2002. The concept of metric four-vector was used first but has since been developed into the tetrad.

Equation (4) is similar to the Einstein equation, but the mixed Latin/Greek indices indicate that tetrads are used, not tensors.

From the basic structure of equation (4) we may obtain three types of field equation:

$$g_{\mu\nu}^{(S)} = q^a_\mu q^b_\nu \eta_{ab} \tag{5}$$

$$g^{ab}_{\mu\nu}{}^{(A)} = q^a_\mu \wedge q^b_\nu \tag{6}$$

$$g^{ab}_{\mu\nu} = q^a_\mu q^b_\nu \tag{7}$$

$g^{ab}_{\mu\nu}$ in equation (7) is the combined asymmetric metric.

Mathematically, the antisymmetric metric $g^{ab}_{\mu\nu}{}^{(A)}$ is the key concept to unification.

By using equation (6) we arrive at the new generally covariant field equation of electrodynamics.

$$q^a_\mu \wedge (R^b_\nu - \tfrac{1}{2} R q^b_\nu) = k\, q^a_\mu \wedge T^b_\nu \tag{8}$$

This gives the turning or spinning of the electromagnetic field.

Evans uses equation (4) and derives the Einstein gravitational equation (5), which is tensor valued. He also uses equation (4) and derives the electromagnetic equation (6) by using the wedge product.

Figure 6-3   The result of using the wedge product is an egg crate or tubes in spacetime that define magnetic field lines.

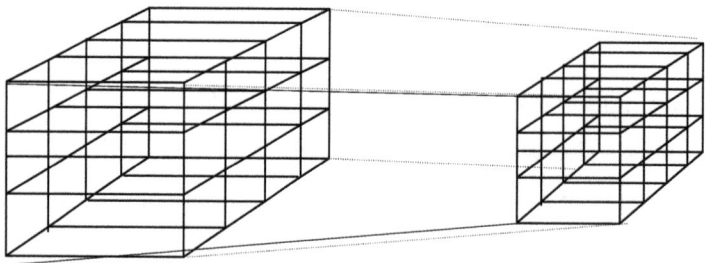

The invariance of the results is seen when the reference frame contracts and a point for point correspondence exists between the first representation and the second.

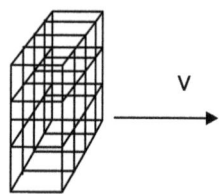

If a reference frame is accelerated, the egg crate of lines is compresed.

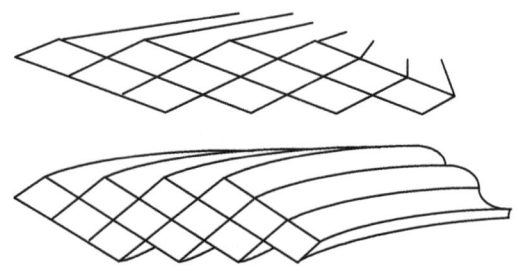

In a spacetime "deformed" - curved - by irregular gravitational fields, the eggcrate force lines can collapse.

A wedge product gives tubes or egg crate grid lines in space. Electromagnetics are described by such lines of force. The torsion can be described as spinning of those lines. The lines are spacetime itself, not something imposed on spacetime. The egg crate deforms due to gravitation. See Figure 6-3.

These are generally covariant <u>field</u> equations of gravitation and electrodynamics:

$$R^a{}_\mu - \tfrac{1}{2} R q^a{}_\mu = k T^a{}_\mu \qquad \text{Evans} \qquad (9)$$

$$R_{\mu\nu} - \tfrac{1}{2} R g_{\mu\nu} = k T_{\mu\nu} \qquad \text{Einstein} \qquad (10)$$

$$q^a{}_\nu \wedge (R^b{}_\nu - \tfrac{1}{2} R q^b{}_\mu) = k q^a{}_\nu \wedge T^b{}_\mu \qquad \text{Torsion (Evans)} \qquad (11)$$

Chapter 6 – The Evans Field Equation

Note that equation (11) is antisymmetric and indicates turning. While it is impossible to go into the details here, there is a symmetric metric and an antisymmetric metric. Electromagnetism is described by the antisymmetric metric, equation (3).

> Using equation (4) and an inner or scalar product of tetrads, one gets Einstein's gravitation.
> Using equation (4) and a wedge product, one gets generally covariant electrodynamics.

A tetrad formulation is simply explained as follows: using $R = -kT$, the inner product of two tetrads gives the gravitational field and the wedge product of two tetrads gives the electromagnetic field.

> The chapter shows the potential field approach. Another approach arrives at the gauge invariant gravitational field.
> $G^a{}_\mu$ is a potential vector tetrad and $T^a{}_\mu$ is the energy momentum vector tetrad.
> $G^a{}_\mu = kT^a{}_\mu$
> The gravitational field is the tetrad $q^b{}_\nu$

Torsion

The concept of torsion in general relativity goes back to Einstein and Cartan.[28] Gravity was established as the curvature of 4-dimensional spacetime.

---

[28] See Letters on Absolute Parallelism, op. cit.

Unification could be achieved if electromagnetism was also a geometrical property of spacetime.

Figure 6-4   Cartan Moving Frames and Evans Rotating Frames

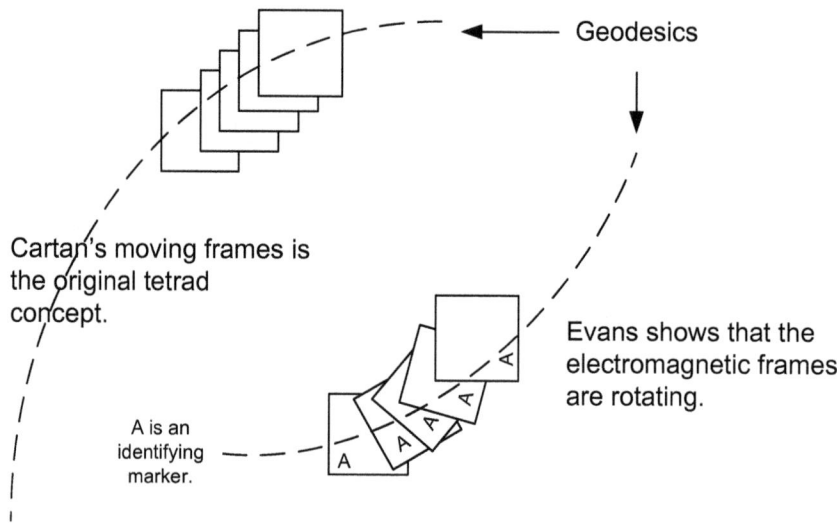

In curved spacetime it is necessary to find a way to move vectors from one location to another while keeping them parallel. In this way we can compare the different spacetimes to one another and determine changes in dimensions. This is done by using Christoffel symbols, $\Gamma^{\kappa}{}_{\mu\nu}$. See Glossary.

In Einstein's relativity, the torsion does not exist. The torsion tensor is $T^{\kappa}{}_{\mu\nu} = 0$. Cartan found that the moving frame tetrad is an alternate way to describe Einstein's relativity. Evans shows the frames are rotating when electromagnetism is present. See Figure 6-4.

Another way to look at this is through the concept of *parallel transport*. Geodesics in Einstein's relativity are the paths of free fall – the straight lines in the curved space. Parallel transport moves vectors (or any object) along the geodesic.

The most familiar form of parallel transport is the use of a Schild's ladder. The vectors are parallel as shown in Figure 6-5. The final vector is parallel to the original vector – its orientation in its own reference frame changes. Note the difference in angles made between the vector and the curved geodesic.

Neither Cartan nor Einstein ever found the method to connect the torsion to Riemann geometry based general relativity. The Evans solution, $R^a{}_\mu - \frac{1}{2} R q^a{}_\mu = kT^a{}_\mu$, is more fundamental than Einstein's equation, $G_{\mu\nu} = R_{\mu\nu} - \frac{1}{2} g_{\mu\nu} R = -kT_{\mu\nu}$. Evans' solution was to show how the metric of the universe is both symmetric and antisymmetric.

A unified field theory requires both curvature of spacetime for gravitation and torsion in the geometry for electromagnetism. The standard symmetric connection has curvature, but no torsion. The Cartan method of absolute parallel transport had torsion, but no curvature.[29]

The equations (4) to (8) are central to unification of gravitation and electromagnetism. The mathematics is involved, particularly for the non-physicist, but the essential concept is quite clear.

---

[29] Lie group manifolds provide both the curvature and torsion connections. See Levi-Civita symbol in Glossary.

Evans Equations of Unified Field Theory

Figure 6-5 Torsion

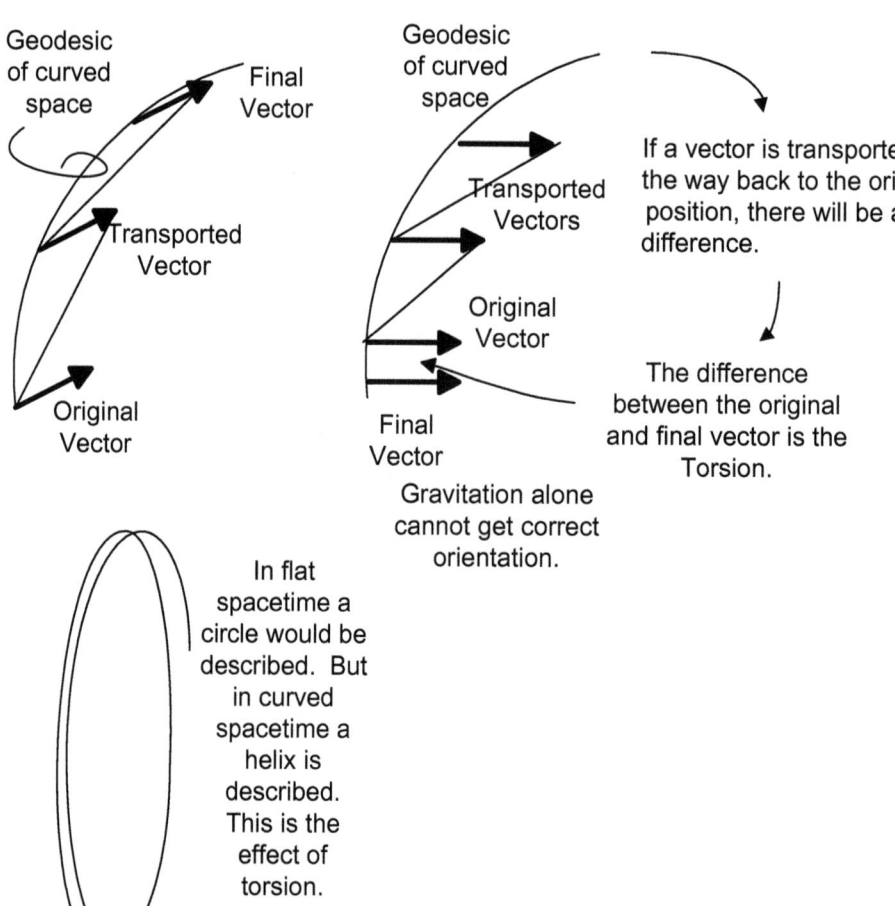

Chapter 6 – The Evans Field Equation

> The development of the Evans Wave Equation in Chapter 7 unites general relativity and quantum theory, completing unification.

Classical and Quantum

The preceding explanation of the Evans equations was in semi-classical terms of Einstein. The equations can be formulated in terms of fields as above or as wave equations as in Chapter 7. When a wave equation is developed, a great deal of new information is found.

Quantum theory then emerges from general relativity.

Figure 6-6

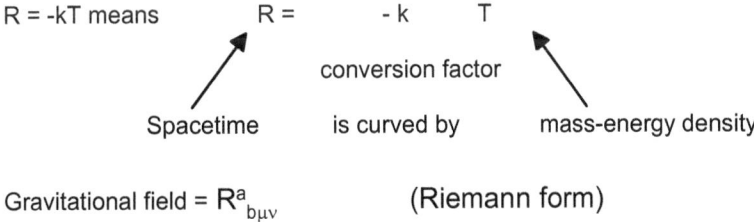

$R = -kT$ means $R =$ $-k$ T

conversion factor

Spacetime   is curved by   mass-energy density

Gravitational field = $R^a{}_{b\mu\nu}$ (Riemann form)

The Tetrad

Depending on what the index "a" represents, the same equation results in equations of gravitation, electromagnetism, the strong force, and the weak force. From the Evans equations can be derived all the equations of physics.

The Evans Wave Equation allows us to interrelate different fields – strong and gravitational for example. All the fields of physics can now be described by the same equation. The electromagnetic, strong, and weak forces are seen as manifestations of curvature or torsion and are explained by general relativity.

In the Evans Wave Equation the tetrad is the proper function for gravitation, O(3) electrodynamics, SU(2) weak force and SU(3) strong force representations. While the present standard model uses SU(2) and SU(3) mathematics, Evans has offered a simpler explanation developed from general relativity.

In the tetrad, $q^a{}_\mu$, the q can symbolize gravitation, gluon strong force, electromagnetism, or the weak force. The a is the index of the tangent spacetime. The μ is the index of the base manifold which is the spacetime of the universe.

The **tetrad** is the gravitational *potential*.

The **Riemann form** is the gravitational *field*.

A(0) is used to denote the electromagnetic potential field. A(0) is referred to as a C negative coefficient. This indicates that it has charge symmetry and e- is the negative electron. A(0) is the fundamental potential = volt-second/meter.

With an electromagnetic factor of A(0) the tetrad is the *electromagnetic potential*. The torsion form of Cartan differential geometry is then the *electromagnetic field*.

The change in perspective that allows unification is that using the tetrad allows all four forces now recognized to exist to be expressed in the same formula. Following chapters go into ramifications and refinements.

The great advantage of the new field and wave equations is that all four fields are tetrads, and all four fields are generally covariant. This means that all forms of energy originate in eigenvalues of the tetrad. That is, all forms of energy originate in real solutions of the tetrad giving scalar curvature.

The wave equation is valid for all differential geometry, irrespective of the details of any connection, so it is valid for a spacetime with torsion. Torsion can now be added to Einstein's theory of general relativity. The weak and strong fields are manifestations of the torsion and curvature tetrad.

> In more mechanical terms, energy and mass are forms of compression or expansion and spinning or antispinning of spacetime vacuum. Compression is the storage or building of potential energy. Expansion builds kinetic energy. Analogously, if spin is positive, then antispin is negative.

The torsion form is written $T^a{}_{\mu\nu} = (D \wedge q^a)_{\mu\nu}$ and is the covariant exterior derivative of the spin connection. Simplified, this is the proper mathematical description of the spinning spacetime itself. It is a vector valued two form. The Riemann form is a tensor valued two form.

In Figure 6-7 the basis vectors as on the left are calculated in one reference frame, then recalculated for another reference frame using the connection coefficients. Each vector influences the others so that they change interactively as a group.

Then in the new reference frame, the new basis vectors are recalculated. They are used to find other vectors, such as four-velocity, in the new frame.

The electromagnetic and weak fields are described by

$$A^a{}_\mu = A^{(0)} q^a{}_\mu \qquad (12)$$

$A^{(0)}$ is a fundamental electromagnetic potential in volts-seconds per meter. It can be written as $\hbar/er_0$ where r_0 is the Compton wavelength. We will discuss this connection to quantum mechanics further. Here we simply see that $A^{(0)}$ is multiplied against each of the elements of the tetrad matrix to produce a new matrix, $A^a{}_\mu$.

Regarding Einstein and Cartan discussion of electromagnetism as torsion, Professor Evans commented, "However they had no **B**$^{(3)}$ field [30] to go on, and no inverse Faraday effect. If they knew of these they would surely have got the answer."

---

[30] Dr. Evans developed the B$^{(3)}$ field in 1991. See Chapter 11.

Evans Equations of Unified Field Theory

At each point in spacetime there is a set of all possible vectors located at that point. There are internal spaces with dual, covariant, etc. spaces and their vectors in addition to higher vector spaces made from these. This is the tangent space which is much larger than the base manifold. The vectors are at a single point and the orthonormal space shown in Figures 6-6 to 6-8 exists at every point. Within the tangent space the vectors can be added or multiplied by real numbers to produce linear solutions.

Figure 6-7   Physical space and the mathematical connections

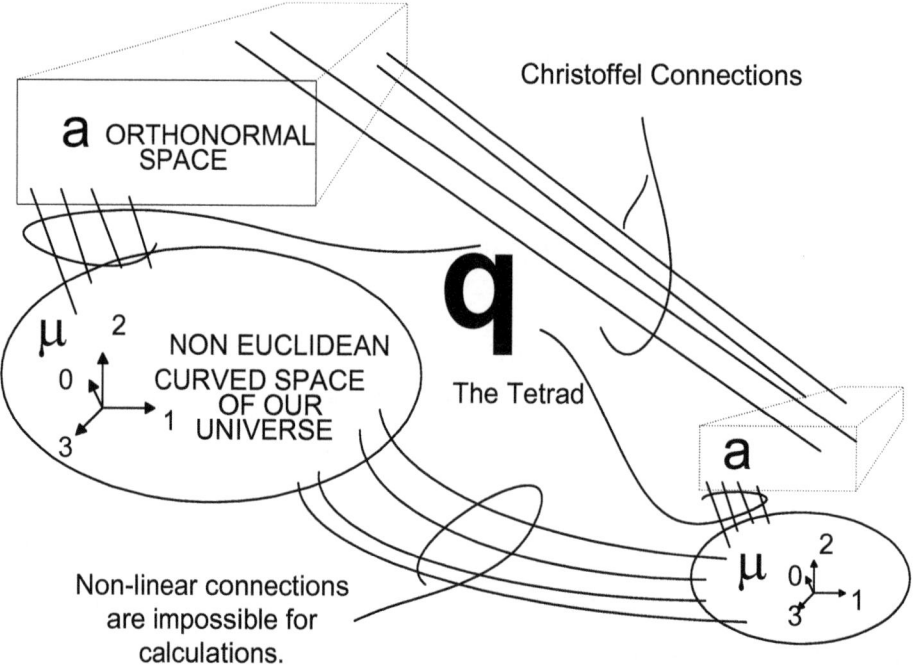

The set of all the tangent spaces is extensive and is called the tangent bundle. In general relativity the spaces are connected to the manifold itself. That is the spaces are real associated geometrical spaces.

In quantum gauge theory there are also internal vector spaces or representation space. These are considered to be abstract mathematical spaces not connected to spacetime. Rather one imagines entering a math space to perform calculations and then leaving the space back to the initial formula.

Chapter 6 – The Evans Field Equation

A vector in the tangent space of general relativity points along a path within the real universe. A vector in the phase space of quantum mechanics is vaguely defined in comparison. There is no tetrad or torsion in the quantum space. Another key to unification is to replace the abstract fiber bundle spacetime of gauge theory by the geometrical tangent bundle spacetime of differential geometry. In Evans' theory, the two are identical. This brings a more real foundation to the mathematics of quantum theory. See Figure 6-8.

Figure 6-8   Abstract Fiber Bundle and Geometrical Tangent Bundle

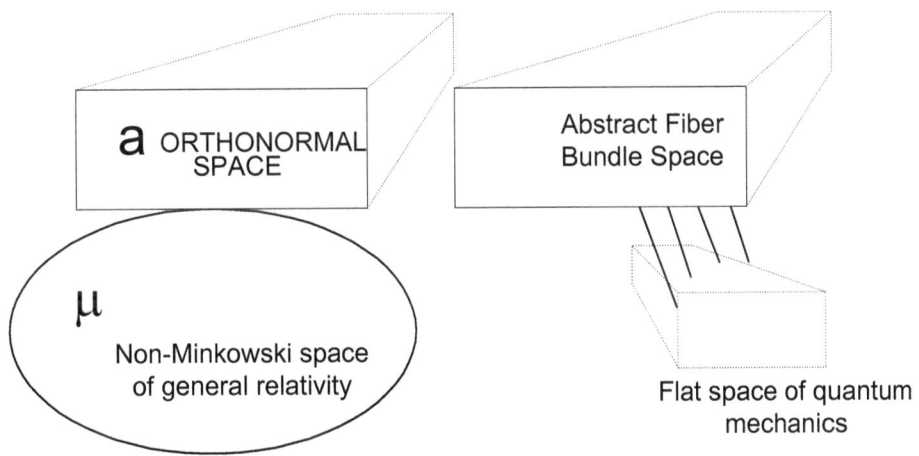

Unification replaces the abstract mathematical space with the physical orthonormal tangent space of general relativity.

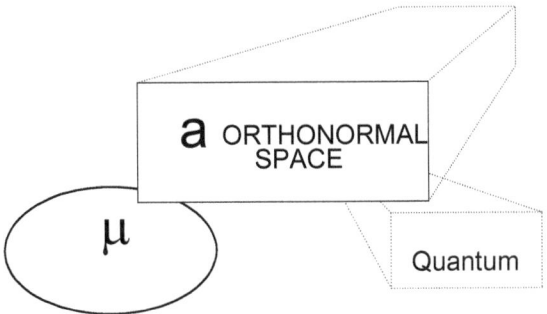

Everything is simplified in terms of differential geometry, using the tangential orthonormal space. The equations of differential geometry are

independent of the details of the base manifold [31] and so one can work conveniently in an orthonormal tangent space for all equations. (Orthogonal and normalized indicates that the dimensions are "perpendicular" and they are multiples, "linearized," with respect to the basis vectors.)

There are 16 connection scalars in the tetrad matrix. Six form an antisymmetric electromagnetic field and 10 form the symmetric gravitational field.

The tetrad is the set of basis scalars comprising the orthonormal basis. The basis vectors are orthogonal and normalized in the space labeled a; the basis vectors of the a space are in general of any type. (**e** is used traditionally in gravitation, **q** is more general and is used to refer to any of the forces.)

As for all matrices, the tetrad matrix can be split into the sum of two matrices – the antisymmetric electromagnetic and the symmetric gravitational. The gravitational field affects the electromagnetic field; that we know already. The tetrad gives us a method to define the vectors necessary to describe the mutual effects of gravitation and electromagnetism. This is done through the fundamental Bianchi identity of differential geometry, a subject beyond the scope of this book.

The gravitational and electromagnetic fields are the same thing in different guises – spatial curvature.

From Einstein we have R describing the curvature due to gravitation and kT describing the energy density that produced it. But from there he went to tensors and $G_{\mu\nu} = R_{\mu\nu} - \frac{1}{2} g_{\mu\nu} R = -(8\pi G/c^2)T_{\mu\nu}$. This formulation does not allow the torsion. From the Evans equations we see that $R^a_{\mu} - \frac{1}{2} Rq^a_{\mu} = kT^a_{\mu}$ which is a tetrad formulation allowing both gravitation and torsion. General covariance must be maintained and it is through the tetrad.

Spatial curvature contains both electromagnetism and gravitation. All forms of energy are interconvertible through spatial curvature. At the particle level, the minimum curvature R_0 is the compressed wave of a stationary particle.

---

[31] See Sean M. Carroll, Lecture Notes on general relativity, pp. 88-98, arXiv:gr-qc/9712019 v1 3 Dec 1997

Chapter 6 – The Evans Field Equation

> The physical meaning of the mathematics is that the electromagnetic field is the reference frame itself, a frame that rotates and translates.
>
> Energy is transmitted from source to receiver by the torsion form in mathematical terms; or in mechanical terms, by spinning and expansion of spacetime vacuum.

Field Descriptions

The Evans Lemma gives quantum field/matter theory from general relativity. Gravitation is described by the Lemma when the field is the tetrad, q^a_μ.

The other three fields – electromagnetism, weak, and strong - are described when the field is the tetrad multiplied by an appropriate scaling factor and is in the appropriate representation space. For example, the fundamental electromagnetic field has the symmetry of equation (6) and is described by equation (12). The strong and weak fields are described respectively as:

$$S^a_\mu = S^{(0)} q^a_\mu \tag{13}$$
$$W^a_\mu = W^{(0)} q^a_\mu \tag{14}$$

The gauge invariant fields, or gauge fields, can be defined as follows:

The gravitational gauge field is the Riemann form of differential geometry. It is curved spacetime.

Electromagnetism is defined by the torsion form. It is spinning of spacetime itself, not an object imposed upon the spacetime.

The weak field is also a torsion form. It is related to electromagnetism. Thus in the neutron's conversion to a proton the torsion form will be involved. We see an electron leave the neutron and a proton remains. Now it is more clear that there was an electrical interaction that has an explanation. See Chapter 12 on the electroweak theory.

The strong field holds the neutron and proton together.

The next chapter deals with the Evans Wave Equation which is the wave equation of unified field theory whose real or eigenoperator is the flat spacetime d'Alembertian, whose eigenvalues or real solutions are kT = -R and whose eigenfunction or real function is the tetrad $q^a{}_\mu$:

$$(\Box + kT) q^a{}_\mu = 0 \qquad (15)$$

This equation gives a new wave mechanical interpretation of all four fields. It is a wave equation because it is an eigenequation with second order differential operator, the d'Alembertian operator. From this wave equation follows the major wave equations of physics, including the Dirac, Poisson, Schrodinger, and Klein Gordon equations. The wave equation also gives a novel view of standard gravitational theory, and has many important properties, only a very few of which have been explored to date.

Summary

There are two expressions of the Evans equations - the classical and the quantum.

The fundamental potential field in grand unified field theory is the tetrad. It represents the components indexed μ of the coordinate basis vectors in terms of the components indexed a of the orthonormal basis defining the vectors of the tangent space in general relativity. In other words, the tetrad connects the tangent space and base manifold.

Differential geometry has only two tensors that characterize any given connection – curvature and torsion. There are only two forms in differential geometry with which to describe non-Minkowski spacetime – the torsion form and the Riemann form.

We can refer to gravitation as symmetricized general relativity and electromagnetism as antisymmetricized general relativity.

There are two new fundamental equations:

1) The Evans Field Equation which is a factorization of Einstein's classical field equation into an equation in a metric vector shown here in tetrad form:

$$G^a_\mu = R^a_\mu - \tfrac{1}{2} R q^a_\mu = k T^a_\mu \qquad (16)$$

From this equation we can obtain the Einstein field equation, describing the gravitational field, and also classical field equations of generally covariant electromagnetism.

2) The Evans Wave Equation

$$(\Box + kT) q^a_\mu = 0 \qquad (17)$$

Different approaches can be taken:

The potential field as indicated in the beginning of this chapter based on tetrads: $R^a_\mu - \tfrac{1}{2} R q^a_\mu = k T^a_\mu$

The gauge invariant field forms are Riemann form and the torsion form. We do not go into detail.

Figure 6-9   The spin connection and gravitation

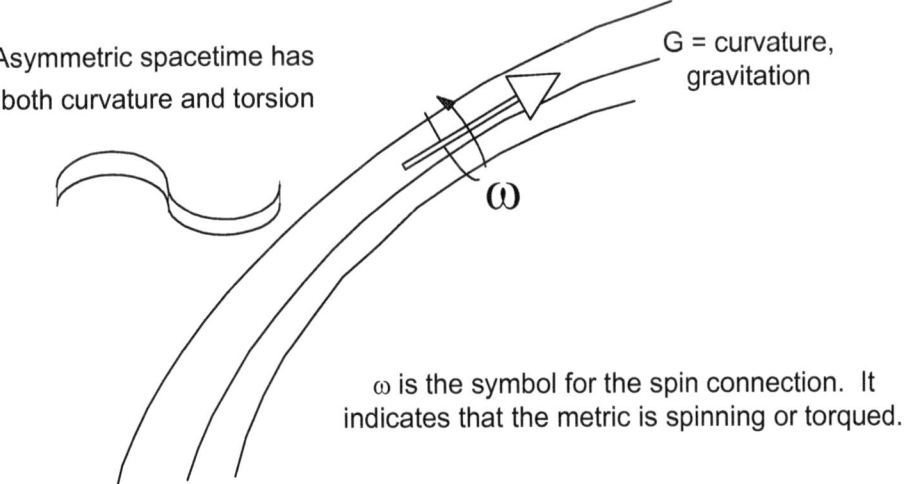

The wave equation, $(\Box + kT) q^a_\mu = 0$, a powerful link to quantum mechanics.

> The field equation can be written in terms of the tetrad:
>
> $$R^a{}_\mu - \tfrac{1}{2} R\, q^a{}_\mu = kT^a{}_\mu$$
>
> with $q^a{}_\mu = q^a{}_\mu{}^{(S)} + q^a{}_\mu{}^{(A)}$

The four fields can be seen to emerge directly from the tetrad itself and are aspects of the tetrad.

While we may not understand all of the intermediate mathematics, we can see the overall concepts.

The Evans metric of spacetime has both curvature and torsion – gravitation and spin. Figures 6-2 and 6-9 depict the two together. They are together an asymmetric metric – neither symmetric nor antisymmetric, but having both contained within. This allows gravitation and electromagnetism to exist in the same equations.

Einstein's basic postulate of general relativity, $R = kT$ means:

| R
Curvature
Mathematics | = | k
A constant
Conversion Factor | x | T
Mass or energy density
Measurable Physics |

Einstein's basic postulate of general relativity defines curvature.

Evans basic postulate of electromagnetic unification is $A^a{}_\mu = A^{(0)} q^a{}_\mu$ and means:

| $A^a{}_\mu$
Physics
Measurable Physics | $A^{(0)}$
A constant
Conversion Factor | $q^a{}_\mu$
Tetrad
Mathematics |

Evans basic postulate of electromagnetism defines spinning.

Evans Field Equation Extensions

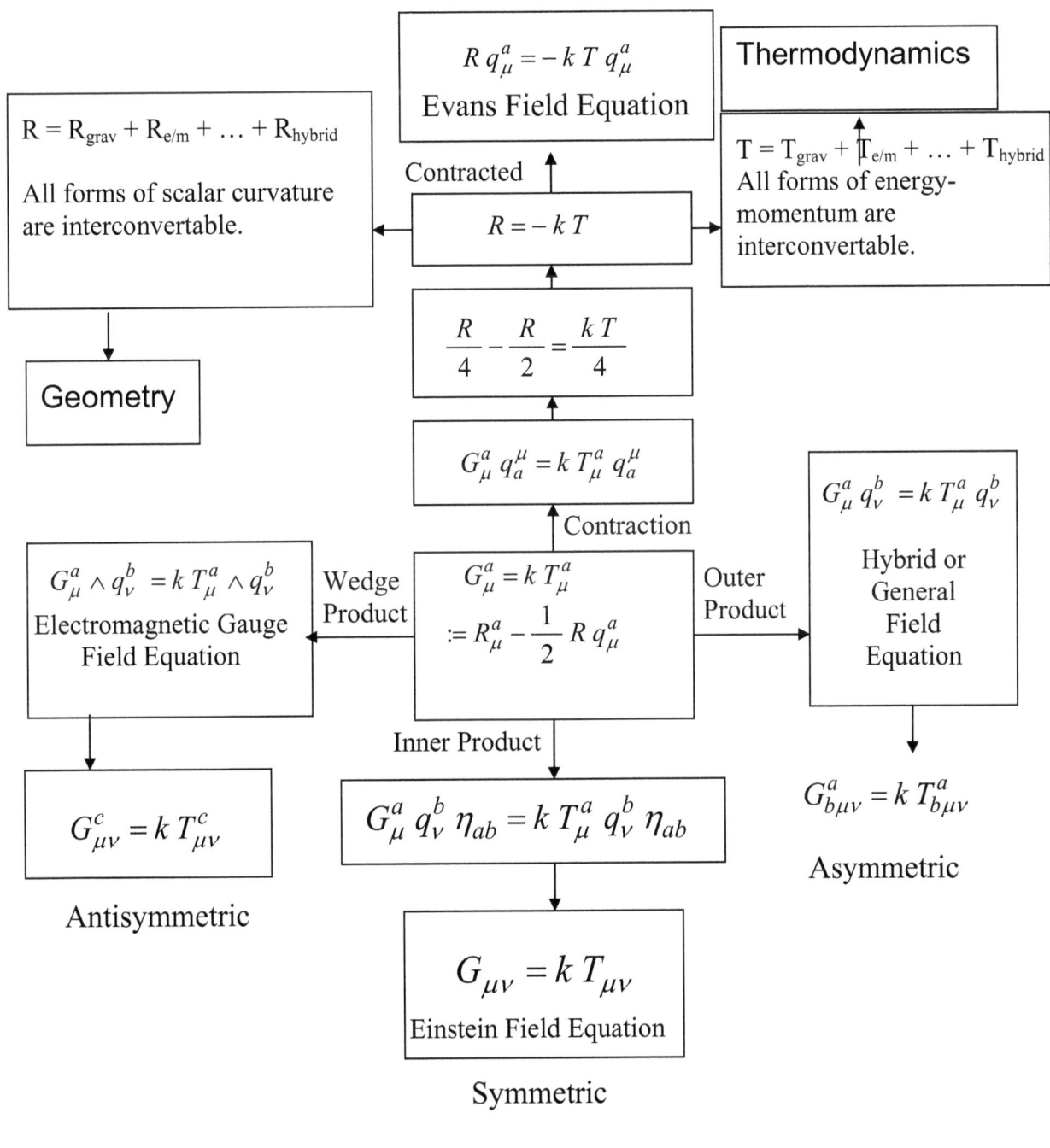

Courtesy ofr. Myron Evans

Chapter 7 The Evans Wave Equation

> Here arises a puzzle that has disturbed scientists of all periods. How is it possible that mathematics, a product of human thought that is independent of experience, fits so excellently the objects of physical reality? Can human reason without experience discover by pure thinking properties of real things?
>
> Albert Einstein

Introduction

In this chapter we will describe the wave equation a number of ways. It is the link between general relativity and quantum mechanics and is the unification equation.

1) The Evans Field Equation in the tetrad which leads to Einstein's classical field equation:

$$G^a{}_\mu := R^a{}_\mu - \tfrac{1}{2} R q^a{}_\mu = k T^a{}_\mu \qquad (1)$$

This states in essence that G, gravitation, is a function of R, curvature, and is derived from the energy density in any region. It is stated in terms of the tetrad, a more complete form than vectors and tensors. From this equation we can obtain the well known classical equations of physics.

2) The Evans Wave Equation of unified field theory:

$$(\Box + kT)\, q^a{}_\mu = 0 \qquad (2)$$

Here $\Box$ is a four dimensional rate of change, kT is the energy density, and $q^a{}_\mu$ is the tetrad - the gravitational potential field. The tetrad adjusts the differences between the base spacetime manifold and the tangent spaces used in calculations. See Chapter 4, equation (11) and Figure 7-1 for a description of the d'Alembertian.

The real physical solutions the wave equation offers, its eigenvalues, will obey:

$$R = -kT \qquad (3)$$

where R is mathematical curvature and T is the stress energy density tensor.

In the introduction to this book, the nature of spacetime was discussed. This subject is critical to understanding the progression in understanding spacetime.

Flat, three dimensional Euclidean geometry defines Newton's space + separate time. This is referred to as the metric. It is the mathematical structure of the space + time.

Minkowski spacetime is flat, but invariant distance is described and showed that space and time are inseparable – that is, they are the same thing.

Electromagnetism can be described in the above metrics using wave equations, however the equations cannot be moved from one gravitational field to another. The theory is incomplete.

Riemann four dimensional curved geometry describes four dimensional curved spacetime. Gravitation is well described. However, again, electromagnetism cannot be described. It is necessary to place equations on top of the metric in order to describe electromagnetism – yet the equations cannot be moved from one gravitational field to another. The theory is incomplete.

Cartan differential geometry allows curving (Riemann) and turning within the metric. The spacetime itself curves and spins. Now the equations of electromagnetism (spinning) are an inherent part of the spacetime. They are not an awkward addition.

It is now possible to build a wave equation that describes the spinning of spacetime while curving. The equations can be moved from one gravitational field to another.

The Evans Wave Equation describes a variety of spinning spacetime states in different gravitational fields.

Figure 7-1 The d'Alembertian and 4-dimensional spacetime

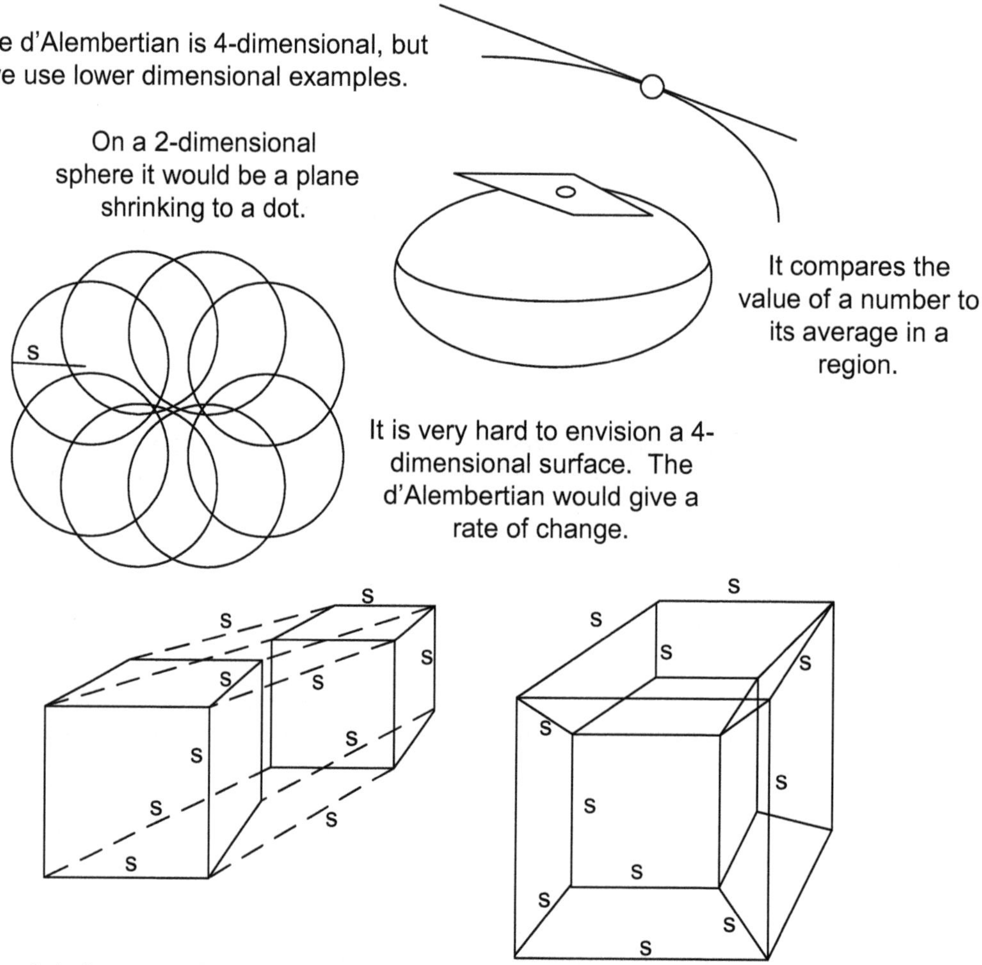

The d'Alembertian is 4-dimensional, but we use lower dimensional examples.

On a 2-dimensional sphere it would be a plane shrinking to a dot.

It compares the value of a number to its average in a region.

It is very hard to envision a 4-dimensional surface. The d'Alembertian would give a rate of change.

It helps to visualize the cube in four dimensions in order to imagine the 4-d sphere. In the left drawing, the front cube is extended through the length of its side in a fourth dimension. There is another three dimensional space where it ends. In the right drawing the sides are labeled s which is their distance. The diagonal lines form cubes with the faces of the easily seen cubes.

The wave equation's real function (eigenfunction) is the tetrad $q^a{}_\mu$. The four fields will be seen to emerge directly from, and are aspects of, the tetrad itself. They are real values of $-kT$. The d'Alembertian operator acts on the tetrad to produce the values of $-kT$.

Chapter 7 – The Evans Wave Equation

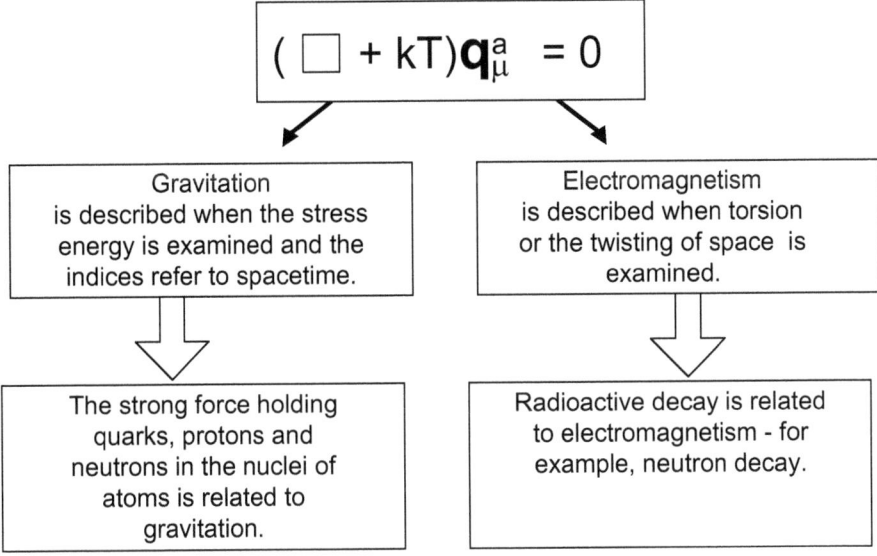

The Wave Equation

The wave equation is a quantized version of the Evans field equation. It is derived from the classical field equation discussed in Chapter 6.[32] See Figures 7-2 and 7-3 for descriptions of the wave equation components.

The d'Alembertian $\square$ is a 4-dimensional measurement of the difference between the value of a real or complex scalar number at a point, typically on a 4-dimensional curve, and its average value in an infinitesimal region near the point. *It gives a type of rate of change of the curvature at a point.*

Another way to look at the equation is to use the Evans Lemma, which is $\square q^a_\mu = -kT\, q^a_\mu$ (or $\square q^a_\mu = R q^a_\mu$).

One can develop Einsteinian gravitational theory using the (in general non-linear) Evans Lemma rather than the Einstein field equation, where gravitation is naturally quantized and radiated from the wave structure of the Lemma. Einstein

---

[32] The derivation can be found at www.aias.us or www.atomicprecision.com

and Evans' field equations are classical – continuously analog in nature whereas the wave equation is quantized.

The wave equation quantizes general relativity showing that although spacetime is not completely analog, the steps of R or kT are very small.

Basic Description of the Evans Wave Equation [33]

The Evans Lemma, or subsidiary proposition of the wave equation is:

$$\Box q^a_\mu = -kT q^a_\mu \qquad (4)$$

This is a geometrical identity based on the tetrad postulate of differential geometry. It states essentially that there exists a tangent spacetime to our universe's spacetime.

This is a very fundamental statement. In three-dimensional Euclidean space, the tetrad postulate and Lemma state essentially that there is a tangent to every curve at some point on the curve (See Figure 7.1, top right). The tangent space is then the Euclidean flat two-dimensional plane containing all possible tangents at the given point.

General relativity is built on geometry and theorems of geometry assume great importance. In the four dimensions of spacetime, it is no longer possible to easily visualize the tangent and curve but the tetrad postulate and Lemma are valid for four dimensions.

April 2003, and in *A Generally Covariant Wave Equation for Grand Unified Field Theory*, "Foundations of Physics Letters" vol. 16, pp. 507 ff., December 2003.
[33] The material in this section is based on an email from Professor Evans to explain the basics. This author has simplified further.

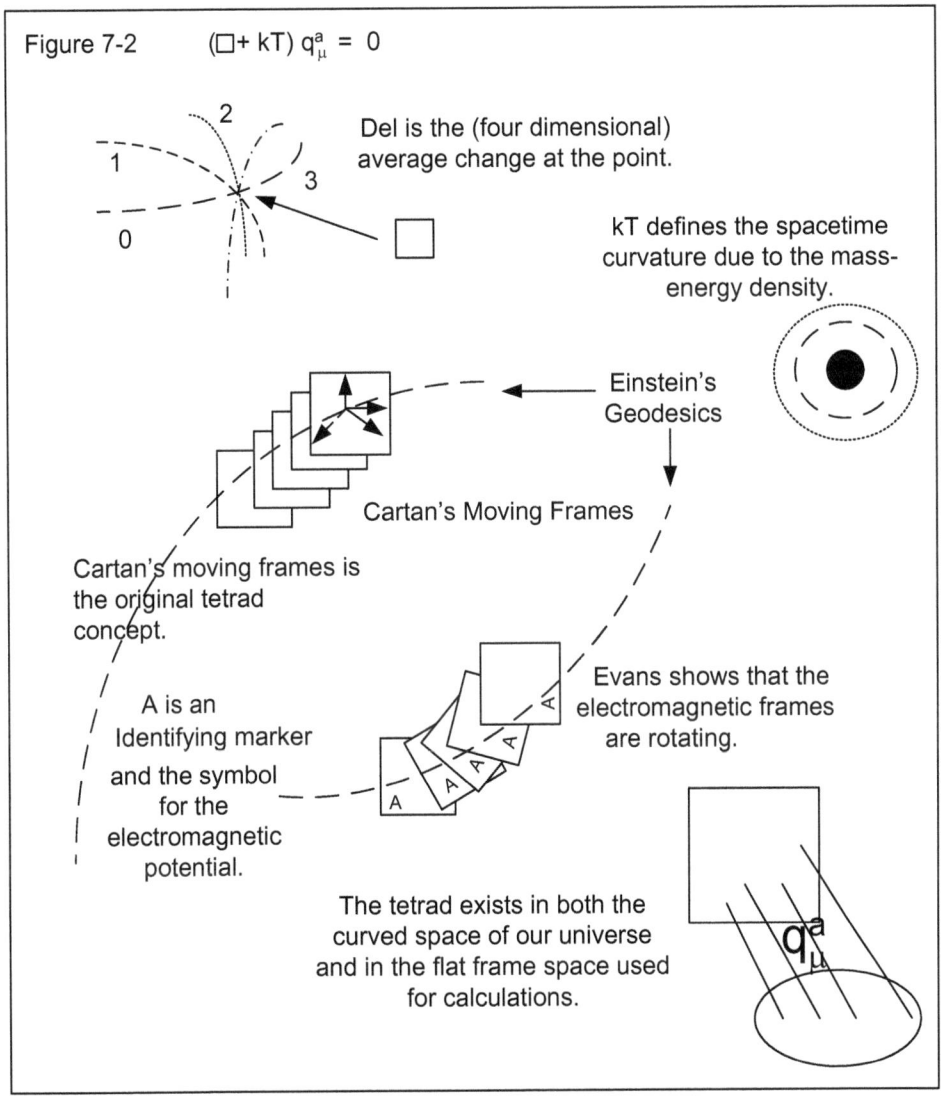

Figure 7-2 $(\Box + kT) q^a_\mu = 0$

The Lemma is an eigenequation. The d'Alembertian *operator*, $\Box$, acts on the tetrad to give a number of values of R. That is the output of the equation is a number of scalar curvatures.

There are many possible eigenvalues, so the eigenfunction is *quantized*. The tetrad is therefore the wave function of quantum mechanics and is

Evans Equations of Unified Field Theory

determined by geometry. This makes the wave equation causal and objective as opposed to the probabilistic interpretation of the Copenhagen school.[34]

The Lemma successfully unifies general relativity with quantum mechanics. Dr. Evans says, "One now sees with great clarity the power that comes from simplicity and fundamentals."

> $(\Box + kT)q^a_\mu = 0$ is derived from the field equation and is the unification equation. It is as rich as in quantum applications as Einstein's field equation is rich in gravitational applications.

The wave equation is a *generalization to all fields* of Einstein's original, famous, but purely gravitational equation, $R = -kT$. Here k is Einstein's constant and T is the index contracted canonical energy momentum density of *any* field (radiated or matter field).

From the Lemma we obtain the Evans Wave Equation:

$$(\Box + kT)\, q^a_\mu = 0 \tag{5}$$

This equation means that the eigenvalues, real results, of the d'Alembertian act on the tetrad two-form eigenfunction to produce a number of eigenvalues which equal real results of $-kT$.

It is most difficult to see the process here unless you are a mathematical physicist. However, these are the results.

T is both quantized *and* generally covariant, something which the Copenhagen School denies. All the main wave equations of physics then follow as in the flow charts at the end of Chapter 6 and this chapter.

The Evans Field Equation $R^a_\mu - \tfrac{1}{2} R q^a_\mu = kT^a_\mu$ is not an eigenequation. It is a "simple" equation in which $R = -kT$ is just multiplied on both sides by the

---

[34] It has been known for some years that the ideas of the Copenhagen School have been refuted experimentally. For example: 1) J. R.Croca "Towards a Nonlinear Quantum Physics" (World Scientific, Singapore, 2003). 2) M. Chown, New Scientist, 183, 30 (2004). This will be discussed later in this book.

tetrad. In this way one deduces a generalization to all fields of the original Einstein / Hilbert field equation (1915 - 1916), $G_{\mu\nu} = kT_{\mu\nu}$.

In the Evans Field Equation there are no operators. In the Evans Wave Equation, the d'Alembertian is the operator.

The material is best understood through differential geometry and tensor analysis, rather than vector analysis. However they are all ultimately equivalent.

The wave function is derived from the metric so the representative space refers to the metric. Wave functions of quantum mechanics can now be interpreted as the metric, not as a probability. That is, they can be interpreted as functions of distance and twisting of spacetime.

The electromagnetic, strong, and weak fields are not entities imposed *upon* spacetime. Just as gravitation is an aspect of spacetime, so the other forces are aspects of spacetime.

See Chart 2 at the end of this chapter for more completed map of ramifications of the wave equation.

The Dirac, Schrodinger and Heisenberg equations have been derived from Evans' wave equation, but Heisenberg is found to be invalid as we will discuss in Chapter 10.

We can now take each of these and describe what is happening.

The tetrad defines the fields presently thought to exist in physics. All four fields are defined mathematically by the tangent space relation to the base manifold.

Gravitation curves spacetime. If a, the tangent space, represents gravitation, then the Evans equation leads to the equations of gravitation.

The matrix defined by the tetrad interrelates two reference frames – the spacetime base manifold and the orthogonal tangent space. For a given upper index, it is generally covariant – that is, it remains invariant as the spacetime reference frame changes due to gravitation.

Evans Equations of Unified Field Theory

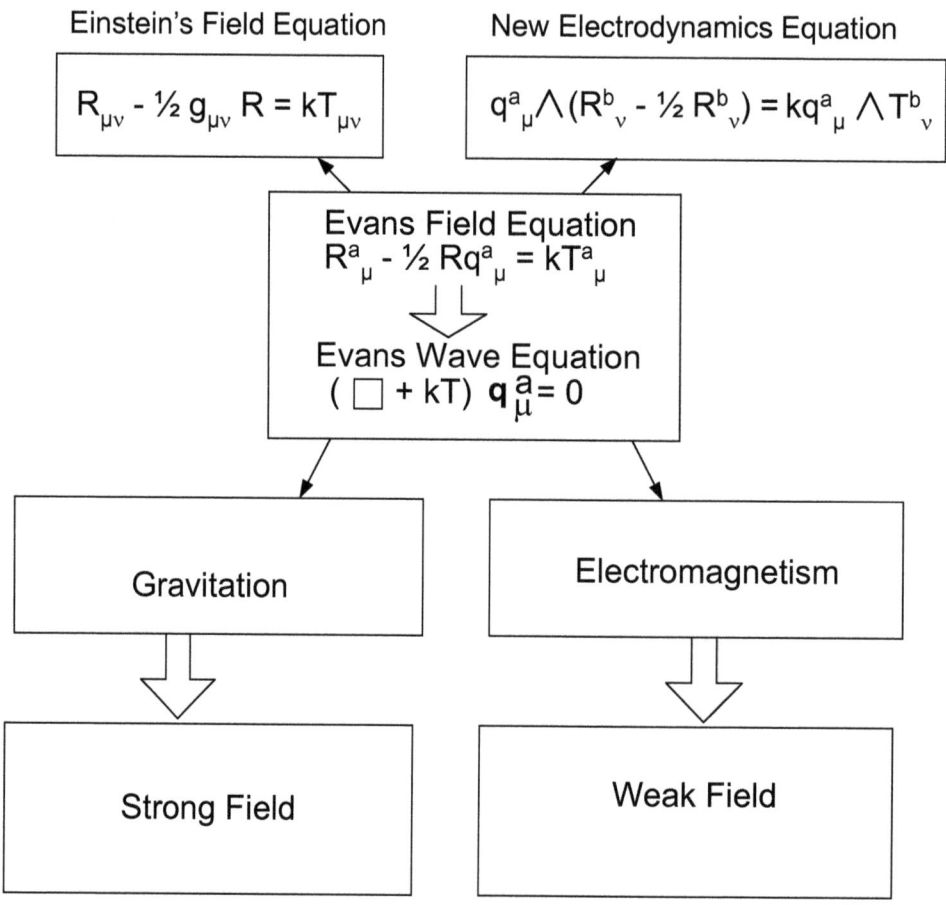

All the fields are defined by the way the tangent space is related to the base manifold. That is the way "a" relates to μ. which is (0,1,2,3) of 4-dimensional spacetime.

Chapter 7 – The Evans Wave Equation

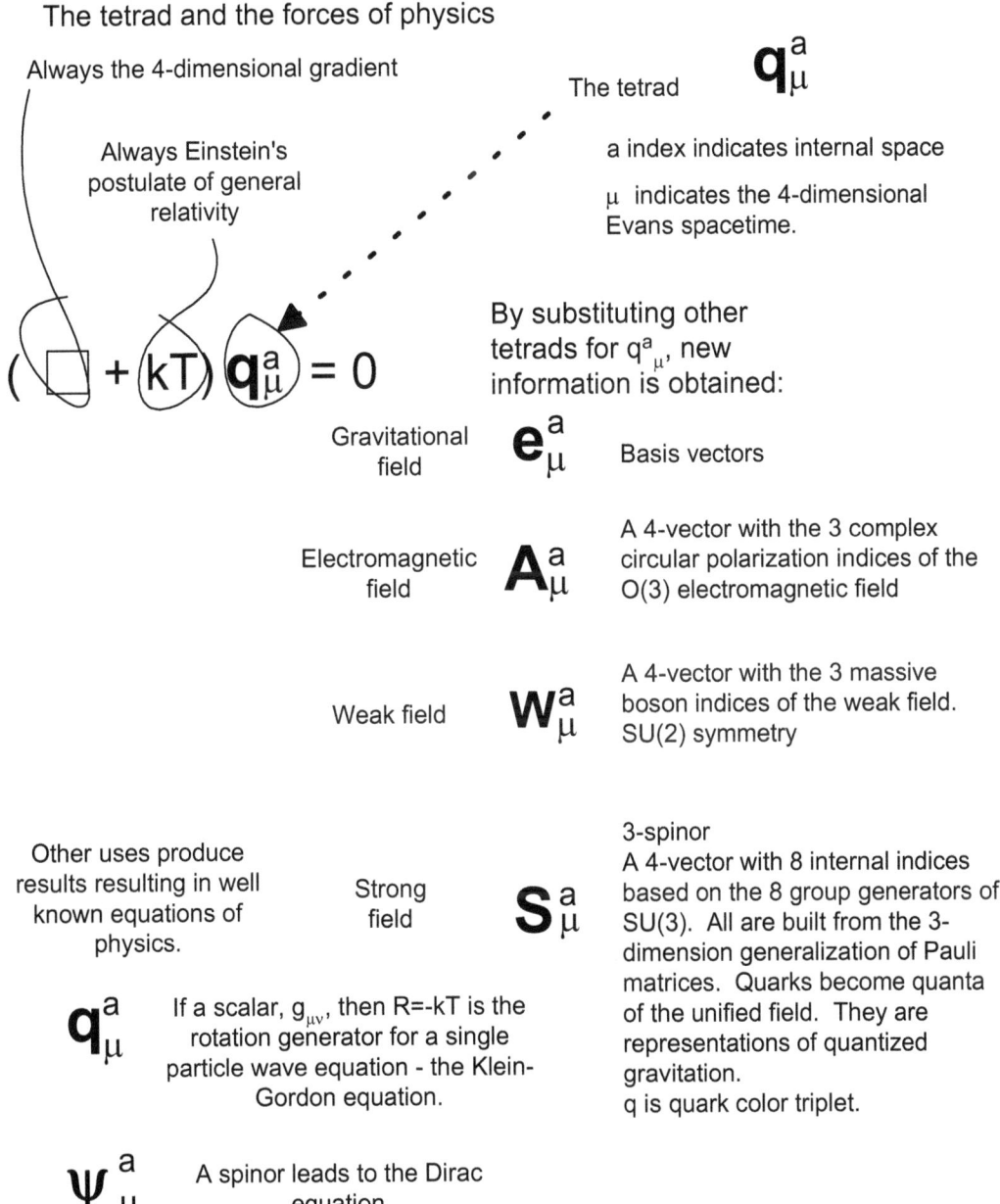

Gravitation is quantized in the wave equation.

Differential geometry is the foundation for general relativity and quantum theory. The tetrad is the potential field itself.

It is also possible to represent both the base manifold and the tangent spacetime in SU(2) representation space, or SU(3) representation space, and the tetrad is then defined accordingly as a matrix in the appropriate representation space. The tetrad thus defined gives the gauge invariant weak and strong fields from general relativity.

> The SU(2) and SU(3) representations are no longer needed, but they were a nice transitional description of the weak and strong force. As of this writing, they are still accepted, but in light of Evans' development of the electroweak force and minimum curvatures, will no longer be necessary. See Chapters 9 through 15.

Any square matrix can be decomposed into three matrices – the traceless symmetric, the trace, and traceless antisymmetric (skew-symmetric).

The Riemann form of gravitation is a square matrix built from the outer product of two tetrads.

The torsion form is built from the cross product of two tetrads.[35]

The tetrad matrix links the two reference frames – the Euclidean tangent spacetime of mathematics and the non-Euclidean base manifold of our universe. There are 16 independent components which are irreducible representations of the Einstein group. Einstein's representation using Riemann geometry had only 10 equations of general relativity. The Evans group has 6 additional equations using Cartan's tetrad.

These 16 simultaneous equations include the mutual effects of gravitation and electromagnetism on one another. The final output at any point are 10 components defining the non-Euclidean spacetime and the 6 components of the magnetic and electric fields, that is B_x, B_y, B_z and E_x, E_y, E_z.

---

[35] Dr. Evans comments: "If the fields form a non-Abelian Lie group with antisymmetric connections, then torsion exists."

Chapter 7 – The Evans Wave Equation

The six electromagnetism equations are the components of the wedge product of the tetrads. These give units of magnetic flux in webers and in units of C negative[36] or $\hbar/e$. The 10 equations resulting from the inner product are the Einstein gravitational field, but now with the effects of electromagnetism included.

The index "a" is the tangent space in general relativity and it is the fiber bundle space in gauge theory. The Riemann spacetime of Einstein's general relativity is replaced by the Evans spacetime as shown in Figure 7-3.

Figure 7-3   Abstract Fiber Bundle and Geometrical Tangent Bundle

These are shown to be equivalent in Evans' unification.

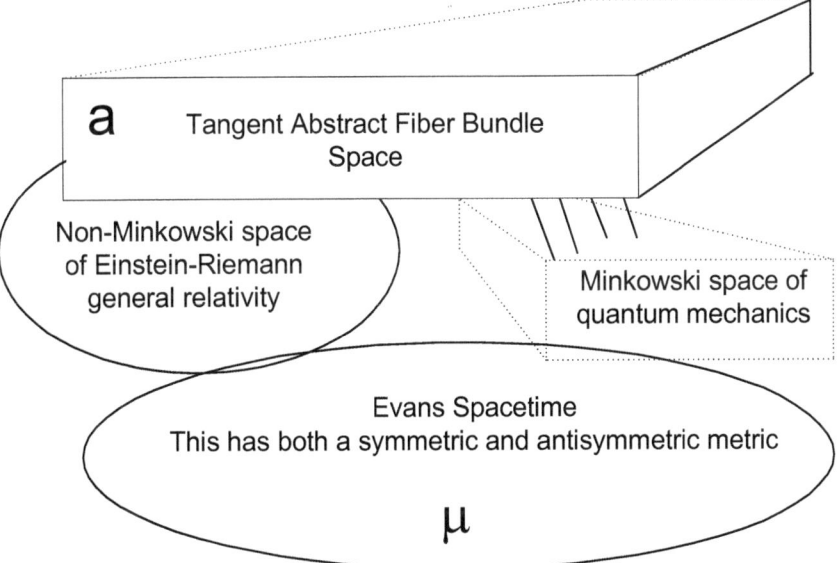

---

[36] See charge conjugation in Glossary.

167

Electromagnetism

The **B**$^{(3)}$ field is covered in its own chapter. This is the essential statement of how it is used within the wave equation. If the index a has O(3) symmetry with bases (1), (2), (3), then the Evans Wave Equation is one of generally covariant higher symmetry electromagnetics. This is $A^a{}_\mu$ which defines the electromagnetic field. $A^a{}_\mu$ is the tetrad multiplied by voltage, $A^{(0)}$.

Figure 7-4   O(3) Depiction of the Unified Field

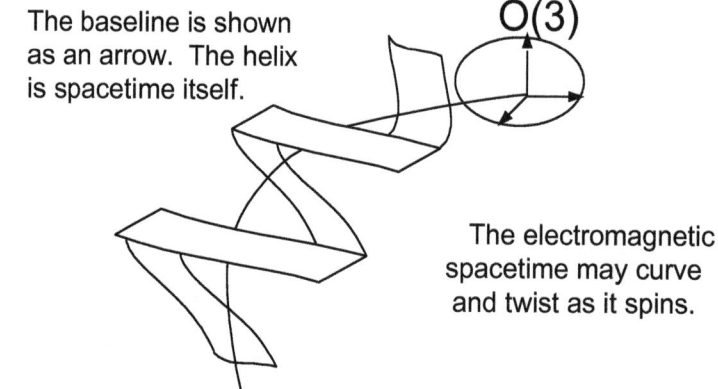

The baseline is shown as an arrow. The helix is spacetime itself.

The electromagnetic spacetime may curve and twist as it spins.

O(3) electrodynamics and the B$^{(3)}$ field

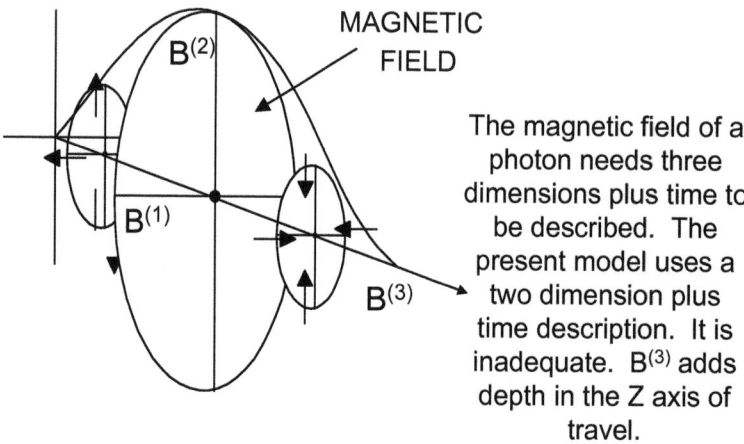

The magnetic field of a photon needs three dimensions plus time to be described. The present model uses a two dimension plus time description. It is inadequate. B$^{(3)}$ adds depth in the Z axis of travel.

O(3) symmetry is the sphere. If spherical shapes are defined and put in the tetrad, then the mathematics leads to electromagnetism. The $\mathbf{B}^{(3)}$ field adds depth to the photon as shown in Figure 7-4 and is described by the O(3) sphere.

The present theory of electromagnetism uses the circle of the Maxwell-Heaviside U(1) mathematics. Electromagnetic waves cannot exist as a circle. A circle is 2-dimensional and would have zero volume. We know electromagnetism has energy, therefore mass, and therefore must have a reality not a flat circular existence. The correct description of waves is the $\mathbf{B}^{(3)}$ field which adds depth to the construct. The Maxwell Heaviside U(1) symmetry is underdetermined. Since spacetime is curved, the U(1) description, which can be viewed as a circle, cannot invariantly describe a four dimensional field.

The $\mathbf{B}^{(3)}$ field is a tetrad component and an element of the torsion form.

Electromagnetism in the unified field theory is described with the three indices (1), (2), and (3) of the complex circular representation of the tangent space superimposed on the four indices of the base manifold. This is O(3) electrodynamics.

The Evans spin field is then the tetrad form: $\mathbf{B}^{(3)*} = -i g \mathbf{A}^{(1)} \wedge \mathbf{A}^{(2)}$. The $\mathbf{B}^{(3)}$ spin field is covered in Chapter 11.

O(3) electromagnetism is an intermediate step between electromagnetism and unified field theory. The equations of O(3) can be derived from the Evans equations in general relativity, another unification link.

Weak force

If the tetrad index a has SU(2) symmetry then the equation leads to the weak field.

The weak field bosons are essentially the eigenvalues of Evans' wave equation in this representation (three indices of the tangent spacetime superimposed on the four indices of the base manifold). The masses of the weak field bosons are minimum curvatures of R. This is explored further in Chapter 12.

Evans Equations of Unified Field Theory

Strong force

If tetrad index "a" has SU(3) symmetry, then the strong field equations result. The standard interpretation of the mathematics is the Gell-Mann quark-gluon concept.

If $q^a{}_\mu$ is the Gell-Mann quark color triplet = $[q_r, q_w, q_b]$, "a" represents the gluon wave function or the proper function of the quantized strong-field. This unifies the strong nuclear field and gravitational field. The point is that the strong force is similar to the gravitational force. See Figure 7-5.

Figure 7-5   Quarks - discrete entities or curvature forms

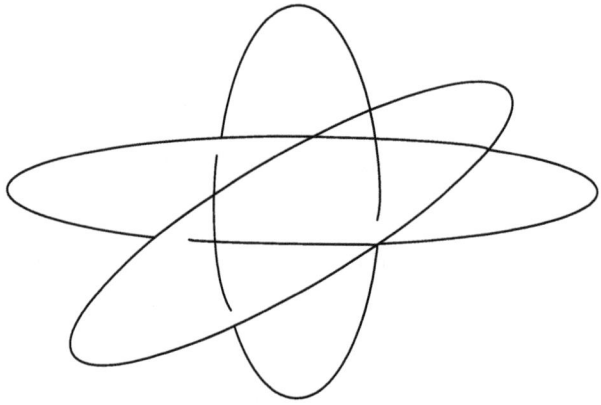

The quark can be described by SU(3) symmetry, but the description of the neutron as three distinct quarks is weak. The neutron decays into proton, electron, and antineutrino. The explanation in Evans' terms as minimum curvature, R, is more clear.

The Gell-Mann color triplet is the three-spinor representation of the metric vector of physical spacetime. The transformation matrices of the representation space of the Gell-Mann triplet are the SU(3) matrices, and are used to describe the strong nuclear field. The representation space of the tangent spacetime has eight indices. These are superimposed on the four indices of the base manifold to give the various tetrads of the strong field. These are the gluons, which are quantized results (eigenfunctions) of the Evans Wave Equation, and the gluon field is therefore derived from general relativity.

In the standard model the gluon field is a construct of special relativity, with an abstract internal index superimposed on the flat spacetime of the base manifold. This abstract internal index is identified in the new unified field theory as the index of the SU(3) representation space of the physical tangent spacetime, which is orthogonal and normalized. The *physical* dimensions of the unified field theory are always ct, x, y and z.

In terms of general relativity, the quark particles are least curvatures. Mechanically, we see spatial compression occurring within the particle.

The standard model uses special relativity which is only an approximation to general relativity. It has some forced explanations that have no experimental proof. Its equations have massless particles that then need another explanation, the Higgs mechanism and spontaneous symmetry breaking, to provide mass.

Therefore if we accept quarks as physical, they emerge from the Evans Wave Equation as eigenfunctions of that equation in an SU(3) symmetry representation space of the orthonormal tangent spacetime of the base manifold.

It now appears that quarks are a mathematical description and there is no physical basis except for rotating curvatures and electrical potential inside the particle. This is unclear at the present stage of research. While we cannot go into details, it is worth noting that either an O(3) or SU(3) representation of space is allowed. Both can be representations of the same manifold and geometrically they are unified. This implies that the strong field is the gravitational field in an SU(3) representation of space. Quarks have not been observed individually, so it remains to be seen whether this SU(3) representation has a physical meaning. (The SU(2) representation is interpreted to be the half integral spin of a fermion. The standard model could imply that quarks are never observable. This would seem to say they are not physical.

The decay of the neutron into antineutrino, electron, and proton from three quarks doesn't add up well. One quark has to change into three entities yet the quark is presumably basic. The transformation of the curvature (and spin) of the neutron into other more stable curvatures adds up better from a simple viewpoint. The flaw in the mathematics is the use of special relativity, not general relativity.

Equations of Physics

A direct derivation of the main equations of physics is given in Evans' work. He starts with the quantized version of $Rq^a_\mu = -kT^a_\mu$, the generally covariant wave equation:

$$(\Box + kT)q^a_\mu = 0$$

Substituting metric vectors, metric tensors, spinors, and symmetries as necessary, the main equations of physics are shown to be derived from the Evans equations.

The equivalence principle states that the laws of physics in small enough regions of spacetime are Lorentzian (flat) and reduce to the equations of special relativity. In special relativity there exist equations such as:

$(\Box - \kappa_0^2)\psi = 0$ where ψ is the four spinor of the Dirac equation,

$(\Box - \kappa_0^2)\phi = 0$ where ϕ is the scalar field of the Klein-Gordon equation, and

$(\Box - \kappa_0^2)A_\mu = 0$ where A_μ is the a four-vector electromagnetic wave function of the Proca equation.

These look like, and actually are, limiting forms of the Evans equation.

κ_0^2 is curvature and is kT_0, that is Einstein's constant times the stress energy tensor in a vacuum.

> The presence of gravitation in any region of space means there is some energy always present – the zero point energy.

Chapter 7 – The Evans Wave Equation

In Chapter 9 we will go into details showing $\kappa_0^2 = 1/\lambda_c^2 = (mc/\hbar)^2$ with λ_c the Compton wavelength, and m is mass of any particle, c is the speed of light in a vacuum, and $\hbar$ is the Dirac constant $h/2\pi$. The form of the equations is evident and ψ, Φ, A_μ can be expressed by the tetrad to become covariant. From this will develop the Dirac equation and the mass-volume relationship of particles. The point here is that the important equations of physics evolve out of the Evans equations.

The following equations, only a few of which we will discuss in this book, have been derived to date from various limiting[37] forms of the Evans' field and wave equations:

1. The Galilean Equivalence of inertial and gravitational mass.
2. Newton's three laws of motion and his universal law of gravitation.
3. The Poisson equations of gravitation and electromagnetism.
4. The equations of O(3) electrodynamics. These have the same structure as the standard Maxwell Heaviside equations for three senses of polarization, (1), (2), (3), but are O(3) gauge field equations.
4. The time dependent and time independent Schrodinger equations.
5. The Klein Gordon equation.
6. The Dirac equation in a 4-spinor derivation.
7. The O(3) Proca equation.
8. The O(3) d'Alembert equation.
9. A replacement for the Heisenberg uncertainty principle.
10. The Noether theorem. All conversion of energy and mass from form to form occurs through spatial curvature, R.

In the chapters that follow we will see that a number of new equations can be found using the Evans' wave equation. The Principle of Least Curvature is developed, the electrogravitic equation gives proportionality between gravity and electromagnetism. O(3) and $\mathbf{B}^{(3)}$ electrodynamics are developed. A simple

---

[37] Limiting means that instead of the real 4-dimensional spacetime, flat space is used in a restricted application. Far from mass-energy, spacetime is nearly flat.

explanation for the Aharonov-Bohm and other optical effects is developed from general relativity.

Of great interest is that the symmetries of the matrix in mathematics are applied to physical quantities:

$$q_\mu^a = q_\mu^{a(S)} + q_\mu^{a(A)}$$

Gravitation can be symmetric or antisymmetric, however it is always curving.

Electromagnetism can be symmetric or antisymmetric, however it is always turning.

Summary

The wave equation of general relativity and unified field theory is:

$$(\Box + kT)\, q_\mu^a = 0$$

By substituting appropriate representations for the tetrad, the various equations of physics can be derived. Gravitation, electromagnetism, the weak force, and the strong force can all be represented. This simple looking equation can be expanded as given in the chart at the end of the chapter.

The wave equation was derived from general relativity using differential geometry.

The factorization of the symmetric and antisymmetric metrics from the asymetric tetrad is basic differential geometry. It gives four forces – a discovery opening new insights into physics. This is $q_\mu^a = q_\mu^{a(S)} + q_\mu^{a(A)}$.

Applied to physics, four potential fields are represented as shown in the chart.

The standard model is not generally covariant – it does not allow calculations of interactions among particles in different gravitational fields, say near a black hole. The standard model does not allow for electromagnetism's and gravitation's mutual effects to be defined.

Chapter 7 – The Evans Wave Equation

The Evans unified equations allow both these processes to be accomplished.

The use of the mathematical representation space for the tangent spacetime "a" is of the essence. This is the Palatini variation of gravitational theory. The tetrad can now be expressed as any of the four fields of physics – G, A, W, and S. Thus four forms of energy can be described within the unified field q^a_μ :

TYPE	POTENTIAL FIELD
Gravitational curvature Symmetric = Centralized Einstein gravitation	$q^{a\,(S)}_\mu$
Gravitational curvature Antisymmetric = turning Unexplored. The strong field, dark matter?	$q^{a\,(A)}_\mu$
Electromagnetic Field Antisymmetric EM Photon, EM waves	$A^{a\,(A)}_\mu = A^{(0)} \, q^{a\,(A)}_\mu$
Electrodynamics Symmetric EM Charge, the electron	$A^{a\,(S)}_\mu = A^{(0)} \, q^{a\,(S)}_\mu$

While the structure is not fully developed yet, the logic is clear. Gravitation is centralized and symmetric. Mass curves space in a spherical shell around it. The electromagnetic wave, the photon, is antisymmetric spinning spacetime. The electron (charge) is centralized spin. Antisymmetric curvature is a matter for study.

BASIC EQUATIONS OF EVANS UNIFIED FIELD THEORY

ORIGIN	TYPE	POTENTIAL FIELD	GAUGE FIELD	FIELD EQUATION (CLASSICAL MECHANICS)	WAVE EQUATION (QUANTUM MECHANICS)	CONTRACTED ENERGY / MOMENTUM	SCALAR CURVATURE
Einstein / Hilbert (1915)	Central Gravitation	$q_\mu^a{}^{(S)}$	$R^a{}_{b\mu\nu}{}^{(A)}$	$R_\mu^{a(S)} - \frac{R}{2} q_\mu^{a(S)} = k T_\mu^{a(S)}$	$(\Box + kT) q_\mu^{a(S)} = 0$	T_{grav} gravitation	R_{grav} gravitation
Evans (2003)	Unified	q_μ^a	$R^a{}_{b\mu\nu}$	$R_\mu^a - \frac{R}{2} q_\mu^a = k T_\mu^a$	$(\Box + kT) q_\mu^a = 0$	T_{total} Hybrid energy	R_{total} Hybrid energy
Evans (2004)	Unknown	$q_\mu^a{}^{(A)}$	$T^c{}_{\mu\nu}$	$R_\mu^{a(A)} - \frac{R}{2} q_\mu^{a(A)} = k T_\mu^{a(A)}$	$(\Box + kT) q_\mu^{a(A)} = 0$	T_{em} symmetric	R_{em} symmetric
Evans (2003) Evans (2004)	Electro-dynamics	$A_\mu^{a(A)} = A^{(0)} q_\mu^{a(A)}$	$A^{(0)} T^c_{\mu\nu}$	$G_\mu^{a(A)} = A^{(0)} k T_\mu^{a(A)}$ $= A^{(0)} \left(R_\mu^{a(A)} - \frac{R}{2} q_\mu^{a(A)} \right)$	$(\Box + kT) A_\mu^{a(A)} = 0$	T_{em} electro-dynamic	R_{em} electro-dynamic
Evans (2003) Evans (2004)	Electro-statics	$A_\mu^{a(S)} = A^{(0)} q_\mu^{a(S)}$	$A^{(0)} R^a{}_{b\mu\nu}{}^{(A)}$	$G_\mu^{a(S)} = A^{(0)} k T_\mu^{a(S)}$ $= A^{(0)} \left(R_\mu^{a(S)} - \frac{R}{2} q_\mu^{a(S)} \right)$	$(\Box + kT) A_\mu^{a(S)} = 0$	T_{es} electrostatic	R_{es} electrostatic

1) Duality: $T^c = \varepsilon^{cb}_a R^{a(A)}_b$

2) Basic Matrix property: $q_\mu^a = q_\mu^{a(S)} + q_\mu^{a(A)}$

3) $A^{(0)}$ is weber / meter = kg·m/(A·s^2) = volt-sec/meter. It converts from mathematics to physics.

COPYRIGHT 2004 by Myron Evans

Thanks to Robert W. Gray for layout of table

Chapter 8 Implications of the Evans Equations

> There is nothing in general relativity that is unknowable because there is nothing in differential geometry which is unknowable.
> Myron Evans, 2004

> Mass and energy are therefore essentially alike; they are only different expressions for the same thing. The mass of a body is not a constant; it varies with changes in its energy.
> Albert Einstein

Introduction

We can now define the term "unified field theory" as the completion of Einstein's general relativity by extending it to all radiated and matter fields in nature. Until Evans derived his equations, we did not have a clear concept of its form nor know the content of a unified theory. The origin of the unified field theory is general relativity and differential geometry. Quantum theory emanates from general relativity and if one looks at the papers that Evans has produced in 2004, a great deal of it is quantum in nature. Differential geometry gives us the abstract foundations and Evans later work in 2004 shows increasingly a geometric approach.

This chapter summarizes and explains material in Chapters 6 and 7 and describes further extensions of the equations.

Geometry

The Evans approach to physics is the same as Einstein's. Physics is geometry. Geometry does not simply describe physics; our universe is geometry. Geometry came first before the universe formed in anything like its present state; then we found it and used it to describe physical processes; then we realized that it actually is physics. Einstein used Riemann geometry which describes curvature, but not torsion. Cartan developed a theory of torsion for electromagnetism. Evans' development using metric vectors and the tetrad is initially pure geometry, but the equations combine curvature and torsion.

The first step to unification is achieved by development of the Evans Field Equation that allows both curvature and torsion to be expressed in the same set of equations. Gravitation and electromagnetism are derived from the same geometric equation. This completes Einstein's goal to show that gravitation and electromagnetism are geometric phenomena.

Figure 8-1   Torsion and Curvature

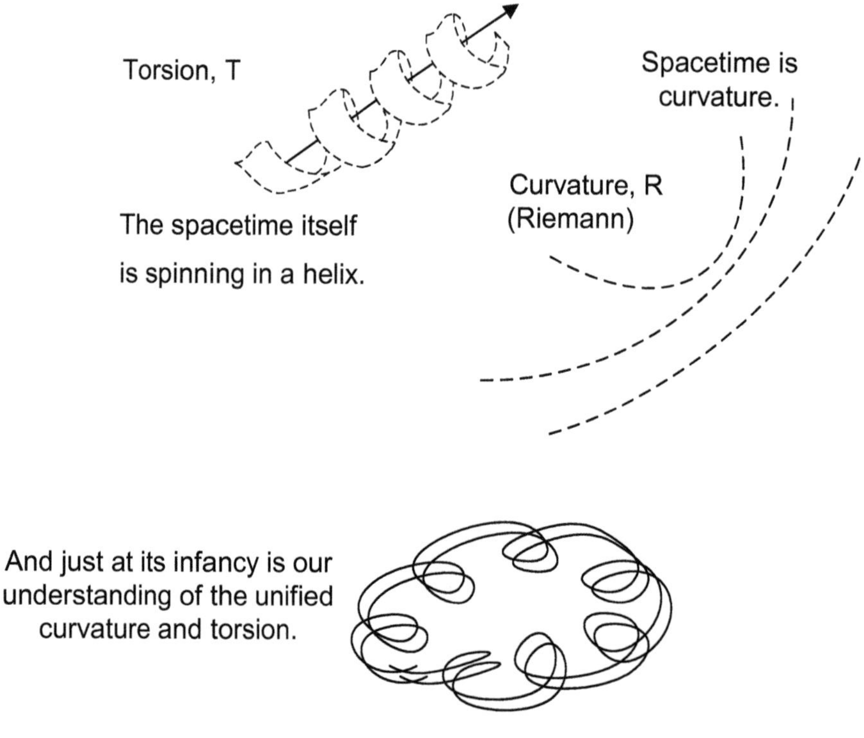

Torsion, T

The spacetime itself is spinning in a helix.

Curvature, R (Riemann)

Spacetime is curvature.

And just at its infancy is our understanding of the unified curvature and torsion.

The second step was the development of the wave equation. The material in Chapter 7 is the beginning, but extensive development followed.

Chapter 8 – Implications of the Evans Equations

There are only two forms in differential geometry and in the real universe that describe connections to spacetime. These are the Riemann curvature form and the torsion form. All forms of mass and energy are derived from these. The Riemann form is symmetrical general relativity. The torsion form is antisymmetrical general relativity. These are depicted in Figure 8-1.

The Evans spacetime is 1) a differentiable 4-dimensional manifold with 2) an asymmetric metric. That is, A) spacetime is geometry that is distinct down to nearly a point.[38] B) Asymmetric = symmetric + antisymmetric. The asymmetric geometry contains both curvature defined by the symmetric portion of the metric and torsion described by the antisymmetric portion of the metric.

The tetrad is a matrix with 16 components that contains information about both fields and their mutual effect on each other. Each of the 16 individual elements of the tetrad matrix are composed of factors that represent curvature and an internal index.[39] For our purposes here it is sufficient to note that the tetrad interrelates the real universe and the various mathematical spaces that have been devised for quantum theory, O(3) electrodynamics, spinor representations, etc. The internal index is used to extend analysis to the weak and strong fields.

The strong force that holds particles together is related to gravitation.

The weak force that holds the neutron together is related to electromagnetism.

All four forms of energy originate in spatial curvature.

> The Evans contention is that if a process is valid in differential geometry, then it is a real physical possibility.

---

[38] Given the minimum curvature that will be indicated in Chapter 9, the manifold is probably differentiable down to orders of magnitude less than the Planck length.
[39] The mathematical methods can be examined in Evans' papers, particularly THE EQUATIONS OF GRAND UNIFIED FIELD THEORY IN TERMS OF THE MAURER-CARTAN STRUCTURE RELATIONS OF DIFFERENTIAL GEOMETRY, June 2003, www.aias.us and published in Foundations of Physics Letters.

Very Strong Equivalence Principle

The *weak equivalence principle* is the equivalence of inertial and gravitational mass: $m_i = m_g$. From calculations on the metric vector the equivalence of inertial and gravitational mass is a geometrical identity in the weak field approximation, showing that the inertial mass and the gravitational mass originate in a geometrical identity.

The mass m enters into the theory from the fact that in the weak field approximation the appropriate component of the contracted energy momentum tensor $T_0 = mc^2/V_0$, where V_0 is the rest volume. So mass enters into the theory from primordial energy. There can be only one m, and so gravitational and inertial mass are the same thing. (This last sentence should produce some twenty books arguing for or against it.)

The *strong equivalence principle* is Einstein's recognition that the laws of physics are the same in all reference frames. In special relativity (non-Euclidean, Minkowski spacetime) equivalence refers to inertial reference frames. In Einstein's general relativity using Riemann curved spacetime, equivalence of accelerated and gravitational reference frames was recognized. Theories retain their form under the general coordinate transformation in all reference frames.

The electro-weak theory states that electromagnetism and the weak force are equivalent at very high energies. Theorists have been working on adding gravitation to the equivalence. In the Evans equations, we see that gravitation comes first - that curvature is primordial. General relativity is the initial theory from which the others can be derived.

The *very strong equivalence principle* simply equates everything.

If we accept the origin of the universe in a big bang, then a near singularity existed and all forms of existence were within a homogeneous region. Initially, there was only curvature. Thus we could say that vacuum = photon = nucleon or space = time = energy = mass. Everything originated from the same one thing.

At high enough energies existence is homogeneous and all forms of matter and energy are identical. They originate in curvature.

Singularities cannot exist according to Noether's Theorem and are only mathematical constructs without physical meaning. They lead to infinite energy in zero volume. Evans has proposed that singularities are impossible with a result derived from the Evans Wave Equation that indicates a particle's mass times its volume is a constant. There is a minimum volume based on minimum curvature. This is covered in Chapter 9.

Figure 8-2 depicts the origin (or beginning of re-expansion after the big crunch) of the universe and the separation of curvature and torsion into the forms we see today.

> The *very strong equivalence principle* equates all the forces with spacetime curvature. The very strong equivalence principle is that all forms of existence are interconvertible and have a primordially identical origin in spatial curvature.

All forms of existence are interconvertible through curvature. For example, the pendulum shows a smooth conversion of potential energy when the arm is lifted to kinetic energy with some potential energy when the arm is moving to kinetic energy when the arm is at the bottom and then back again. Gravitational potential is turned into electricity in hydroelectric power stations.

Einstein's $E = mc^2$ already stated that a type of equivalence exists between mass and energy. This is not controversial; Evans simply shows that the equivalence is more extensive than previously shown.

Figure 8-2 Separation of Forces

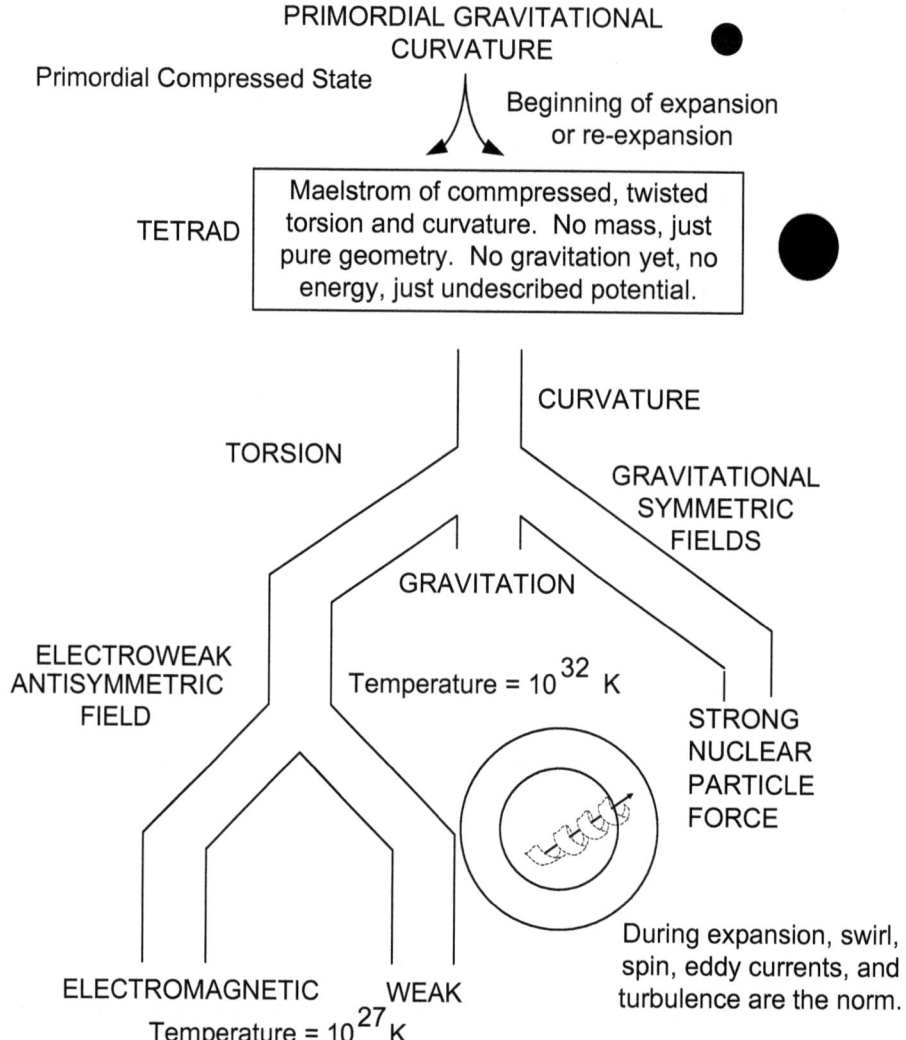

A Mechanical Example

In a particle accelerator, we see kinetic energy added to a particle during acceleration to near c. The particle's spatial reference frame Lorentz compresses and more of the energy is held in that compression than goes into velocity. When the particle collides with another, the velocity drops to near zero, the reference frame expands with an explosive decompression, and particle creation can occur. The momentum and velocity are conserved, but the curvature (E_k, kinetic energy) can be converted to particles. Momentum = mass x velocity and it is conserved. E_k = mass x velocity squared. A large amount of kinetic energy (curvature) is available for particle creation.

Curvature (compression) transmits the energy. Force is expansion, which creates particles under the proper circumstances.

Evans uses the term curvature, the correct mathematical physics term, but compression is the equivalent engineers' term. That compression is potential energy is an engineering basic. Expansion is application of force. This is quite clear at the macro and micro levels. We can state that energy is curvature.

The insight that electromagnetism is twist, spin, or torque - collectively torsion - is not new since Einstein and Cartan knew that torsion would explain electromagnetism. However, they did not put it into a formulation. The Evans addition to general relativity is significant. The **B**$^{(3)}$ field is circularly polarized radiation and gives us this twist.

In a straightforward analysis, it can be shown that the rest energy of a particle is kinetic energy. That all forms of curvature are interchangeable has already been established, but putting the principle into practice can give some surprising results. It can be shown that:

$$T = mc^2 (\gamma - 1). \tag{1}$$

where $\gamma = (1 - (v/c)^2)^{-1}$. Thus the rest energy is

$$E_0 = mc^2 = T (\gamma - 1)^{-1}. \tag{2}$$

This implies that potential, rest energy is kinetic energy. The particle accelerator example above indicates this also.

> We have seen that all forms of energy are interconvertable. The indication here is that regardless of form, they are the essentially the same thing.

Implications of the Matrix Symmetries

The different products of the tetrads give different physically significant results:

Wedge product gives antisymmetric torsion fields – electromagnetism.

Outer product gives asymmetric fields and as suggested below are indicative of hybrid matter fields.

The inner or dot product gives well known symmetric gravitation.

Contraction leads to geometry and thermodynamics via $R = -kT$.

We have seen that $G\, q^a_\mu = kT\, q^a_\mu$. This says that curvature, G, and stress energy, T, are related through the tetrad matrix. The unified field is the tetrad q^a_μ.

The tetrad is an asymmetric square matrix. This can be broken into its symmetric and antisymmetric parts:

$$q^a_\mu = q^{a\,(S)}_\mu + q^{a\,(A)}_\mu \tag{3}$$

The gravitational potential field is the tetrad q^a_μ and the electromagnetic potential field is $A^a_\mu = A^{(0)} q^a_\mu$ where $A^{(0)}$ has units of volts times seconds per meter or $\hbar/e$. $\hbar/e$ has the units of weber, the magnetic fluxon. It is the conversion factor from mathematics to physics.

We can then expand or redefine $R = -kT$ to:

$$R_1\, q^{a\,(S)}_\mu = -kT_1\, q^{a\,(S)}_\mu \tag{4}$$

$$R_2 \, q^a_\mu{}^{(A)} = -kT_2 \, q^a_\mu{}^{(A)} \tag{5}$$
$$R_3 \, A^a_\mu{}^{(S)} = -kT_3 \, A^a_\mu{}^{(S)} \tag{6}$$
$$R_4 \, A^a_\mu{}^{(A)} = -kT_4 \, A^a_\mu{}^{(A)} \tag{7}$$

where T is the energy momentum tensor. T_1 is symmetric and gives gravitation. T_2 is antisymmetric gravitation as of the time of this writing not well understood. T_3 is symmetric torsion indicating electrostatics, and T_4 is antisymmetric giving electromagnetism. See Figure 8-3.

It is concluded that $q^a_\mu{}^{(S)}$ is the central gravitational potential field and that $A^a_\mu{}^{(S)}$ is the central electrostatic potential field. These both obey the Newtonian inverse square law with strength of field proportional to $1/r$ with r being the radius or distance from the center. Symmetry leads to centralized $1/r^2$ force laws for both gravitation and electromagnetism.

The Newton and Coulomb laws thus have a common origin.

Figure 8-3   Geometry and Applications in General Relativity

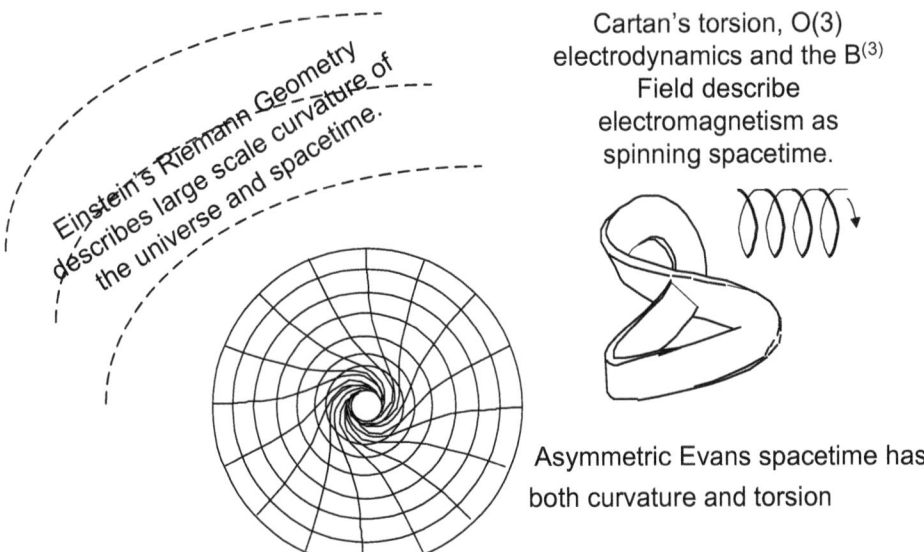

$q^a_\mu{}^{(A)}$ is a spinning gravitational potential field. This field is not centrally directed so does not obey the inverse square law. $q^a_\mu{}^{(A)}$ appears to meet the

criteria for dark matter. We know that $A^a_\mu{}^{(A)}$ turns and moves forward and we generalize its counterpart to antisymmetric gravitation. So antisymmetric gravitation will turn and travel. Given what we know of the normal particle, it may be that it too has these aspects. Evans has speculated on this, but has not published anything yet.

Electrodynamics

$A^a_\mu{}^{(A)}$ is the rotating and translating electrodynamic potential field.

The scalar curvature R operates on A^a_μ. We know that gravitation affects electromagnetic fields, but now the interaction between both radiated fields and matter fields can be considered. We assume without seeing any calculations that the spinning electromagnetic field is also curved in spacetime near a gravitational field. See Figures 8-4 and 8-5.

Figure 8-4

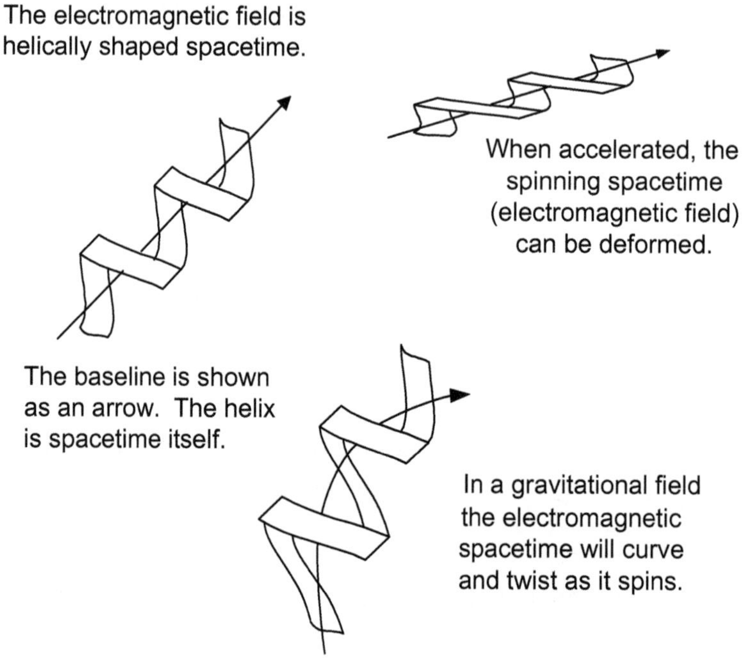

The electromagnetic field is helically shaped spacetime.

When accelerated, the spinning spacetime (electromagnetic field) can be deformed.

The baseline is shown as an arrow. The helix is spacetime itself.

In a gravitational field the electromagnetic spacetime will curve and twist as it spins.

Chapter 8 – Implications of the Evans Equations

In deriving O(3) electrodynamics (see Glossary) from the Evans Wave Equation, we can see that the electromagnetic field is the reference frame itself. The frame translates and rotates. The antisymmetric tensor describes rotation and gives the connection to general relativity. Spin and angular momentum are seen as antisymmetric torsion forms of general relativity.

The wave functions of quantum mechanics can be interpreted as the metric of Einstein, not a probability. The wave equation function is derived from the metric and the representative space of quantum mechanics refers to the Einstein metric.

The tangent space of general relativity is the internal space of gauge theory. The equivalence is further unification.

Nature operates in waves, giving us quantum mechanics. However causal logic and geometry are the foundations of general relativity and geometry is not random. It is not possible for something to happen without prior cause.[40] The Heisenberg uncertainty principle is unnecessary as shown in Chapter 10.

Figure 8-5   Spinning Spacetime

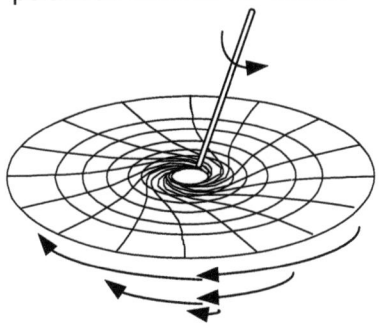

The stirring rod is the magnetic field or the circularly polarized laser or RF beam.

Like a twirling whirlpool in water, the electromagnetic field is spinning spacetime. In neither case is there some entity superimposed on the medium. Rather the energy is the medium.

---

[40] Professor Evans commented, "..it amazes me that physicists have accepted the Copenhagen voodoo for so long." Einstein felt the same way.

R and T in the Evans equations are quantized. They are expressed as wave equations but the are also equations in general relativity and are therefore causal. Evans refers to this as "causal quantization" to distinguish his unification from Heisenberg's quantization which was probabilistic. Physics retains the Einstein deterministic approach.

While quantum mechanics emerges from the Evans equations, the probabilistic nature of the Schrodinger equation and the Heisenberg uncertainty principle are modified. Conflicts between general relativity and quantum mechanics can be resolved.

String theory and its relatives are not necessary. The Evans equations show quantum mechanics emerging from general relativity and in following chapters, we will see more of the power of the wave equation. String theory appears to be no more than beautiful mathematics, but it has never produced any physical results. It explains nothing and it predicts nothing.

Reality has four dimensions, those of Einstein. The need for 9,10,11, or 26 dimensions is unnecessary. There is a rich mathematical research field coming from string theory back to general relativity and quantum theory, but a return to basic geometry is necessary to find physically meaningful results.

We see now that the electric field strength E and the acceleration due to gravity g are both generated by Evans spacetime. Therefore, it is possible in principle to obtain an electric field from curved spacetime. It has been known since 1915 that it is possible to obtain gravitation from spacetime.

The hydroelectric generator is an indirect example of this. The generator uses the gravitational potential of water mass (curvature) to turn (spin) a magnetic field (rotor and armature) to produce electrical potential and current flow. Direct conversion has a theoretical foundation in the Evans equations. We would gain a source of electric power if we could directly extract power from the curvature of spacetime. Unknown as of the time of this writing is if the earth's gravitational field is strong enough to gain efficiencies needed to make this practical.

Particle physics

Quarks have not been observed - they are inferred indirectly by particle scattering. In the weak field limit, the Gell-Mann quark color triplet emerges from the Evans equations – but as energy fields, not particles. The concept of quark confinement is an attempt to explain why they are not observed. Usually in science we explain things that *are* observed. The strong force in Evans' theory is similar to the gravitational force using SU(3) geometry. If quarks exist, then an explanation of quark confinement using Evans' general relativity equations may be attainable. However, at this juncture, research seems more likely to show that quarks and gluons are not discrete particles but are rather energy states with primarily internal curvature and torsion.

Mass-energy is the source of curvature in spacetime and the result of curvature in spacetime. Curvature = spacetime.

Gravitation and electromagnetism both originate in $q^a{}_\mu$, the tetrad, so the influence of one on the other depends on the dynamics of the tetrad or the metric vectors that define it. This influence is given quantitatively by identities of differential geometry.

In the first descriptions of the Evans equations, vectors were used instead of the tetrad. Using the metric vector rather than the metric tensor achieves a simplification in gravitational theory and allows electromagnetism to be described covariantly in general relativity. The tetrad is more advanced. Paraphrasing Professor Evans:

The mathematical representation space used for the tangent spacetime is of greatest importance. Its index is "a" and indicates the Minkowski spacetime in the Palatini variation of the generally covariant theory of gravitation. The electroweak field has been produced using the minimal prescription in the Dirac equation, and this gives the influence of gravitation on radioactivity without using the Higgs mechanism. So there may only be two fundamental fields in nature, the gravitational and electromagnetic, the existence of non-observable confined quarks being an unphysical assertion of the standard model.

Charge

Charge in general relativity is a result of and a source of torsion in spacetime. It is symmetric torqued or spinning spacetime.

Evans defines the electromagnetic field as the torsion form in differential geometry (in physics, the torsion tensor) within a factor $A^{(0)}$ which is electromagnetic potential. This factor is negative under charge conjugation symmetry. This means that if one changes the sign of e then the sign of $A^{(0)}$ changes. e is e^+ the proton charge and is positive, the electron charge is e^-.

Instead of regarding $A^{(0)}$ as intrinsically "positive" or "negative" from some arbitrary symmetry of the scalar field, components of the tetrad itself $q^a{}_\mu$ may be positive or negative.

> The origin of charge is to be found in the direction of the spinning spacetime.

The tetrad can be positive or negative – spinning one direction or the opposite. Note that the tetrad is the reference frame itself. It is the spacetime. When spinning symmetrically, it is an electric field. When simply curved, it is gravitational. When spinning antisymmetrically, it is the magnetic field. (And when spinning curvature, we are not yet sure what it is.)

Particle spin is definable through the tetrad. Antiparticle spin is the tetrad with reversed signs in all its components.

The curvature R may be positive or negative, depending on the direction of spacetime curvature (one assumes "inward" or "outward" in a 4th dimension), and so the particle mass energy is positive or negative valued. The Evans Wave Equation can be reduced to the Dirac equation (Chapter 9) and so is capable of describing antiparticles, the Dirac sea, and negative energies. We see a picture emerge of particle as curvature and antiparticle as anti-curvature or curvature in the other direction – the "outward."

Chapter 8 – Implications of the Evans Equations

Evans reduces everything to geometry and attempts to remove concepts which do not appear in Einsteinian general relativity, (and therefore are not objective and have no place in physics).

The Symmetries of the Evans Wave Equation

$$(\Box + kT_g^{(S)})\, q^a{}_\mu{}^{(S)} = 0 \qquad (8)$$

where $T_g^{(S)}$ is the symmetric gravitational energy momentum. This is Einstein's symmetric gravitation and leads to Newtonian gravitation and laws of motion in low limit mass-energy situations.

$$(\Box + kT_g^{(A)})\, q^a{}_\mu{}^{(A)} = 0 \qquad (9)$$

where $T_g^{(A)}$ is antisymmetric gravitational energy momentum. This leads to any gravitational effect with spin such as centripetal acceleration.

$$(\Box + kT_e^{(S)})\, A^a{}_\mu{}^{(S)} = 0 \qquad (10)$$

where $T_e^{(S)}$ is symmetric electromagnetic energy momentum and $A^a{}_\mu{}^{(S)}$ is the symmetric component of the electromagnetic potential tetrad. This leads to electrostatic charge and the Poisson equation.

$$(\Box + kT_e^{(A)})\, A^a{}_\mu{}^{(A)} = 0 \qquad (11)$$

where $T_e^{(A)}$ is antisymmetric electromagnetic energy momentum and $A^a{}_\mu{}^{(A)}$ is the antisymmetric component of the electromagnetic potential tetrad. This leads to O(3) and $\mathbf{B}^{(3)}$ electrodynamics. With no matter field interaction, the Proca equation results and when there is field-matter interaction it leads to the d'Alembert equation.

In all the forms of the tetrad, the four-dimensional spacetime is "μ" and is Riemann non-Euclidean curved spacetime with gravitation. It represents the real universe we inhabit. In all forms of the tetrad, "a" is a mathematical Minkowski spacetime with the indices of special relativity used in quantum mechanics. These are the familiar ct, x, y, and z dimensions.

Figure 8-6 shows the equations of the Evans wave equation and the physical properties of spacetime that they represent.

Evans Equations of Unified Field Theory

Figure 8-6   Approximate dimensions of atom and constituants in terms of the tetrad symbols.   Not to scale.

$$(\Box + kT)\, q^a_\mu = 0$$

Wave equation governs all processes.

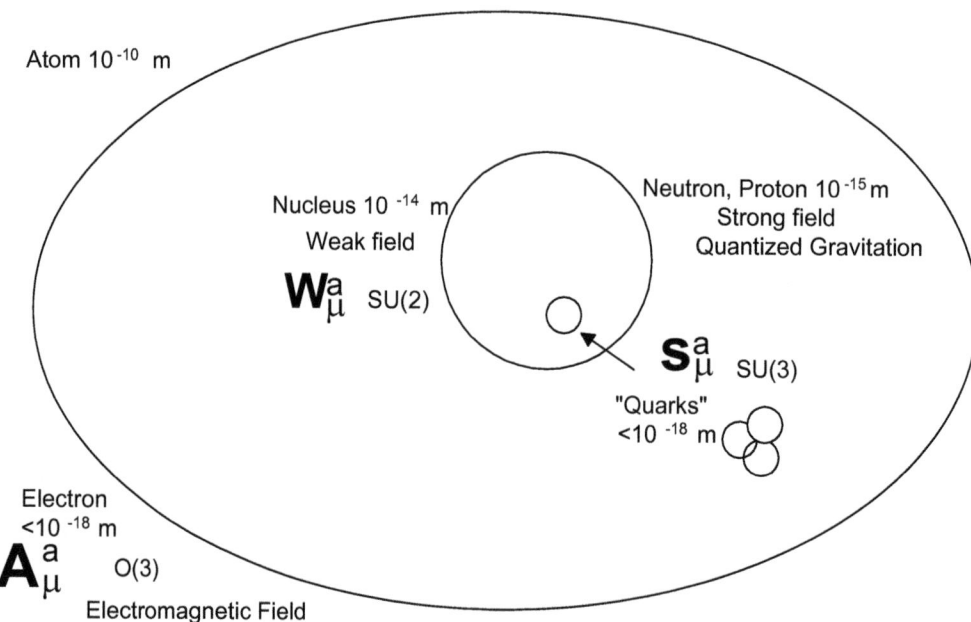

Each constituent is probably a standing wave, not a discrete entitiy as pictured here. The electron is a wave that surrounds the nucleus and stretches out to infinity. The quarks are energy levels described by SU(3) symmetry, but the physical make-up is questionable.

e^a_μ Weakest of the fields is the gravitational

R = -kT

The source of the fields in physics is kT. Energy is transmitted from source to receiver by the scalar curvature, R. This is true for all four fields. It is also true for velocity. Accelerated spacetime is compressed and therefore carries energy.

Chapter 8 – Implications of the Evans Equations

R = -kT is consistent with the fact that energy is primordial and that all forms of energy are interconvertible, so all forms of curvature are also interconvertible. Therefore, gravitation and electromagnetism emerge from the fact that energy is both primordial and interconvertible . We may reverse the argument and state from the beginning that from non-Euclidean geometry all forms of curvature are primordial and interconvertible and that the existence of primordial energy is the existence of primordial and non-zero R.

> All forms of energy are either compression (curvature) or spinning (torsion) of the spacetime vacuum.

Everything originates in the geometry of spacetime. The originator is the "vacuum" and its structure. If R is identically zero then there is no energy in the region or the universe. As soon as R departs from zero, energy is formed and from that energy emerges gravitation and electromagnetism, and their aspects - the weak and strong forces, and matter fields. Spacetime is all these aspects of existence.

The gravitational and electromagnetic fields are both states of spacetime.

Mathematically, we can say "spacetime" and envision a completely differentiable manifold upon which to establish physics.

A classical mechanical interpretation is that spacetime is our experienced reality. It has a variety of aspects like a table, this book, the moon, the sun, and the vast spacetime in which the planetary systems and galaxy exist. It ranges in size and composition from particle to nearly empty space.

Spacetime curvature in and around particles exists. One assumes that particles are a form of compressed spacetime, probably with some torsion.

We will find that a redefinition of "fully differentiable manifold" is necessary. In the next chapter on the Dirac and Klein Gordon equations, we see that there is a minimum curvature. It is many times smaller than the Planck length, but it is a quantized limit. One has either zero curvature or the first step of curvature. Although physics operates down to a very small size, a point cannot be achieved.

If zero curvature were to occur, there is no existence.

> The physical description here is that of a helical and curved spacetime with the electromagnetic field as the spacetime itself. In special relativity the spacetime is fixed and the electromagnetic field is visualized as rotating. In general relativity the field is seen as part of spacetime itself. The electromagnetic field is the spacetime and the baseline is a helical curve.

Summary

Physics is geometry and at the present time, differential geometry appears to be sufficient for explanations. There are more complicated geometries, but their use is probably unnecessary. (Of course, the mere fact of their existence within this physics we live in may imply that they are necessary to explain something.)

The Evans Wave Equation quantizes curvature. The quanta of R are the real physical steps of change that can occur. The Evans wave equation is a quantization of the contracted Einstein field equation $R = -kT$ with the operator replacement R goes to d'Alembertian. This unification occurs by showing that quantum theory emerges from general relativity.

The very strong equivalence principle is that all forms of existence are interconvertible and have a primordially identical origin in spatial curvature.

The square tetrad matrix is asymmetric. It can be decomposed into its symmetric and antisymmetric parts: $q^a_\mu = q^a_\mu{}^{(S)} + q^a_\mu{}^{(A)}$. This gives us deeper insight into the meaning of the equations and how they relate to physical reality. Spacetime can be symmetric or antisymmetric gravitation or it can be symmetric or antisymmetric electromagnetism. Unified, the components are asymmetric.

Chapter 8 – Implications of the Evans Equations

Figure 8-7   The magnetic field is spinning spacetime.

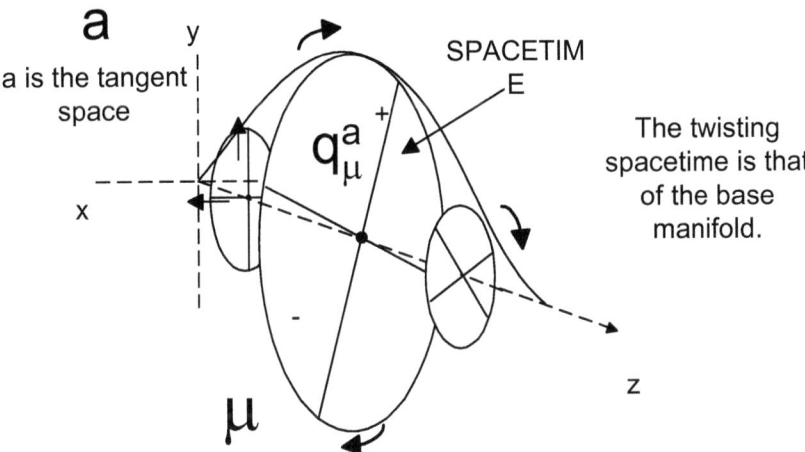

All forms of energy are either compression (curvature) or twisting (torsion) of spacetime. Einstein's postulate of general relativity R = -kT is the beginning of understanding. R is curvature or geometry; kT is physics.

The Evans Wave Equation

$$(\Box + kT) q^a{}_\mu = 0$$

$\Box$ or "del" is the d'Alembertian or $\dfrac{1}{c^2}\dfrac{\partial^2}{\partial^2 t} - \nabla^2$

$\Box$ is a Lorentz invariant description of the change in the curvature at a point in spacetime. k is Einstein's constant and T is the stress energy tensor. $q^a{}_\mu$ is the tetrad described below.

This equation is part wave equation – quantum mechanics, $\Box$, part Einstein general relativity, kT, and part Cartan differential geometry, $q^a{}_\mu$, the tetrad. The solutions are quantized and a variety of meanings can be given to the tetrad.

You can't go out and buy a stenciled T-shirt yet, but at least $E=mc^2$ has a companion, $(\Box + kT) q^a{}_\mu = 0$.

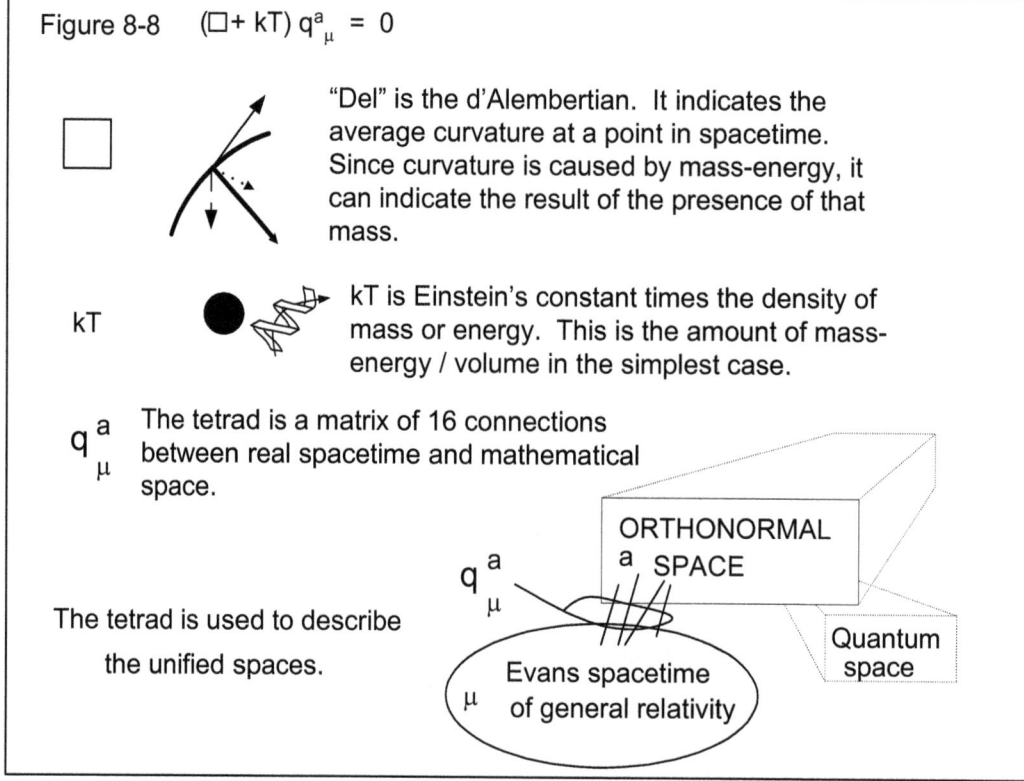

Chapter 9 The Dirac, Klein-Gordon, and Evans Equations

> Determination of the stable motion of electrons in the atom introduces integers and up to this point the only phenomena involving integers in physics were those of interference and of normal modes of vibration. This fact suggested to me the idea that electrons too could not be considered simply as particles, but that frequency - wave properties - must be assigned to them also.
> Louis de Broglie, 1929

Introduction

We will arrive at two new important equations:

$$E = \hbar c \sqrt{|R_0|} \qquad (1)$$

Here E is total energy, $\hbar$ is $h/2\pi$ or the Dirac constant, c is the velocity of electromagnetic waves, and R_0 is the spacetime curvature in the low limit. This gives us the Principle of Least Curvature.

We also are given a new fundamental relationship:

$$mV_0 = k \hbar^2 / c^2 \qquad (2)$$

where m is mass, V_0 is the volume of a particle in the low limit, and k is Einstein's constant. The implications are that there is a minimum particle volume, that it can be defined from basic constants – G, c and $\hbar$, and that there are no singularities in physics.

A link is made with these equations between general relativity and quantum mechanics at the most basic level.

Before looking at the Dirac equation derived from the Evans Wave Equation, we review some basics.

The wave number can be defined two ways:

$$\kappa = 2\pi/\lambda \text{ or } \kappa = 1/\lambda \qquad (3)$$

where λ is the wavelength. The latter is used here. It can be applied to electromagnetic waves or to particle waves.

ω is angular frequency measured in rotations per second. It is defined:

$$\omega = 2\pi f = 2\pi/t \tag{4}$$

where f = 1/t, f is frequency and t is time. Then ω =1 when there is one rotation per second. See Figures 9-1 and 9-2.

Figure 9-1   Wave Number

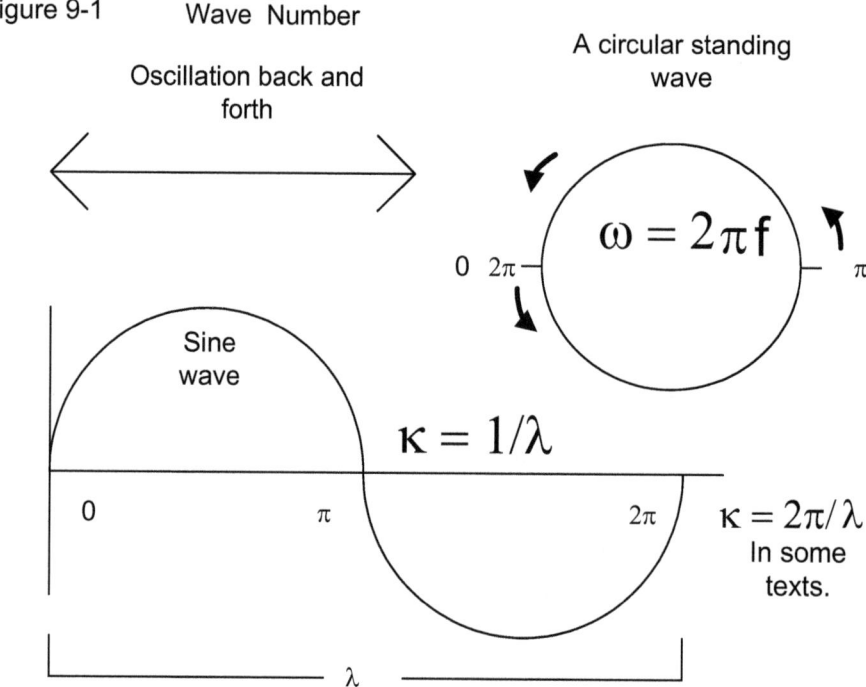

0 and 2π are the same point. In an oscillation or circular motion this is clear, but in the sine wave it is not at first glance. A high value of κ implies energetic, short wave lengths.

Figure 9-2   Equivalence of oscillation and circular motion.

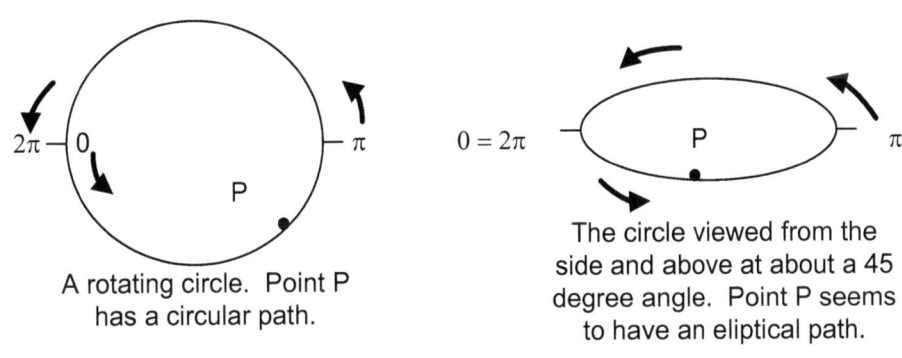

A rotating circle. Point P has a circular path.

The circle viewed from the side and above at about a 45 degree angle. Point P seems to have an eliptical path.

The circle viewed from the side and just above. Point P has a very flat eliptical curve, almost back and forth.

From the exact side the motion of P appears to be totally oscillatory. Its apparent speed is slower at the ends than in the middle.

The Compton wavelength is that of a photon having the same energy as the mass of a particle. Another description is that if the mass of an electron were converted to a photon, that photon would have the frequency associated with the Compton wavelength. In other words, if a given photon of frequency x were converted to mass, that particle would have the same frequency. The Compton wavelength is:

$$\lambda_c = \hbar / mc \qquad (5)$$

The Compton wavelength is a measure of energy.

The electron moves in complex standing wave patterns which are resonantly stable around the nucleus of an atom. When free, the electron behaves more like a particle wave traveling through spacetime.

The de Broglie wavelength describes the free electron as a standing wave with angular frequency as any other wave. This is the basis of the wave-particle duality of quantum mechanics.

$$\lambda_{de\,B} = \hbar / p = \hbar / mv \tag{6}$$

where $\lambda_{de\,B}$ is the de Broglie wavelength, $\hbar$ the Dirac constant, and $p = mv$ is the momentum.

We can associate the de Broglie and Compton wavelengths together as describing two aspects of the same process. The Compton and de Broglie wavelengths are descriptions of energy.

We also recall the basic quantum and Einstein energy equations:

$$E = nhf \tag{7}$$

$$E = mc^2 \tag{8}$$

Here E is energy, n the quantum number, h is Planck's constant, f is frequency, m is mass, and c is velocity of electromagnetic waves. These equations link relativity and quantum mechanics giving $nhf = mc^2$. That is that Planck's action times frequency equals mass or energy.

The basic equations of curvature are:

$$\kappa = 1/r \tag{9}$$

where r is the radius of an osculating circle that touches a point on a curve. And

$$R := \kappa^2 \tag{10}$$

where κ is the curvature and R is the scalar curvature.

In some texts Latin k is used for wave number and curvature. Here we use only kappa κ to describe them.

We know that spacetime curvature increases as mass-energy increases. Simple examples are the Lorentz contraction as a particle approaches c and the shrinking of a black hole to near a dot as mass increases. High energy density reference frame volumes are highly compressed compared to, and as viewed from, low energy density regions like our spacetime on Earth.

Dirac and Klein-Gordon Equations

The Dirac equation in its original form is:

$$(i\gamma^\mu \partial_\mu - mc/\hbar)\psi = 0 \tag{11}$$

where i is $\sqrt{-1}$ and γ^μ designates the Dirac matrices.

The Dirac equation is a wave equation valid in special relativity whereas the Schrodinger equation applies to Euclidean space only. The Evans Wave Equation is valid in general relativity as well and therefore supplants the Dirac equation in theoretical physics.

The Evans Wave Equation is

$$(\Box + kT)q^a{}_\mu = 0 \tag{12}$$

Where $\Box$ is the d'Alembertian operator, k is Einstein's constant, T is the standard contracted form of the canonical energy momentum tensor, and $q^a{}_\mu$ is the tetrad. This could be written:

$$(\Box + kT)\gamma^\mu = 0 \tag{13}$$

where γ^μ is the Dirac matrix generalized to non-Euclidean spacetime. It is a 4 x 4 matrix related to the tetrad.

In special relativity the Dirac and Klein-Gordon equations are of the same form as the Evans Wave Equation.[41] The Dirac equation can be written:

$$(\Box + 1/\lambda_c^2)\psi = 0 \tag{14}$$

---

[41] $(\Box + \kappa_0^2)\phi = 0$ is the general form. Note that $\kappa_0^2 = 1/\lambda_c^2$ and the Compton wavelength is derived form the scalar curvature in the Evans unified field theory in the special relativistic limit. This allows equating (unifying) curvature with wavelength.

where λ_c is the *Compton wavelength* and ψ is a four-spinor. The Klein-Gordon equation can be written:

$$(\Box + (mc/\hbar^2))\phi = 0 \qquad (15)$$

This is equivalent to:

$$(\Box + kT)\psi^a{}_\mu = 0 \text{ or } (\Box + (mc/\hbar^2))\psi \qquad (16)$$

where $\psi^a{}_\mu$ is a tetrad.

Therefore we can see that $kT \to (mc/\hbar)^2 = 1/\lambda_c^2$ in the weak field limit of flat spacetime. That is kT approaches $(mc/\hbar)^2$ and $1/\lambda_c^2$. The stress energy density of Einstein and the Dirac energy are seen to be equivalent in the weak field limit. The energy of the Klein-Gordon equation is reinterpreted to be equal to kT and has no negative solutions. The Klein-Gordon equation, as originally interpreted, had negative solutions. Since solutions were considered probabilities, and probabilities cannot be negative, the equation was considered defective. Now we see this is not so.

In general relativity the equation becomes:

$$(\Box + kT)\psi = 0 \qquad (17)$$

where

$$\psi = \begin{pmatrix} q_1^{(R)} \\ q_2^{(R)} \\ q_1^{(L)} \\ q_2^{(L)} \end{pmatrix} = \begin{pmatrix} q^{(R)} \\ q^{(L)} \end{pmatrix}$$

with the superscripts denoting right and left handed two-spinors.

We have not discussed spinors in this book. They are sort of like the square root of a vector. When looking at these equations, picture that; it may help to understand.

Chapter 9 – The Dirac, Klein-Gordon, and Evans Equations

In the derivation of the Dirac equation, one begins with the wave equation, and the first step is to transform the metric four-vector into a metric two-spinor. The two-spinor is then developed into a four-spinor with the application of parity.

The spinor analysis is necessary to show that ψ can be expressed as a 2 x 2 tetrad, $\psi^a{}_\mu$.

Then the flat spacetime limit is approached by recognizing:

$$kT \rightarrow (mc/\hbar)^2 \qquad (18)$$

We arrive at the Dirac equation with dimensionless metric four spinor.[42]

Using equations (5) to (10) we can now see that:

$$1/\lambda_c^2 = (mc/\hbar)^2 = kT_0 = R_0 = \kappa_0^2 \qquad (19)$$

Any of these equivalencies can be used as necessary, for example:

$$\lambda_c = R_0^{-1/2} \qquad (20)$$

This allows us to equate elements of curvature and general relativity with quantum mechanics and wave equations. We can substitute any of the above for one another. The Compton wavelength is the rest curvature in units of $1/m^2$. It is the inverse of the curvature squared. It is possible to define any scalar curvature as the square of a wave number.

The Evans derivation of the Dirac wave equation shows that the particle must have positive or negative (right or left) helicity, and gives more information

---

The Dirac equation is well known in quantum mechanics. It has now been deduced from general relativity using geometry.

No probabilistic assumptions were made in the derivation and a reinterpretation of the meaning of quantum theory is needed.

---

[42] Professor Evans stated, "The Dirac equation can then be expressed with Dirac matrices. The initial equation is obtained from covariant differentiation of the new metric compatibility condition of the tetrad postulate, $D_\nu q^a{}_\mu = 0$ on the metric four vector."

than the original Dirac equation.

From the Dirac equation one can deduce the Schrodinger equation.

Thus we see several basic equations of physics derived from the Evans equations. The Einstein principle says that the equations of physics must be geometry and this is fulfilled.

The components of $R = -kT$ originating in rest energy are the Compton wavelength in special relativity. R in the low limit is the least curvature and is an example of Evans principle of least curvature.

R cannot become 0. If it were 0 then spacetime would be flat and empty. There would be no universe in that region.

Compton Wavelength and Rest Curvature

Rest curvature is the minimum curvature associated with any particle. The rest curvature is the inverse of the square of the Compton wavelength.

$$|R_0| = 1/\lambda_c^2 = (mc/\hbar)^2 \tag{21}$$

Then:

$$\sqrt{|R_0|} = mc/\hbar \tag{22}$$

$$\hbar \sqrt{|R_0|} = mc \tag{23}$$

$$\hbar c \sqrt{|R_0|} = mc^2 \tag{24}$$

and since mc^2 is energy, rearranging:

$$E = mc^2 = \hbar c \sqrt{|R_0|} \tag{25}$$

Since $E = hf = \hbar \omega$, the rest curvature is related to the rest energy.

In general relativity, the quantum of energy is now known to exist and is for any particle or the photon:

$$E = \hbar c \sqrt{|R_0|} \tag{26}$$

This relates the Planck law and de Broglie wave-particle dualism.

The Compton wavelength is the rest curvature in units of inverse meters squared. The scalar curvature is related to the Compton wavelength. The Compton wavelength is characteristic of the mass of each particle, and thus scalar curvature of Einstein is joined with quantum theory in the Evans unified field theory.

$E = \hbar c \sqrt{|R_0|}$

This is an equation of general relativity, not quantum mechanics. Quantum mechanics emerges from general relativity.

The significance of this is that this term is a scalar curvature that is characteristic of the wave nature of a particle. The wavelength is related to rest curvature, which is R_0 and is defined by mass. This is an expression of particle wave duality.

The wave function is derived from the metric, an important new procedure that defines the Dirac equation in general relativity. The wave function is deduced to be a property of Evans spacetime.

The relationship $R_0 = -(mc/\hbar)^2$ can be made. In other words, if a particle has mass of m, then the curvature is defined in terms of c and $\hbar$, which are fundamental constants.

The Principle of Least Curvature:

The minimum curvature defining rest energy of a particle is, in the limit of special relativity:

$R_0 = -(mc/\hbar)^2$

Relationship between r and λ

Professor Evans points out the relationship between curvature and wavelength. We can look at it in a bit more detail. The speculative material here is primarily by this author. The equations are those of Professor Evans.

Figure 9-3 shows the two-dimensional geometry of $r = \lambda$, curvature = wavelength relationship. We can imagine in three or four dimensions that λ = AB length is a volume. As spacetime compresses like an accordion, the length AB is constant and becomes the distance along the wave while λ decreases. r is a measure of λ and is used to define the curvature.

The spacetime compression is the wave. A high-energy photon has a very short wavelength. An electromagnetic wave, a photon, is compressed spacetime itself spinning. That is, it does not "have" curvature and torsion, rather it is curvature and torsion of spacetime.

λ is physics and r is mathematics. From r we get κ, $|R_0|$, and kT. From λ we get $(\hbar/mc)^2$. Mixing them we get unified field theory.

One wonders if we have been looking for the origin of the universe at the wrong end of the universe. The high energy density near singularity we call the precursor to the big bang is as likely the end result of the geometry after it compresses.

Curvature and wavelength originate together in the scenario in Figure 9-3. The electromagnetic wave is space folded or compressed – the accordion. Curvature originates simultaneously. The mathematics is the first indication that this occurs and the pictorial description is not at odds with mechanical logic. In Figure 9-4 some more description is given.

It is only in the limit of special relativity that $r = \lambda$.

Chapter 9 – The Dirac, Klein-Gordon, and Evans Equations

Figure 9-3   Relationship between curvature and wavelength

Since $\kappa_0^2 = 1/\lambda_c^2$ and $1/r = \kappa_0 = 1/\lambda_c$ then $r = \lambda_c$

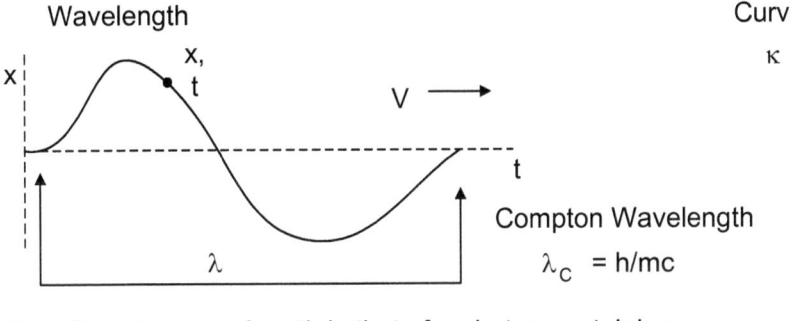

Wavelength

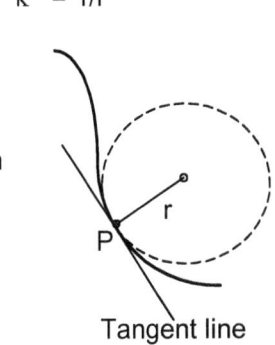

Curvature

$\kappa = 1/r$

Compton Wavelength

$\lambda_C = h/mc$

The Compton wavelength is that of a photon containing the rest energy of a particle, that is, its mass equivalence.

Tangent line

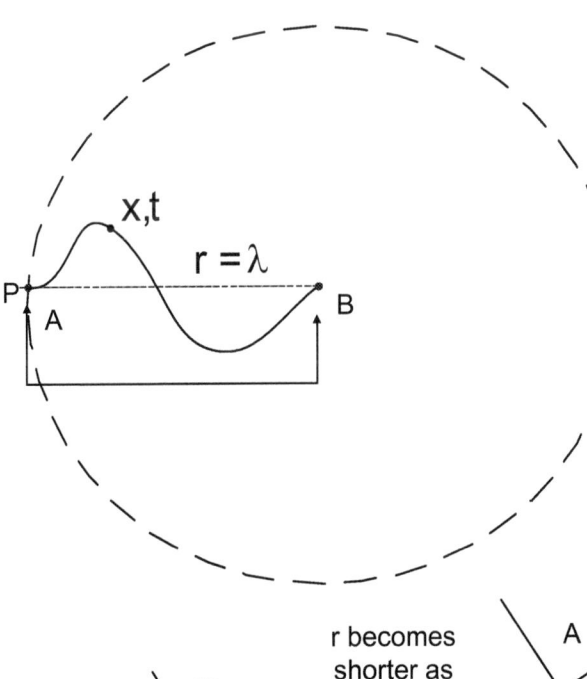

The arc length of the wavelength = λ when space is flat. It is a plane line.

Then as space scrunches like an accordion, the wave forms. The space is compressed (curved) and the wavelength defines r the radius of curvature.

λ is physics, r is mathematics

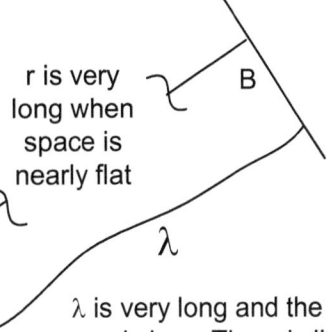

r is very long when space is nearly flat

λ is very long and the energy is low. There is little curvature of the wave.

r becomes shorter as energy increases and space is more compressed = curved.

207

Figure 9-4   Curvature and Wavelength

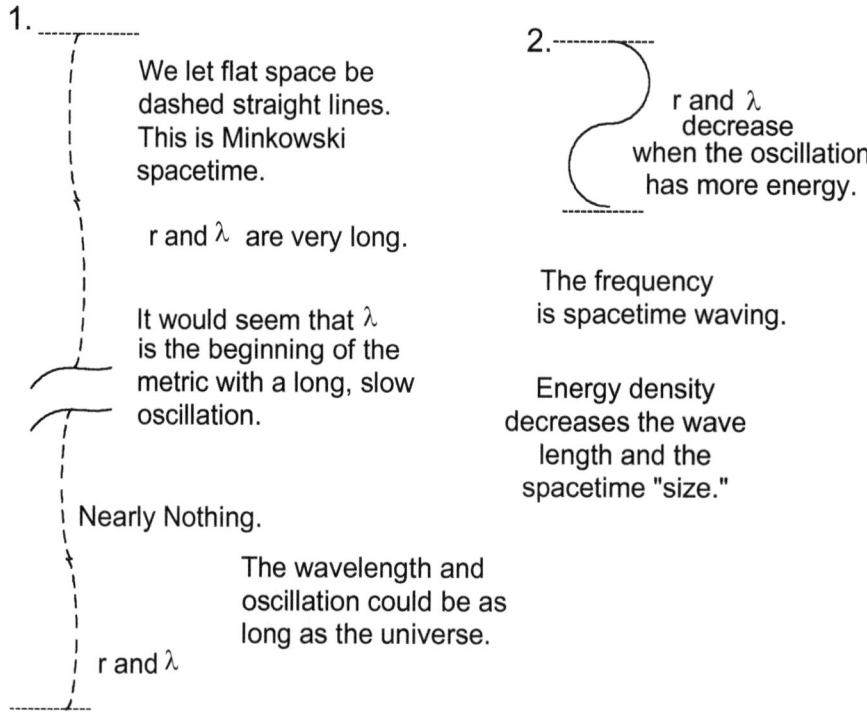

Derivation of the Inherent Particle Mass-Volume relationship

We have $kT_0 = |R_0| = \kappa_0^2 = 1/\lambda_c^2 = (mc/\hbar)^2$. Here $k = 8\pi G/c^2$ which is Einstein's constant, T_0 is the stress energy in the low limit.

In the weak field limit we know that $T_0 = m/V_0$ or simple mass density. Substituting we get

$$kT_0 = k\, m/V_0 \tag{27}$$

and from above $kT_0 = (mc/\hbar)^2$ so:

$$k\, m/V_0 = (mc/\hbar)^2 \tag{28}$$

After rearranging and canceling m, we get:

$$mV_0 = k\hbar^2/c^2 \tag{29}$$
$$V_0 = k\hbar^2/mc^2$$

Chapter 9 – The Dirac, Klein-Gordon, and Evans Equations

and since $E = mc^2$,

$$V_0 = k \hbar^2 / E \qquad (30)$$

> This means that the product of the rest mass and the rest volume of a particle is a universal constant in terms of Einstein's general relativity and Planck's constant.

We cannot have point particles. This has been a problem in quantum theory in the past. A particle has energy. If it also has no volume, then the energy density is infinite. Renormalization used an arbitrary minimum volume to avoid this. A similar but more accurate solution is clear – there is a minimum volume and it is given by the mV_0 equation. Renormalization is not necessary if the volume is given. Renormalization has been very successful and now the actual volume can be used.

That mass-energy and volume are inversely related is well recognized.

The wavelength of a high-energy photon is small. As energy increases, wave length decreases. More is packed into a smaller region – as viewed from our low energy reference frame.

In special relativity we have Lorentz-Fitzgerald contraction. As kinetic energy builds, the reference frame contracts in the directions in which energy is increased, as seen from a low energy density reference frame. Those directions are one space and one time in special relativity. In general relativity they are the spatial directions.

As mass increases beyond certain limits, the volume decreases. The black hole is the common example. Beyond a certain threshold, spacetime collapses into a dot. A Kerr spinning black hole is smaller than a Schwarzschild non-spinning hole. The spin energy compresses the spacetime.

One may tentatively assume that the Evans curvature equation above can be applied to a black hole as well as the particle. There is no singularity at the center of a black hole. There is a minimum volume since mV_0 is always > 0.

Particles

It can be said that particles are discrete regions of spacetime occupied by a standing wave. It may be more accurate to say that particles are standing spacetime waves. The field concept describes them mathematically. A zero rest mass particle is frequency only - that is a photon. Its momentum is wave number. No zero volume points can exist. As long as there is energy or mass, there is some curvature of spacetime present.

As of the time of this writing there is not yet a precise definition of the particle in terms of curvature and torsion, but it is obvious that it will be obtained.

The Shape of the Electron[43]

We can expect interesting discoveries in the future using the principle of least curvature and the mV_0 equation. This is an example.

One immediate application is to rearrange the equation to:

$$V_0 = (k/m)(\hbar^2/c^2) \tag{31}$$

We can then solve for the volume of a particle based on the mass. The electron would not be a sphere, but a flattened disk or ring possibly rotating to

---

[43] The speculation here is by this author, not by professor Evans. A point charge is hypothetical and the electron size can be characterized by a radius. The classical electron radius r_0, the Compton radius, is defined by equating the electrostatic potential energy of a sphere of charge e and radius r_0 with the rest energy of the electron,
$U = e^2/r_0 = m_e c^2$.

Chapter 9 – The Dirac, Klein-Gordon, and Evans Equations

take up a spherical region and giving rise to spin. For an electron with a mass of 9.11 x 10^{-31} kg, V_0 is about 1 x 10^{-79} m^3, and has a radius about 4 x 10^{-13} m.

Radius of the electron is variously estimated to be from 3.86 x10^{-13} m based on the magnetic moment and 1/2 the Compton radius or 1.21 x 10^{-12} m. The analysis below uses the value 3.86 x10^{-13} m for the radius.

A spherical shape does not fit since $V_0 = 4/3 \pi r^3$ = 4/3 x 3.14 x 1.21 x 10^{-12} m = 9 x 10^{-36} m^3 is much too large.

A flattened disk with a radius of 1.21 x 10^{-12} and a thickness of 2.18 x 10^{-56} fits the volume better. $V_0 = \pi r^2$ x thickness = 3.14 x $(1.21 \times 10^{-12})^2$ x 2.18 x 10^{-56} = 1.00 x 10^{-79}. A ring shape could also exist and the Compton radius may be more appropriate but the result in any event would be in the same neighborhood. Is spin then the turning of the plane of the disk in the 3rd and 4th dimensions? This is purely speculative, but there are other indications that the electron may be a ring. Then we can look at the proton also.

Figure 9-5   Possible Electron Shape

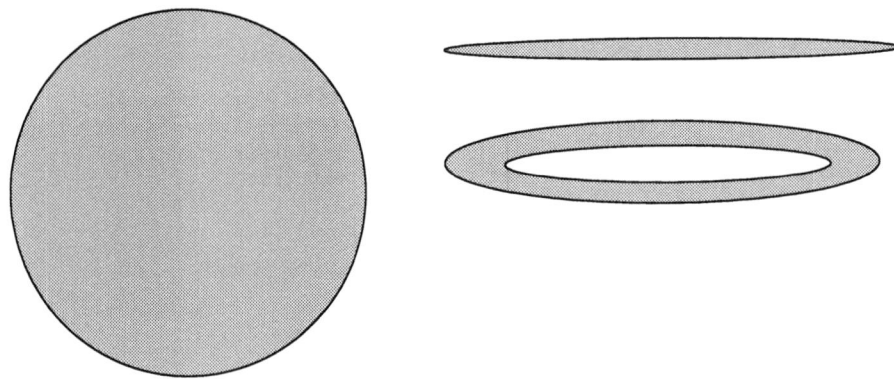

The radius, mass and volume relations do not allow a spherical shape for the electron.

A very flat disk or a ring shape would allow the known dimensions.

Summary

The two new equations introduced here are:

$$E = \hbar c \sqrt{|R_0|} \tag{1}$$
$$mV_0 = k\hbar^2/c^2 \tag{2}$$

The first gives us the Principle of Least Curvature and the second gives us the minimum volume of a particle. Figure 9-6 shows implications. We do not know exactly why large mass contracts spacetime with respect to our low energy density reference frame, but we can now describe another part of the puzzle mathematically.

Figure 9-6 Inverse Mass Energy to Volume Relationship

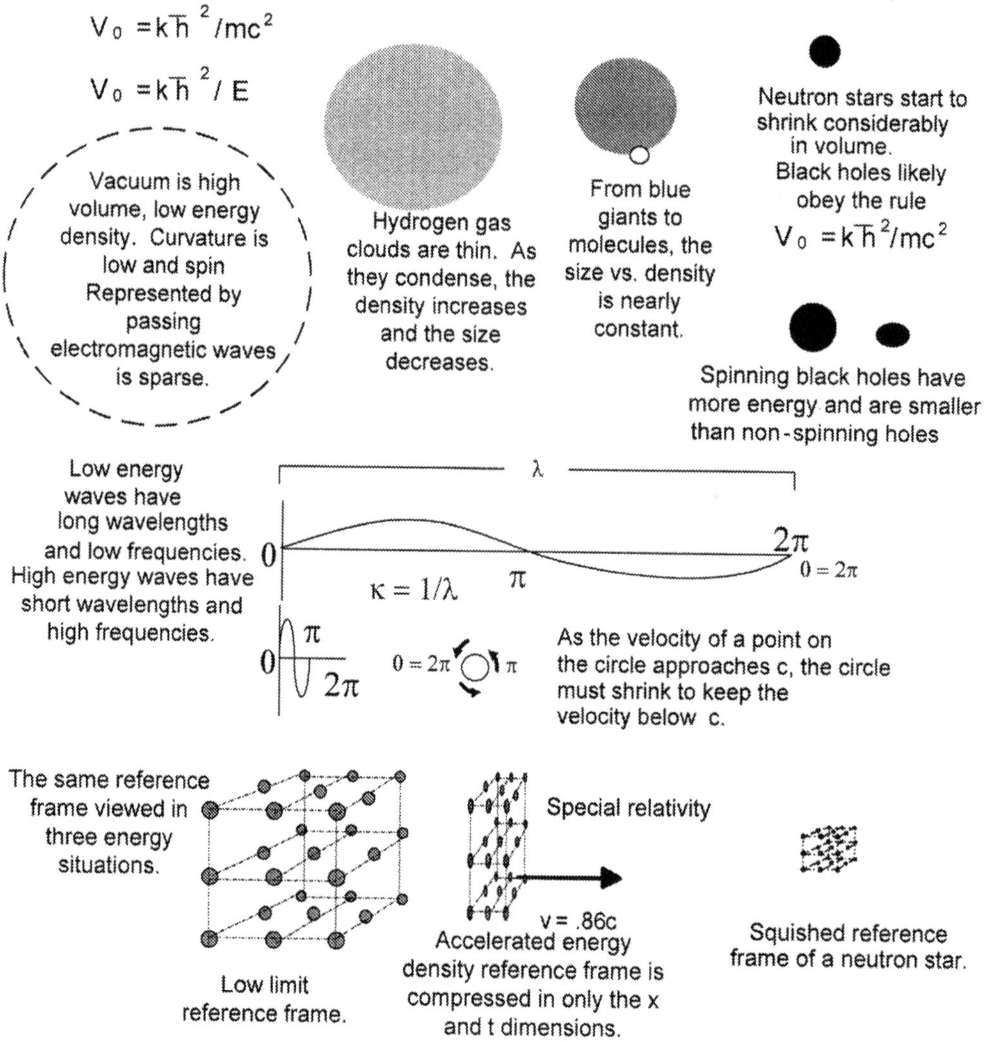

$V_0 = k\bar{h}^2/mc^2$

$V_0 = k\bar{h}^2/E$

Vacuum is high volume, low energy density. Curvature is low and spin Represented by passing electromagnetic waves is sparse.

Hydrogen gas clouds are thin. As they condense, the density increases and the size decreases.

From blue giants to molecules, the size vs. density is nearly constant.

Neutron stars start to shrink considerably in volume. Black holes likely obey the rule
$V_0 = k\bar{h}^2/mc^2$

Spinning black holes have more energy and are smaller than non-spinning holes

Low energy waves have long wavelengths and low frequencies. High energy waves have short wavelengths and high frequencies.

$\kappa = 1/\lambda$

As the velocity of a point on the circle approaches c, the circle must shrink to keep the velocity below c.

The same reference frame viewed in three energy situations.

Low limit reference frame.

Special relativity

$v = .86c$
Accelerated energy density reference frame is compressed in only the x and t dimensions.

Squished reference frame of a neutron star.

The formulations mix general relativity and quantum mechanics and are equations of unified field theory.

Never before have a least curvature or minimum particle volume been defined.

$\hbar = h/2\pi$ so in terms of a circle or oscillation, h is the circumference and $\hbar$ is the radius. Alternatively, we can see $\hbar$ as the frequency and h as ω in Figures 1 and 2. Then in equation (1) $E = \hbar c \sqrt{|R_0|}$, energy equals frequency or oscillation times c times κ or $1/r$. We have a unified quantum and relativistic equation.

$$mV_0 = k\hbar^2/c^2 \text{ with } k = 8\pi G/c^2 \Rightarrow mV_0 = 8\pi G\hbar^2/c^4 \quad (31)$$

Rearrangement several ways is possible, but all are still somewhat enigmatic.

$$1/r^2 = \kappa_0^2 = R_0 = kT_0 \quad \text{General Relativity}$$

$$\kappa_0^2 = 1/\lambda_c^2 = (mc/\hbar)^2 \quad \text{Wave Mechanics}$$

$$kT_0 = (mc/\hbar)^2 \quad \text{Unified Theory}$$

$$V_0 = k\hbar^2/E$$

Chapter 10 Replacement of the Heisenberg Uncertainty Principle

> Things should be made as simple as possible, but not any simpler.
> Albert Einstein

Basic Concepts[44]

While quantum mechanics emerges from the Evans equations, the probabilistic nature of the Schrodinger equation and the Heisenberg uncertainty principle are modified. The Klein-Gordon equation, which was abandoned, is resurrected and correctly interpreted. Conflicts between general relativity and quantum mechanics are resolved.

During the last few years new experiments have cast serious doubt on the Heisenberg uncertainty concept and the Bohr complementarity concept.[45]

Very high resolution microscopes have been developed that can measure position accurately down to 1/50th of a wavelength while keeping a simultaneous measurement of momentum at $2h/\lambda$. That is, $\Delta x = \lambda/50$ and $\Delta p = 2h/\lambda$. Thus

$$\Delta x \, \Delta p = h/25$$

---

[44] The full derivation of the material covered here can be found in "New Concepts from the Evans Unified Field Theory, Part Two: Derivation of the Heisenberg Equation and Replacement of the Heisenberg Uncertainty Principle" (Feb, 2004)and "Generally Covariant Quantum Mechanics" (May 2005). This chapter is a simplification based on the above and on handwritten notes titled "The Heisenberg Equation of Motion from the Evans Wave Equation of Motion," M.W. Evans, Feb 13, 2004, www.aias.us.

[45] Beyond Heisenberg's Uncertainty Limits, J.R. Croca, *Gravitation and Cosmology From Hubble Radius to Planck Scale*, edited by R.I. Amoroso, G. Hunter, and M. Kafatos, J-P. Vigier, Kluwer Academic Publishers (now Springer), Dordrecht, 2002. Available at http://cfcul.fc.ul.pt/Equipa/croca/berkeley%20-%20paper.pdf

Chapter 10   Replacement of the Heisenberg Uncertainty Principle

instead of h/2 as Heisenberg predicted.

Shariar Afshar while at Harvard performed a new type of double slit experiment in which was shown definite simultaneous wave and particle properties of a photon.[46] This disproves the complementarity concept. The photon or electron wave or particle aspects are simultaneously observable.

The Copenhagen probabilistic interpretation of quantum mechanics cannot hold and the Einstein-de Broglie viewpoint of causal physics must be accepted. Quantum mechanics is accepted, but the Copenhagen interpretation is rejected in the unified field theory and now by reproducible experimental evidence which is of course the strongest argument.

The Heisenberg equation can be derived from both the Evans wave and field equations. General relativity is objective. It does not accept uncertainty. The uncertainty equation can be derived from Evans' equations, but the interpretation is causal. A new law, derived using differential geometry and the Evans Wave Equation, replaces the Heisenberg uncertainty principle.

There are a few equations that we can look at in preparation for Evans' analysis.

$$p = \hbar \kappa \tag{1}$$

where p is momentum, $\hbar$ is Dirac's constant, and κ is the wave number. This is derived from the definition for wave number, $\kappa = 2\pi / \lambda$ and the de Broglie wavelength of a particle, $\lambda = \hbar / p$.

The momentum and wave number can be expressed as tetrads in the unified field theory. They are related by:

$$p^b{}_\mu = \hbar \kappa^b{}_\mu \tag{2}$$

---

[46] See http://users.rowan.edu/~afshar.

The position can also be expressed as a tetrad, $x^b{}_\mu$.

Figure 10-1 Heisenberg Uncertainty

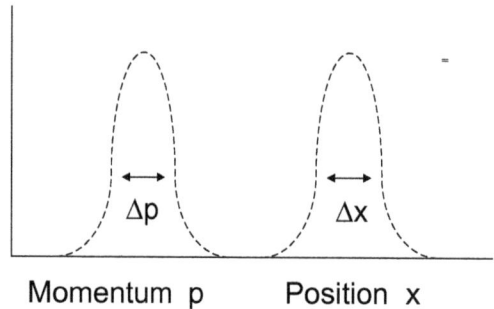

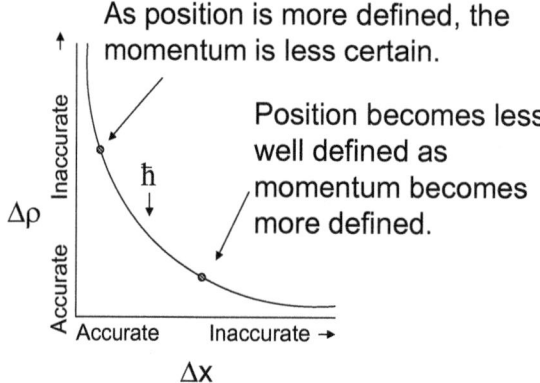

The Heisenberg uncertainty principle is often stated as:

$$\Delta x \, \Delta p \geq \hbar/2 \qquad (3)$$

See Figure 10-1. Here x is position and p is momentum = mv or = γmv. ℏ is h/2π. That is, if the position and momentum of a particle are measured simultaneously, there will be an error greater than or equal to Planck's constant. In the original paper, Heisenberg stated it as:

$$pq - qp = \frac{h}{2\pi i} \qquad (4)$$

He stated that he believed that the existence of the classical "path" came into existence only when we observe it.[47] Various explanations and

---

[47] Heisenberg, in the Uncertainty Principle paper, 1927.

interpretations have come out over the years since. The predicted error is not due to *inaccuracy* in measurement. It is due to uncertainty in actual physical reality according to Heisenberg and the Copenhagen school of thought.

The principle is now more formally stated as:

$$(xp - px)\psi = i\hbar\psi \tag{5}$$

where momentum $p = -i\hbar\, \partial/\partial x$.

Replacement of the Heisenberg Uncertainty Principle

As a consequence of the Evans equations:

$$(\Box + kT)\psi^a{}_\mu = 0 \quad \text{and} \quad (R + kT)\psi^a{}_\mu = 0 \tag{6}$$

and dependent on the Evans Lemma:

$$\Box q^a{}_\mu = R\, q^a{}_\mu \tag{7}$$

the Heisenberg uncertainty principle can be reinterpreted.

R and T in the Evans equations are quantized. They are contained within wave equations but the are also equations in general relativity and are therefore causal. Evans refers to this as "causal quantization" to distinguish his unification from Heisenberg's quantization which was probabilistic.

Evans first derives the Schrodinger equation from a non-relativistic limit of the Evans Wave Equation, then derives the Heisenberg equation from the Schrodinger equation. Evans shows that the Schrodinger equation is a restatement of the Heisenberg equation with a operator equivalence derived from general relativity. Again we see equations of quantum mechanics derived from the geometry of Evans' general relativity.

Instead of using p, momentum, and ℏ, angular momentum, their densities are defined using bold italics to indicate density:

p = p/V and **ℏ** = ℏ/V_0

The volumes here are derived from $V = \hbar^2 k / mc^2$ shown in Chapter 9.

The Heisenberg uncertainty principle is replaced by:

$$x_a \wedge p_b \to \hbar \qquad (8)$$

where p_b is a volume differential forms. This states loosely that the wedge product of position and momentum density approaches $\hbar$, angular momentum density as a limit. That this should be is completely logical. If $\hbar$ is the smallest observable action in the universe, any less must be zero – not existence. The least observable action is $\hbar$. However now volume density is used.

Figure 10-2 Wedge Product of Position and Momentum (Wave Number) Tetrads

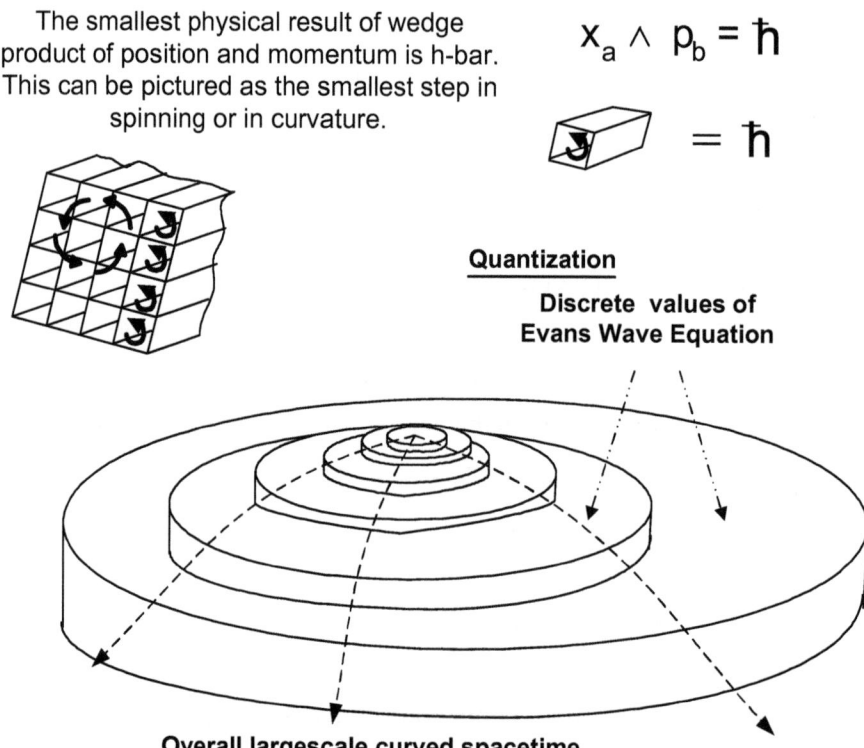

This is the law that governs the behavior of the least amount of angular momentum *density*, $\hbar$, in the limit of special relativity. $\hbar$ is the quantum of angular momentum, $\hbar$ is the quantum of *angular momentum density*.

Chapter 10 Replacement of the Heisenberg Uncertainty Principle

The equations are derived from general relativity and are generally covariant. They replace the Heisenberg equation which was a flat spacetime equation.

Born concept that the product of a wave function with its complex conjugate is a probability density is replaced with a volume density. Pictorially, the product of position volume density and momentum or wave number is shown in Figure 10-2. The smallest product possible is $\hbar$, that is $\hbar/V_0$.

Angular momentum is the wedge product of tetrads. In general relativity the uncertainty principle is replaced by a wedge product of tetrads. The principle of least curvature is used to interpret $\hbar$ as the least possible *angular momentum density* or *action density*.

This indicates that the quantum of energy originates in scalar curvature.

In Evans' general relativity we can measure down to the smallest connected dot of the vacuum. Evans shows that this smallest dot is the least curvature that can exist. Rather than approaching it from the idea that an inherent error always exists, Evans states that we can always be accurate to the least amount of differentiation in the spacetime. And density must be considered, not simply the total energy.

No longer do we think in terms of probability, but rather in terms of the smallest possible quantum step in our spacetime manifold. The least curvature is defined by the least action $\hbar$, and so $\hbar$ is the causally determined least possible action *density* or least possible angular momentum *density* for the photon and for any particle.[48]

The Heisenberg uncertainty principle becomes a causal and geometrical formula. This conserves the least action according to the least curvature

---

[48] In an email Professor Evans says, "There is no room in any of this for an "acausal physics" or "unknowable physics", so I reject the Copenhagen School as being unphysical (outside natural philosophy). It is a solid fact of observation that the Uncertainty Principle has been refuted experimentally by the Croca group in Lisbon Portugal, but the protagonists of the Copenhagen School stubbornly refuse to accept the evidence of the data, a classic blunder in physics. No doubt, I will get into an awful lot of more trouble for these common sense remarks and my sanity once more questioned in a derisive manner. This is another example of the least action principle, to make fun." No one has made fun up to the date of this writing. LGF

principle. When momentum is high, the wave length is small – that is, R is small. The limit is ℏ. This applies to the known complimentary pairs. For example, if E, energy, is large, measurable time is small.

Evans tells us that ℏ is causally determined by minimizing action to define a particle of any given type. It is a constant because it is not possible to have zero action for a particle, and not possible to go below a certain minimum. This is the quantum limit.

The use of densities solves the Croca observation puzzle.

The Klein Gordon Equation

The Klein Gordon equation is a limiting form of the Evans wave equation:

$$(\Box + m^2c^2/\hbar^2) q^a{}_\mu = 0 \tag{12}$$

where $kT = m^2c^2/\hbar^2 = 1/\lambda_c^2$. See Chapter 9. The Compton wavelength, $\lambda_c = \hbar/mc$, is similar to the de Broglie wavelength, $\lambda_{de\,B} = \hbar/mv$.

Evans identifies

$$R_0 = -(m^2c^2/\hbar^2) \tag{13}$$

as the least curvature needed to define the rest energy of a particle.

Figure 10-3   Minimum Curvature

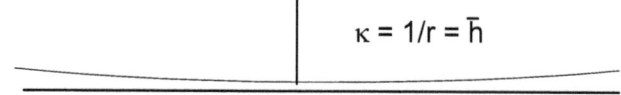

$\kappa = 1/r = \bar{\hbar}$

Spacetime cannot be flat, however it can be nearly so. The minimum curvature possible is measurable as ℏ.

Any less is non-existence = flat spacetime.

The Klein Gordon equation is a scalar component of the tetrad. It is not a probability. A probability cannot be negative whereas a tetrad can have a negative component. The Klein Gordon equation is valid.

We have seen that all forms of energy are interconvertible. The indication here is that regardless of form, they are the same thing. The minimum curvature, R_0 is in the limit of special relativity, a function of mass and basic constants c and $\hbar$. See Figure 10-3.

This simplifies the quantum uncertainty approach and agrees with Einstein that physics is causal.

Summary

$x_a \wedge p/V \rightarrow \hbar/V_0$ is the replacement for the Heisenberg equation. The wedge product of position and momentum density approaches *$\hbar$* density as a limit. The least observable action density is $\hbar$. This is not a statement of uncertainty, rather it gives the border between existence and non-existence.

Another way to look at this is that quantum physics was unable to determine the position and momentum of a particle at the same time and saw this as due to an inherent probabilistic nature of reality. In actuality, they can be determined to a much smaller size – right down to existence or non-existence. Use of the density allows going to the limit of that existence.

With quantum physics emerging from general relativity, Evans has showed that the interpretation is incorrect. Any measurement is limited at that point where a quantum jump takes place. The change can only occur in steps of $\hbar$. This was Planck's original quantum hypothesis. This interpretation is causal and is the argument that Einstein made.

In the unified field theory, the matter becomes clear. In general relativity kT, density gives a volume density which when used allows theory to agree with experiment. Not $\hbar$ or momentum, but *$\hbar$ density* and momentum *density* are the correct terms to apply.

Chapter 11   The Evans $B^{(3)}$ Spin Field

> It is only the circumstances that we have insufficient knowledge of the electromagnetic field of concentrated charges that compels us provisionally to leave undetermined the true form of this tensor.
>   Albert Einstein

> $B^{(3)}$ is a triumph of Einstein's own general relativity applied to electromagnetism and unified field theory.
>   Myron Evans

Introduction

As always, feel free to skip the math if you fear brain damage. Just get the essential concepts down.

The electromagnetic field is the twisting and turning of spacetime itself.

Figure 11-1 shows the older U(1) Maxwell-Heaviside concept of the electromagnetic B field and the electric E field. The depiction on the right takes slices through the traveling field while it expands, reaches maximum size, and then contracts again. (U)1 is a circle. This is not completely logical since the field moves forward and has existence in all dimensions. The Evans concept is that the B and E fields are spacetime, not an entity superimposed on spacetime. The fields turn as they propagate. Spacetime itself is turning. See Figure 11-2. Figures 11-1 and 11-2 look alike at first, but there is a significant difference.

Photon spin can produce magnetism in materials.[49]

---

[49] There has been experimental confirmation the existence of a $B^{(3)}$ field in the inverse Faraday effect of an order of a million gauss with pulsed circularly polarized lasers in under-dense plasma. So there is no doubt that the $B^{(3)}$ field exists. The various types of experimental and theoretical evidence for $B^{(3)}$ is reviewed in Volume 119 of Advances in Chemical Physics. This evidence has been collected and analyzed over about a decade

Chapter 11 – The Evans B$^{(3)}$ Spin Field

This is observed in and explains the Inverse Faraday Effect which cannot be explained using the older two dimensional U(1) concept. Magnetic spin causes magnetization of materials by radiation at all frequencies. The photon must have mass, however small, for this to take place.

Figure 11-1   Depiction of the Magnetic and Electric Fields

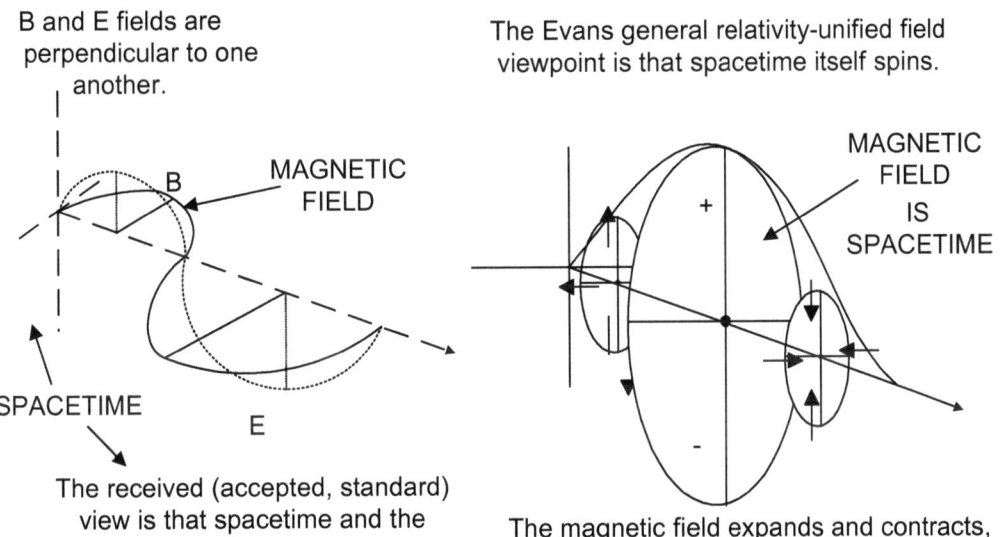

B and E fields are perpendicular to one another.

The Evans general relativity-unified field viewpoint is that spacetime itself spins.

The received (accepted, standard) view is that spacetime and the magnetic and electric fields are different entities.

The magnetic field expands and contracts, changing polarity (direction) every half cycle.

The Evans B$^{(3)}$ Spin Field

B$^{(3)}$ is the angular momentum multiplied by a constant. It is the longitudinal component of the photon.

$$\mathbf{B}^{(3)*} = i g \, \mathbf{A}^{(1)} \wedge \mathbf{A}^{(2)} \tag{1}$$

where $g = e/\hbar$ for one photon; e is the charge on an electron. This means that the B$^{(3)}$ field is described by the turning of its components in three dimensions.

---

of research, initiated at Cornell Theory Center in 1991 (M. W. Evans, Physica B, 182, 227, 237, 1992. See also list of publications on www.aias.us).

Evans Equations of Unified Field Theory

The actual mathematical description is beyond that with which we can deal in this book.

The spin can be right (negative) or left (positive). Therefore the angular momentum can be right or left. If $\mathbf{B}^{(3)}$ is negative there is right circular polarization and if $\mathbf{B}^{(3)}$ is positive there is left circular polarization. Polarization of light, of any electromagnetic field or photon, can be left, right or linear. Linear polarization is the superposition of two circularly (or elliptically) polarized photons.

Figure 11-2   The Evans Spin Field

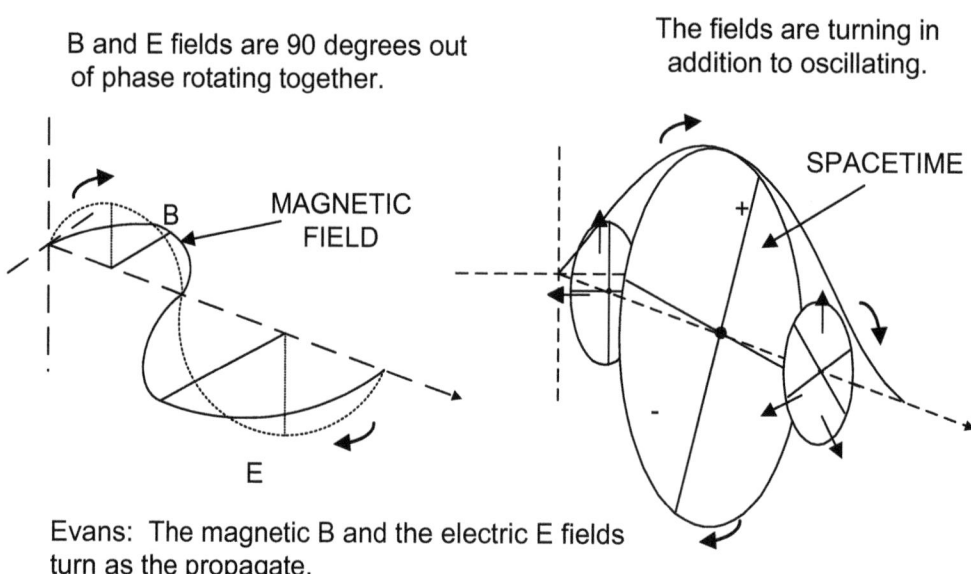

Evans: The magnetic B and the electric E fields turn as the propagate.

In the unified field theory Evans came to the realization that the spin field is spacetime itself.

The actual mathematical description is beyond that with which we can deal in this book.

If $\mathbf{B}^{(3)}$ is negative there is right circular polarization and if $\mathbf{B}^{(3)}$ is positive there is left circular polarization. Polarization of light, of any electromagnetic field or photon, can be left, right or linear. Linear polarization is the superposition of two circularly (or elliptically) polarized photons.

The $\mathbf{B}^{(3)}$ field is a component of the torsion form. It is an essential concept in taking electromagnetism from special relativity to general relativity and thus discovering the unified field theory. The Planck constant is the least amount of angular momentum possible and is the origin of the $\mathbf{B}^{(3)}$ field. The $\mathbf{B}^{(3)}$ field is a vector field directly proportional to the spin angular momentum of the radiated electromagnetic field. This occurs in the Z or (3) axis of propagation. The mathematics that describes $\mathbf{B}^{(3)}$ is non-linear and non-Abelian (rotating).

Figure 11-3

As the circle rotates and moves forward, a helix is drawn.

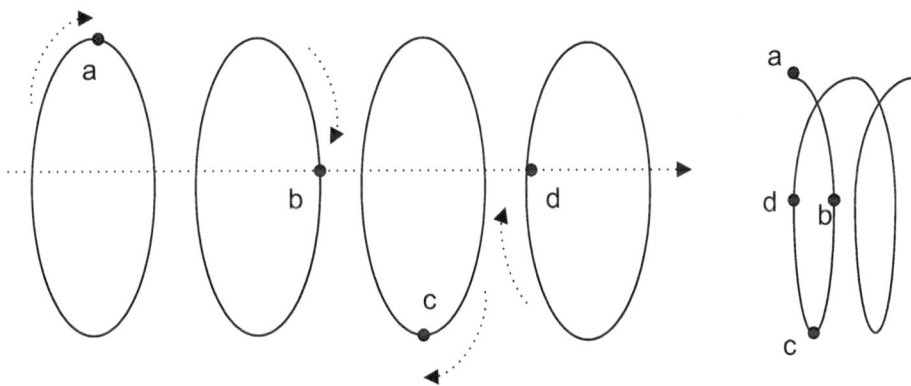

The photon exists in three dimensions. The U(1) geometry only describes two. $\mathbf{B}^{(3)}$ describes the third.

Basic geometric description

The $\mathbf{B}^{(3)}$ component of the B field was the discovery of Professor Evans in 1991. Up until then, only the $\mathbf{B}^{(1)}$ and $\mathbf{B}^{(2)}$ portions were used.[50]

---

[50] The spin field is now (2005) well established. For excellent visuals see: http://departments.colgate.edu/physics/research/optics/oamgp/gp.htm

Figure 11-3 shows the circle turning as it propagates forward. Point "a" on the circle forms a helix as it moves forward to points b, c, and d. In U(1) Maxwell Heaviside theory, a, b, c, and d are identical points. In Evans formulation, d has moved forward and has not rotated all the way to the a position. The arc length is a function of the rotation and the movement forward.

U(1) electrodynamics is a circle with $\mathbf{B}^{(1)}$ and $\mathbf{B}^{(2)}$ components. Evans considers the angular momentum about the Z-axis as the fields move forward. This is the $\mathbf{B}^{(3)}$ component.

$\mathbf{B}^{(3)}$ electrodynamics is a sphere with O(3) symmetry. The helix describes its movement and the curvature of the helix is $R = \kappa^2$.

We saw a version of this in Chapter 9 with curvature R and the Compton wavelength.

The $\mathbf{B}^{(3)}$ field is a longitudinal photon, called the Evans photomagneton. It is a phaseless magnetic field. It can be described by

$$\mathbf{B}^{(3)*} := i g \mathbf{A}^{(1)} \times \mathbf{A}^{(2)} \tag{2}$$

where $g = e/\hbar$ for one photon and $\mathbf{A}^{(1)} = \mathbf{A}^{(2)*}$ is the transverse vector potential. This can be developed further. Evans gives the magnetic field components:

$$B^c{}_{ij} = B^{(0)} (q^a{}_i q^b{}_j - q^a{}_j q^b{}_i) \tag{2}$$

and the electric field components:

$$E^c{}_{oj} = E^{(0)} (q^a{}_o q^b{}_i - q^a{}_i q^b{}_o) \tag{3}$$

Note that the structure is that of the wedge product.

In the older view of gravitation, the electromagnetic field was a structure imposed on flat spacetime. This viewpoint is known to be incorrect since spacetime curves and twists, but until the Evans equations, a method for describing electromagnetics in four-dimensional curved spacetime was unknown.

Now the accepted viewpoint must be that the electromagnetic field is also a manifestation of spacetime described by an equation similar to Einstein's symmetric field equation, $R_{\mu\nu} - \frac{1}{2} R q_{\mu\nu} = k T_{\mu\nu}$.

Chapter 11 – The Evans $B^{(3)}$ Spin Field

Figure 11-4   Basic Geometry

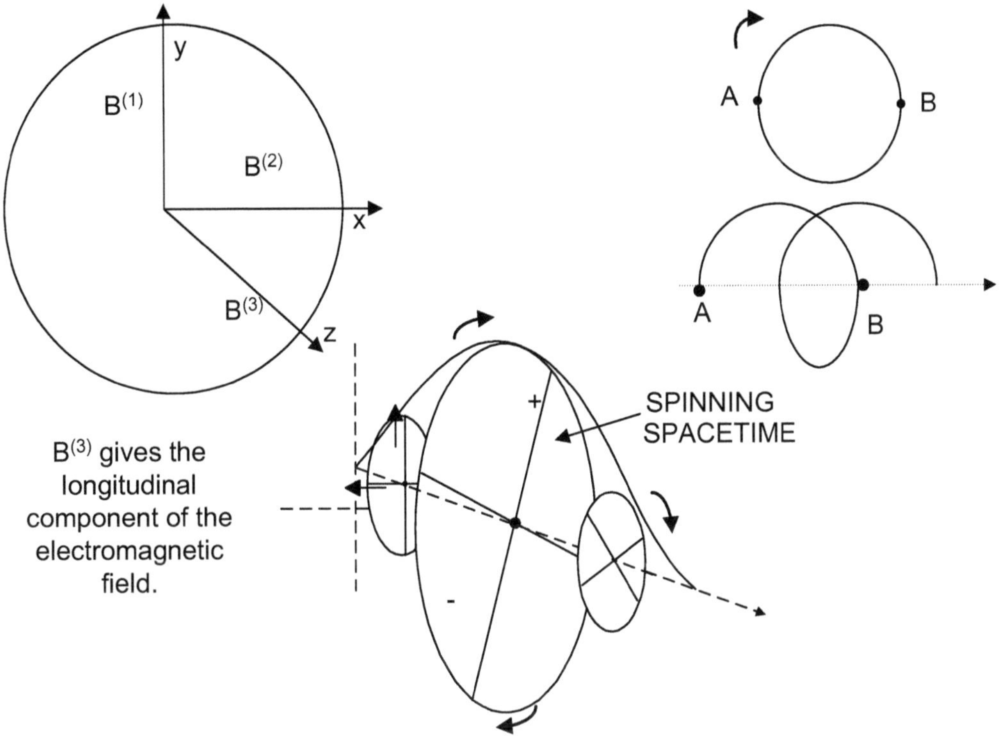

$B^{(3)}$ gives the longitudinal component of the electromagnetic field.

SPINNING SPACETIME

While the gravitational field is described by the symmetric Einstein field equation, the electromagnetic field is described by the antisymmetric metric dual to the symmetric metric. The existence of a symmetric metric and an antisymmetric metric – curvature and torsion – is the concept that allows this description.

> The electromagnetic field is spinning spacetime.

The magnetic and electric fields in special relativity have circular U(1) symmetry and are seen as entities on flat spacetime. In general relativity we know spacetime is curved. In the unified field theory we see that electromagnetism is turning spacetime. This is depicted in Figure 11-5.

Einstein's principle of general relativity says that all equations of physics must be generally covariant. This includes electromagnetism. Professor Evans

Figure 11-5 Gravitation and Electromagnetism

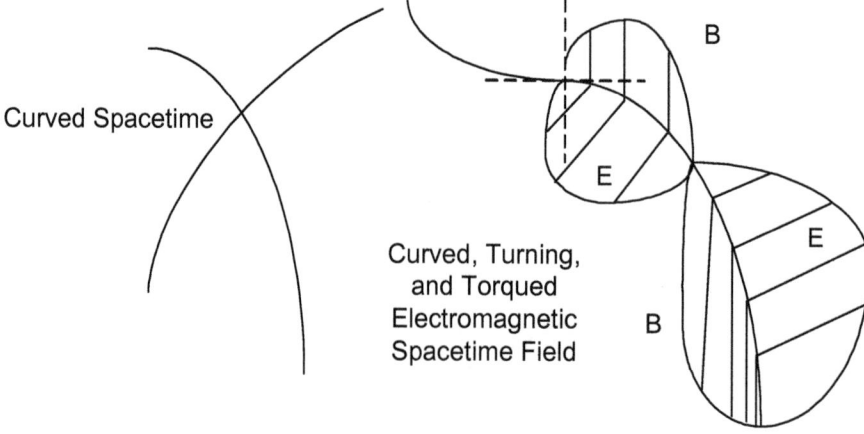

as taken Einstein seriously and shows that all physics can be developed from Einstein's basic postulate, $R = -kT$. The helix is $R = \kappa^2$, a clear connection between electromagnetics and general relativity geometry.

The Metric

The metric tensor is the tensor product of two four dimensional vectors whose components are scale factors:

$q^\mu = (h^0, h^1, h^2, h^3)$ and $q^\nu = (h_0, h_1, h_2, h_3)$

These give the symmetric metric tensor in curved spacetime with curvilinear coordinates.

The symmetric metric tensor is:

$$q^{\mu\nu(S)} = \begin{bmatrix} h_0^2 & h_0h_1 & h_0h_2 & h_0h_3 \\ h_1h_0 & h_1^2 & h_1h_2 & h_1h_3 \\ h_2h_0 & h_2h_1 & h_2^2 & h_2h_3 \\ h_3h_0 & h_3h_1 & h_3h_2 & h_3^2 \end{bmatrix}$$

The antisymmetric metric tensor is:

$$q^{\mu\nu(A)} = \begin{bmatrix} 0 & -h_0h_1 & -h_0h_2 & -h_0h_3 \\ h_1h_0 & 0 & -h_1h_2 & h_1h_3 \\ h_2h_0 & h_2h_1 & 0 & -h_2h_3 \\ h_3h_0 & -h_3h_1 & h_3h_2 & 0 \end{bmatrix}$$

The metric must have a factor that is negative under charge conjugation for the electric field to exist.

Then the complete electromagnetic field tensor is the well-known tensor:

$$G^{\mu\nu} = \begin{bmatrix} 0 & -E^1 & -E^2 & -E^3 \\ E^1 & 0 & -B^3 & B^2 \\ E^2 & B^3 & 0 & -B^1 \\ E^3 & -B^2 & B^1 & 0 \end{bmatrix}$$

The equations showing the results of inner, outer, contraction, and wedge products of the metric are in the chart titled "Map of the Evans Field Equation Extensions " at the end of Chapter 6.

The B field components are a rotating metric with antisymmetric metric components.

Comparison of the two matrices $q^{\mu\nu(A)}$ and $G^{\mu\nu}$ show that the signs are the same and if one were to work them all the way out, they describe the same rotation.

The Principle of general relativity has been applied in the equations of the $B^{(3)}$ field. The magnetic field is a metric (an antisymmetric metric tensor) that rotates and translates.

There are three components of the B field. Two are circularly polarized phase dependent complex conjugates perpendicular to the direction of propagation. These are the $B^{(1)}$ and $B^{(2)}$ fields. There is one phase independent component in the direction of propagation. This is the $B^{(3)}$ field.

The magnetic field components are related by the B Cyclic Theorem of O(3) electrodynamics. The electric field components can also be defined in terms of the rotation generators of the O(3) group. A metric describes the electric components that also rotate and translate and is perpendicular to the magnetic field. The metric is used to define the magnetic fields, $\mathbf{B}^{(1)}$, $\mathbf{B}^{(2)}$ and $\mathbf{B}^{(3)}$ of O(3) electrodynamics. The O(3) electromagnetic field has been derived from curved spacetime and geometry.

Professor Evans' analysis shows that his formulations lead to the Coulomb, Gauss, Faraday, and Ampere-Maxwell equations. The $\mathbf{B}^{(3)}$ field of O(3) electrodynamics is responsible for interferometry, the Aharonov-Bohm and Sagnac Effects, the Berry Phase and all topological and optical effects.

The concepts are depicted in Figure 11-6.

Figure 11-6

Evans: the electromagnetic field is spinning spacetime itself - the $B^{(3)}$ spin field

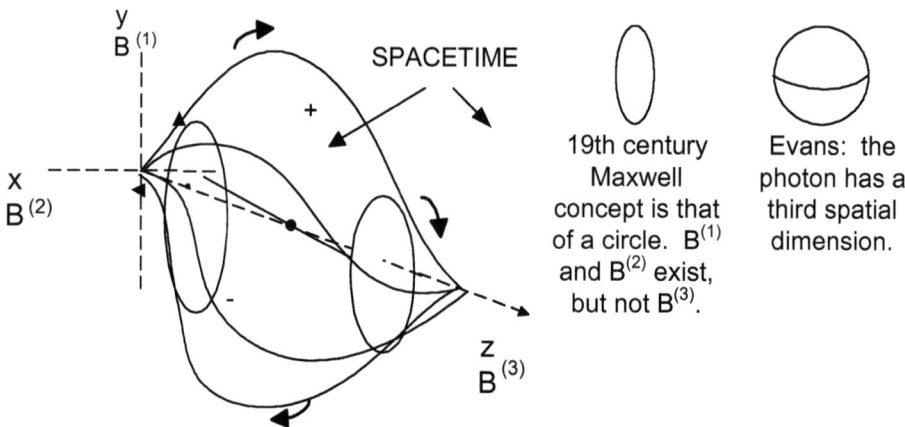

The photon has volume and some (very small) mass.

Mass and spin characterize all particles and fields. The origin of the Planck constant in general relativity is explained and is shown to be the least amount possible of action, angular momentum or spin in the universe.

In Professor Evans' words, "We can also now identify 'higher symmetry electrodynamics' as 'generally covariant electrodynamics', i.e. electrodynamics that is indicated by the Principle of general relativity, that all theories of natural philosophy (including electrodynamics) must be generally covariant, i.e. theories of general relativity."

Given that gravitational curvature is related to electromagnetic spin, we may be able to directly convert one form of energy to the other. This needs to be determined in experiments to see if the effect is large enough.

Electromagnetic energy resides in the Evans spacetime (Riemannian vacuum with Cartan torsion) without the presence of radiating electrons. We may be able to tap this energy, which is a manifestation of curvature. Vacuum electromagnetic energy is zero if and only if R is zero, and this occurs only in flat spacetime, in which there are no fields of any kind (gravitational, electromagnetic, weak or strong).

Summary

The development of the $\mathbf{B}^{(3)}$ field is highly mathematical. A good source for further study is Professor Evans' Lecture Notes 1 and 2 that are available on www.aias.us and his volumes on *Generally Covariant Unified Field Theory.*

Starting with geometry Evans has taken the metric vectors which are written as antisymmetric rank two tensors and then developed tensors written in the complex circular basis to arrive at electromagnetism. He does this in tetrad formulation which allows complete analysis.

More simply stated, starting with Einstein's geometry of our spacetime it has been shown that electromagnetism is also described.

The $\mathbf{B}^{(3)}$ field is the fundamental longitudinal and phaseless magnetic field component of electromagnetic radiation in general relativity. The $\mathbf{B}^{(3)}$ field can be derived from general relativity and it is necessary to use the $\mathbf{B}^{(3)}$ concept to obtain generally covariant – reference frame free – electromagnetics.

The photon has mass as inferred by Einstein and de Broglie, and their followers.

The metric that we label Evans spacetime has Einstein's curvature and Cartan's torsion. The effects could be large enough to produce electromagnetic power from gravitational curvature.

Chapter 12   Electro-Weak Theory

> Problems cannot be solved at the same level of awareness that created them.
> Albert Einstein

Introduction

In this chapter the Evans Lemma is used to understand the masses of the weak field bosons and to develop electro-weak theory without the concepts used in the standard model. Electro-weak refers to the unification of the electromagnetic and weak forces. The weak force holds some particles together. For example, the isolated neutron and the muon transform after short times into other particles. The neutron becomes a proton plus an electron plus an anti-neutrino. The muon becomes a muon-neutrino.

The present standard model uses concepts like spontaneous symmetry breaking and the Higgs mechanism, and it compensates for infinities due to zero volume particles using renormalization. The path integral method is used for calculating probabilities. The standard model uses the Higgs mechanism to adjust the masses of the particle, but after many experiments, the Higgs particle has not been found. The standard model uses adjustable, ad hoc parameters to gain agreement between theory and experiment. These can accurately predict results of most experiments however they are flawed at the theoretical level.

The W and Z bosons mediate the weak nuclear force.

Use of the Evans equations gives solutions from first principles in terms of basic constants of physics and is to be preferred over the present standard theory. It is simpler and explains more.

General relativity is generally covariant while the existing standard model is not generally covariant – it is a theory of special relativity that approximates general relativity and is incomplete since the electromagnetic, strong, and weak forces are not generally covariant – effects of gravitation cannot be predicted. General covariance is a fundamental requirement of physics. Without covariance, the laws of physics would vary in different reference frames.

The Evans Lemma is:

$$\Box q^a{}_\mu = R\, q^a{}_\mu \tag{1}$$

where the eigenfunction is the tetrad $q^a{}_\mu$. The values of R will be real observables in our universe.

The Principle of Least Curvature is:

$$R_0 = -(mc/\hbar)^2 \tag{2}$$

This is valid in the limit of special relativity with flat spacetime. This is the limit of the smallest value of R, scalar curvature, for a given mass. Mass m is adjustable in equation (2) for c and $\hbar$ are, as far as we know, fundamental constants. R the curvature is then fixed for any given mass.

The Compton wavelength of a particle is:

$$\lambda_0 = \hbar/mc \tag{3}$$

This is the wavelength of a photon with the same energy as the mass of a particle with mass m. Using equations (2) and (3) one sees that the least curvature of a particle or spacetime is $R_0 = (1/\lambda_0{}^2)$. The de Broglie equation is the same as the Compton equation with v the velocity of a particle instead of c. See Chapter 9 for discussion.

Using Einstein's index contracted field equation, $R = -kT$ and the Evans Lemma above, leads to the Evans Wave Equation, $(\Box + kT)\, q^a{}_\mu = 0$.

The wave equation in terms of the least curvature is:

$$(\Box + (mc/\hbar)^2)\, q^a{}_\mu = 0 \tag{4}$$

where $q^a{}_\mu$ is a tetrad, spinor, matrix, vector, or other as necessary.

Chapter 12 – Electro Weak Theory

Figure 12-1 Electroweak Theory

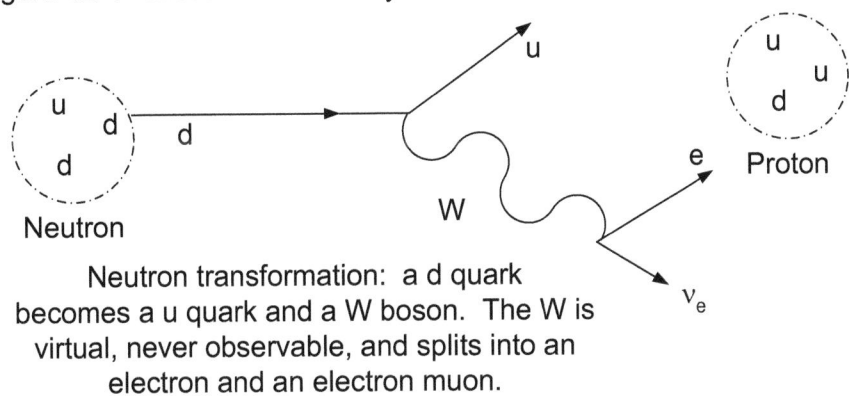

Neutron transformation: a d quark becomes a u quark and a W boson. The W is virtual, never observable, and splits into an electron and an electron muon.

For present purposes the following definitions (that are not universally agreed upon) will suffice. void is true nothing. Vacuum is the special relativistic Minkowski flat spacetime that approaches zero curvature and zero torsion. Real spacetime can never become perfectly flat since where there is energy, there is curvature. Non-Minkowski or Riemann spacetime is the curved universe of Einstein's general relativity. It has only a symmetric metric giving distances. Evans spacetime is curved and torqued. It is the real spacetime of our universe with symmetric and antisymmetric metric. The antisymmetric metric allows for electromagnetism. It could be as tenuous as the space far between galaxies or it could be as dense and turbulent as the region in a particle or near a black hole. Curvature and torsion are clearly recognized as present.

Eventually it may be that the terms twisting, turning, spinning, rotating will have differences in connotation. For now, they are essentially the same thing.

Spin is defined by the tetrad with and without the presence of gravitational curvature. The real spacetime base manifold and the tangent bundle are spinning with respect to one another. If gravitation is not present then the base and the bundle are both Minkowski spacetimes. There is compressibility due to the energy density of velocity, but the spacetime is flat – gravitation is ignored. See Figures 12-1 and 12-2.

Figure 12-2   Tangent Gauge Space and Normal Curved Space

The index space is like a snapshot at any one set of conditions.

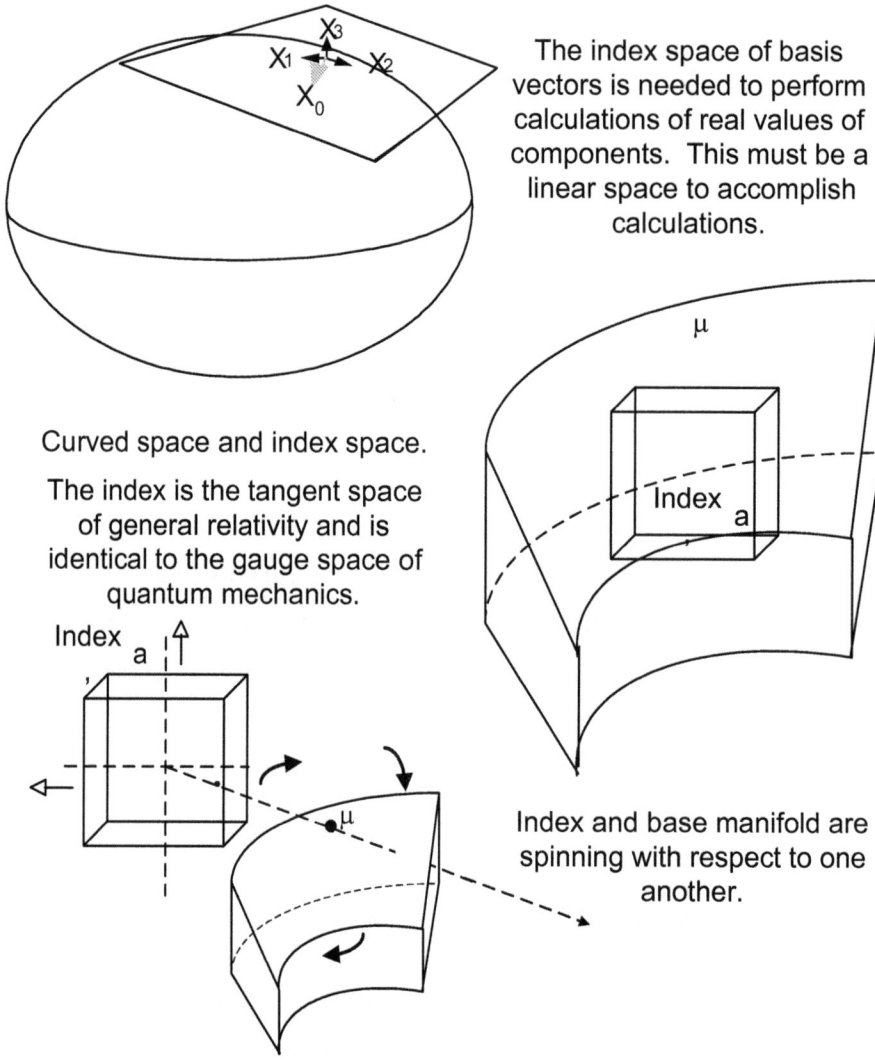

The index space of basis vectors is needed to perform calculations of real values of components. This must be a linear space to accomplish calculations.

Curved space and index space.

The index is the tangent space of general relativity and is identical to the gauge space of quantum mechanics.

Index and base manifold are spinning with respect to one another.

Chapter 12 – Electro Weak Theory

> We see here that the rvature of a particle in general relativity is expressed in quantum mechanics terms. The curvature is proportional to the mass. The electro-weak interactions are described by curvature in the Evans' generally covariant approach. That is, quantum (electro-weak) interactions are described in terms of general relativity (curvature) in the unified field theory.

Derivation of the Boson Masses

In the Evans development of the electro-weak theory, wave functions are tetrads governed by the Evans Lemma, equation (1), and the particle masses are eigenvalues of the lemma – the values of R. This indicates that the masses have real physical occurrence. In this way, the equations derive from physical geometry as required by Einstein's general relativity.

In the standard model, particles are assumed initially massless and then "spontaneous symmetry breaking" occurs and the Higgs field-particle gives them mass. The neutrino has no mass in the standard theory. The complexity is unnecessary and it has been shown that the neutrino does have mass.[51] Any patch to the standard model to explain this is another unnecessary complication.

Equation (2), the Evans Principle of Least Curvature, is simpler and is based on known constants. In addition, the Evans development is covariant and can predict actions in different gravitational fields.

The Evans equations show that mass is defined by spacetime geometry – the curvature R. And the masses of the electron and all particles are minimum curvatures or minimum eigenvalues of equation (2).

$$R^L = -(m^L c/\hbar)^2 \quad (5)$$

---

[51] See http://neutrino.phys.washington.edu/~superk/sk_release.html The Super-Kamiokande experiment found evidence for non-zero neutrino mass.

Where the superscript L indicates the electron.[52]

$$\Box W^a{}_\mu = R\, W^a{}_\mu \qquad (6)$$

Here $W^a{}_\mu$ is a tetrad of the Weak field and

$$W^a{}_\mu = W^{(0)}\, q^a{}_\mu \qquad (7)$$

with $W^{(0)}$ a scaling factor.

The mass of the three electroweak bosons, the Z and two W particles, can be derived from the Evans equations without the use of the Higgs mechanism using the Evans Lemma and the Dirac equation. Past experiments have given energies (masses) of 78.6 GeV/c² and 89.3 GeV/c² for the bosons. If the Evans equations can arrive at these masses, proof of its validity is obtained.

In the standard theory the bosons get mass from an equation with η, the Higgs mechanism, which is adjusted to find the experimental results using equations of the form:

$$L_1 = g^2\, \eta^2 ((W^1{}_\mu)^2 + ((W^2{}_\mu)^2)/4 \qquad (8)$$

Where g is a coupling constant and W is the boson. This leads to:

$$m^2 = g^2\, \eta^2 /2 \qquad (9)$$

Ignore all the terms but m the mass of the boson and η the Higgs "particle." Higgs has not been found in nature, rather it is "predicted" based on the other known values. As it happens Higgs can be replaced by Evans' minimum curvature value. Evans equations of the form:

$$m^2 = \hbar^2/4c^2\, ((W^1{}_\mu)^2 + ((W^2{}_\mu)^2) \qquad (10)$$

are found where use of the Higgs is unnecessary since the minimum curvature is the same thing and replaces Higgs. However the minimum curvature is based on fundamental constants of physics, $\hbar$ and c in equation (2). Boson masses are replaced by spacetime curvatures. This is a technically superior method.

The results of the calculations are the experimentally observed masses of 78.6 GeV/c² (W) and 89.3 GeV/c² (Z).

---

[52] Actually the left electron but we are leaving out a lot of detail in this explanation.

It is virtually impossible for this calculation to arrive at the boson masses if the equations used were not correct. From fundamental constants the boson masses are found.

Particle Scattering

Figure 12-3 shows the basic process of particle scattering when two fermion particles collide. Conservation of energy and momentum must be preserved. In Evans' formulation, it is the conservation of curvature which must be maintained, since energy and momentum are forms of curvature. p indicates a particle's momentum where $p = mv$ or relativistic $p = \gamma mv$.

The law of conservation of momentum applies in all collisions. While total kinetic energy, E_k, can be converted to new particles, the momentum is conserved separately. Figure 12-3 shows the basic equations. Note that p_1 is not typically equal to p_3 nor p_2 to p_4. It is the sums that are equal. Total energy in equals total energy out. Total momentum before collision equals total momentum afterwards.

See Particle collisions in the glossary for more information.

Figure 12-3   In Special Relativity and the Standard Model,
Tangent Space and Base Manifold are Flat

In the weak field limit, the index space and the base manifold are both flat. They do not curve or rotate with respect to one another. This is Minkowski spacetime.

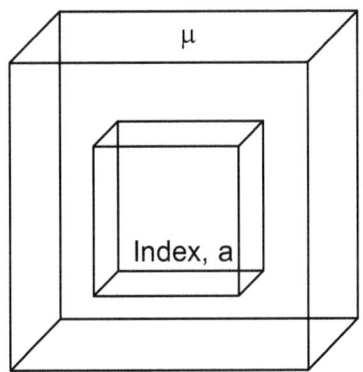

The equations here are over-simplifications of Evans' equations, but the process can be written essentially in the form:

$$f(k, m_1) p^b{}_1 = f(m_3) p^b{}_3 = 0 \tag{11}$$

$$f(k, m_2) p^b{}_2 = f(m_4) p^b{}_4 = 0 \tag{12}$$

where f is some function, k is the boson (energy exchanged), $m_{1\text{ to }4}$ are the masses of the particles in Figure 12-3, and $p^b{}_{1\text{ to }4}$ are wave functions which are tetrads in differential geometry. The wave functions $p^b{}_3$ (and $p^b{}_4$) are that of the particles after picking up (and losing) momentum mediated by the boson k.

Equation (11) and (12) state that there is a function of the boson energy k and m the mass of the particle times the wave function which is conserved. There is no creation or disappearance of momentum. The boson is also governed by an Evans equation of the form:

$$f(m_k) k^b{}_i = 0 \tag{13}$$

where m_k is the mass of the boson and $k^b{}_i$ is the initial wave function of the boson before colliding with the fermion. Equation (13) is derived from the Evans Wave Equation:

$$(\Box + (m_k c/\hbar)^2) k^b{}_t = 0 \tag{14}$$

This is seen as a form of the now familiar Evans Wave Equation.

Figure 12-4   Momentum Exchange

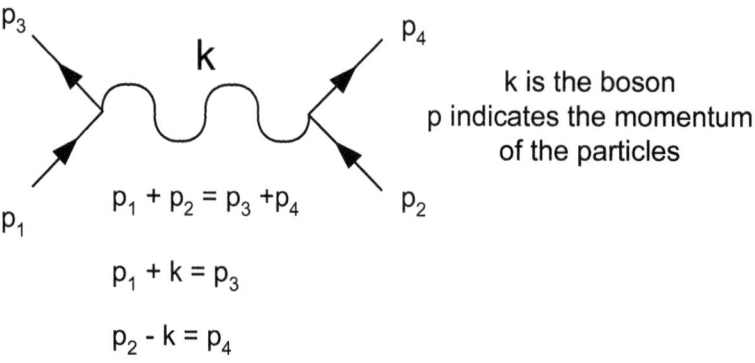

k is the boson
p indicates the momentum of the particles

$p_1 + p_2 = p_3 + p_4$

$p_1 + k = p_3$

$p_2 - k = p_4$

Chapter 12 – Electro Weak Theory

Collisions between particles are governed by two laws: one, conservation of momentum, and two, conservation of mass-energy. Conservation of the mass energy is separate from the momentum; each is conserved separately. Equations (10) to (14) replace standard model equations producing the same results, but the Evans equations use fundamental constants. Two simultaneous equations can describe any scattering process. These are covariant and transfer to any gravitational system. See Figure 12-4.

Figure 12-5 Particle scattering seen as curvatures

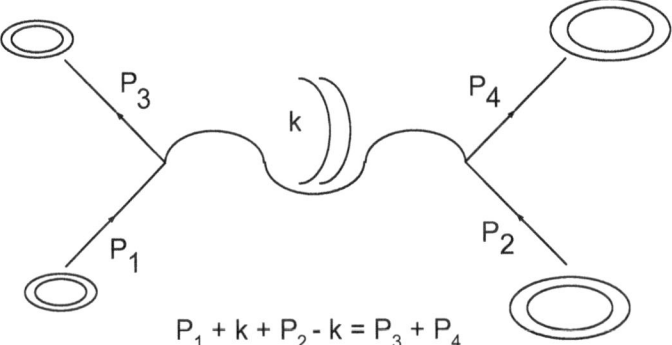

$P_1 + k + P_2 - k = P_3 + P_4$

The test of any theory is not that it is renormalizable, but rather that it is generally covariant as Einstein stated. See Figure 12-5.

The Neutrino Oscillation Mass

The neutrino has been "observed" to change from the muon type to the tau type.[53] This can only occur if the three neutrinos have mass and those masses are different. The standard not explain the phenomenon, however the Evans model can and does explain it.

---

[53] Actually muon neutrinos disappear and it is hypothesized that they turn into tau neutrinos in the atmosphere. So rather than being observed, they are not observed when they should be.

Evans Equations of Unified Field Theory

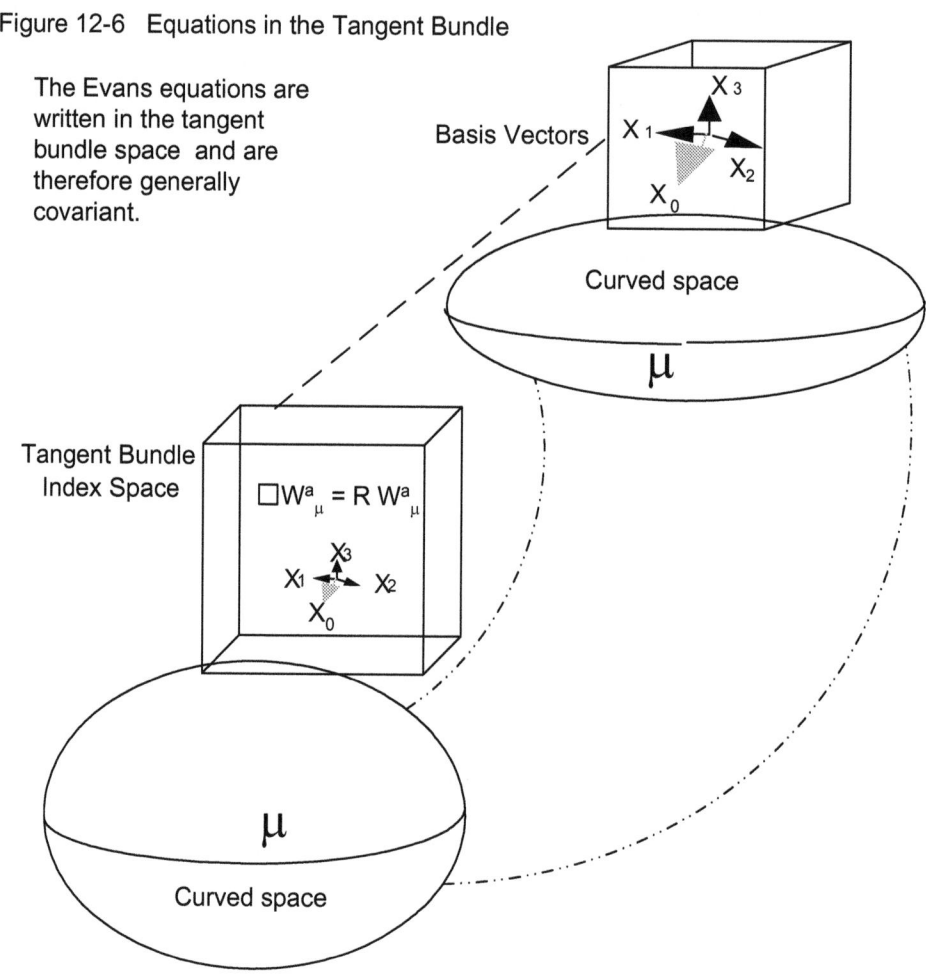

Figure 12-6   Equations in the Tangent Bundle

The Evans equations are written in the tangent bundle space and are therefore generally covariant.

One neutrino, say an electron neutrino, can be initially in two quantum states with two different mass energies. In the Evans equations, this means two different scalar curvatures or eigenvalues of the Evans Lemma. That is two different valid solutions to the equations.

A mixture can be parameterized by an angle θ which we will not go into here. See www.aias.us for several detailed explanations.

Simpler is to note that the Evans oscillation hypothesis allows the muon and tau neutrinos to be mixtures of x and y in the equation:

$$\nu_\mu + i\nu_\tau = 2x$$

Chapter 12 – Electro Weak Theory

$$\nu_\mu - i\nu_\tau = 2y \qquad (17)$$

where ν_μ is the muon neutrino and ν_τ is the tau neutrino. After several lines of equations that could cause severe emotional trauma in us normal humans, Evans shows that the neutrino oscillation can be governed by the Evans Lemma:

$$\Box\, v^a{}_\mu = R v^a{}_\mu \qquad (18)$$

We see that $v^a{}_\mu$ is a tetrad and real values of R of the neutrino oscillations are scalar curvatures in general relativity.

The standard model is the name given to the collected accepted theories of physics. Quantum mechanics lies at the center. Gravitation as described by Einstein, while known to be accurate, has not been connected to quantum descriptions in the standard model.

There are some theoretical ideas that have entered the standard model without any proof. According to Evans well documented concepts, these are incorrect ideas.

Quantum mechanics has proven to be highly accurate in its predictions of experimental results. However, the theory underlying the experiments is sometimes fuzzy at best. Unproven concepts do not belong in the standard model.

The mathematics of quantum theory uses abstract spaces, "gauge space," to perform calculations. After calculation, the numbers are brought back into Minkowski spacetime where they gain real values and can be tested.

General relativity uses similar spaces, the "tangent space," for calculation. What Evans indicates is that the index in his equations is the same as the two mathematical spaces. Thus the tangent space of general relativity is the same as the gauge space of quantum mechanics.

This unites quantum mechanics and general relativity. The two previously separate theories are mathematically joined.

Standard Model with Higgs versus the Evans method

1. The standard model is not objective because it is not a generally covariant theory of physics. The Evans equations do give a covariant formulation.
2. The gauge space used in the standard theory is an abstract mathematical device with no physical meaning. A number of concepts are ad hoc constructions. The Evans method uses differential geometry and Einstein's concept of curvature.
3. The Higgs mechanism is a loose parameter introduced by an abstract mathematical model of the Minkowski vacuum with no physical meaning. The Higgs particle-field-mechanism is a mathematical parameter found by fitting data to known energies. The Evans equations use the concept of minimum curvature which is based on fundamental constants.
4. Renormalization is used in the standard model. This sets an arbitrary minimum volume of a particle to avoid infinities. This is untestable. The formula based on the Evans equations is a concrete method. That is $V_0 = k/m \, (\hbar/c)^2$ as given in Chapter 9. Both Einstein and Dirac rejected renormalization. It was a clever method to set minimum volumes while they were still unknown. However it is only an approximation that was necessary only because of the incomplete standard model.

Generally Covariant Description

The first generally covariant description of the transmutation of the muon into the muon-neutrino was given in notes placed on the www.aias.us website. This was:

$$(i\hbar\gamma^a \, (\partial_a + igW_a) - m_\mu c)\mu^b + (i\hbar\gamma^a \, \partial_a - m_\mu c)\nu^b) = 0 \qquad (19)$$

This is a landmark equation although unlikely to become quite as famous as $E = mc^2$.

Chapter 13   The Aharonov-Bohm (AB) Effect

> General relativity as extended in the Evans unified field theory is needed for a correct understanding of all phase effects in physics, an understanding that is forged through the Evans phase law, the origin of the Berry phase and the geometrical phase of electrodynamics observed in the Sagnac and Tomita Chiao effects.
> Myron W. Evans

Phase Effects

A simple explanation of a phase effect is shown in Figure 13-1. The expected wave has no phase shift. However in certain experiments, the phase is shifted. Explanations to date of why they occur have been awkward and questionable. Evans shows that his generally covariant equations are simpler and demonstrate the causes of phase effects. The mathematics involved in phase effects and descriptions of phenomena like the AB effect are difficult and beyond what we can deal with in any detail here. However, the effects can be discussed without mathematics details.

Figure 13-1   Phase Effects

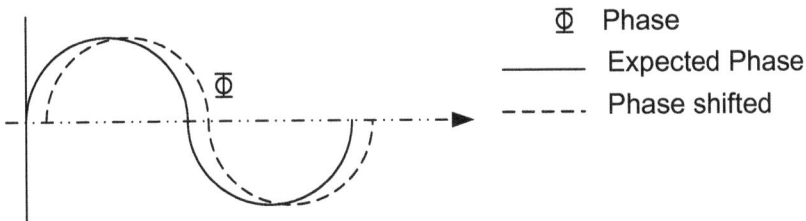

Consider an experiment where a beam of photons or electrons is shot at a screen. An interference pattern will occur as shown in Figure 13-3. For an electromagnetic field the phase law tells us that the phase is a function of potential voltage and the area of the beam or the magnetic field strength and the area. That is:

$$\Phi = \text{function of } (\mathbf{A} \cdot d\mathbf{r}) = f(\mathbf{B}^{(3)} \cdot \mathbf{k}\, dAr) \qquad (1)$$

where A is potential voltage, $\mathbf{B}^{(3)}$ is the magnetic field (both are directed in the Z axis of propagation) and Ar is the area of a circle enclosed by the beam.

(Actually, Evans gives: $\Phi = \exp(ig \oint \mathbf{A}^{(3)} \cdot d\mathbf{r}) = \exp(ig \oint \mathbf{B}^{(3)} \cdot \mathbf{k}\, dAr) := \exp(i\,\Phi_E)$. For matter fields, $\Phi = \text{function of } (\kappa \cdot d\mathbf{r}) = f(\kappa^2 \cdot dAr)^1$ where κ is the wave number (inverse wavelength). This author is simplifying significantly in order to explain.)

Figure 13-2

The sine wave describes the circular motion shown below.

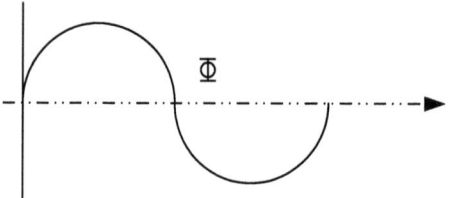

The diameter of a circle equals $2\pi r$. The distance that this describes when moving forward is not a full circle. Rather the arc length of the helix is drawn.

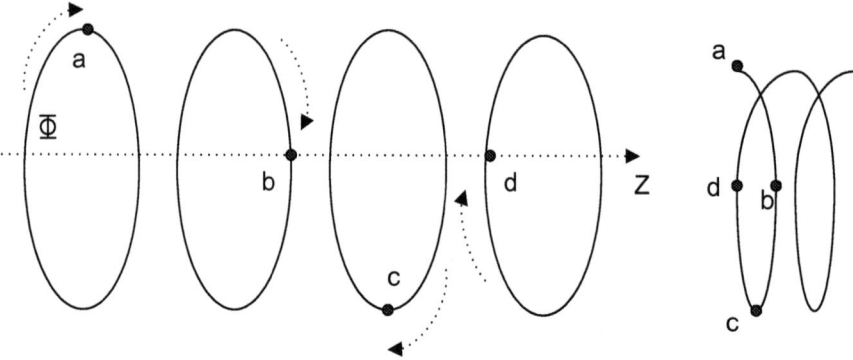

The arc length of the circle as it turns becomes that of a helix as shown in Figure 13-2.

The curvature of the helix is defined as $R = \kappa^2$. We thus see a geometric curvature of relativity expressed in the description of phase. The helix in Figure 13-2 is the baseline of the electromagnetic field.

Spacetime itself spins. Both electromagnetic and matter waves are manifestations of spacetime itself.

The angle Θ through which light is rotated out of phase originates in the Evans phase of unified theory. This is:

$$\Theta = \kappa \oint ds = R \int dAr \qquad (1)$$

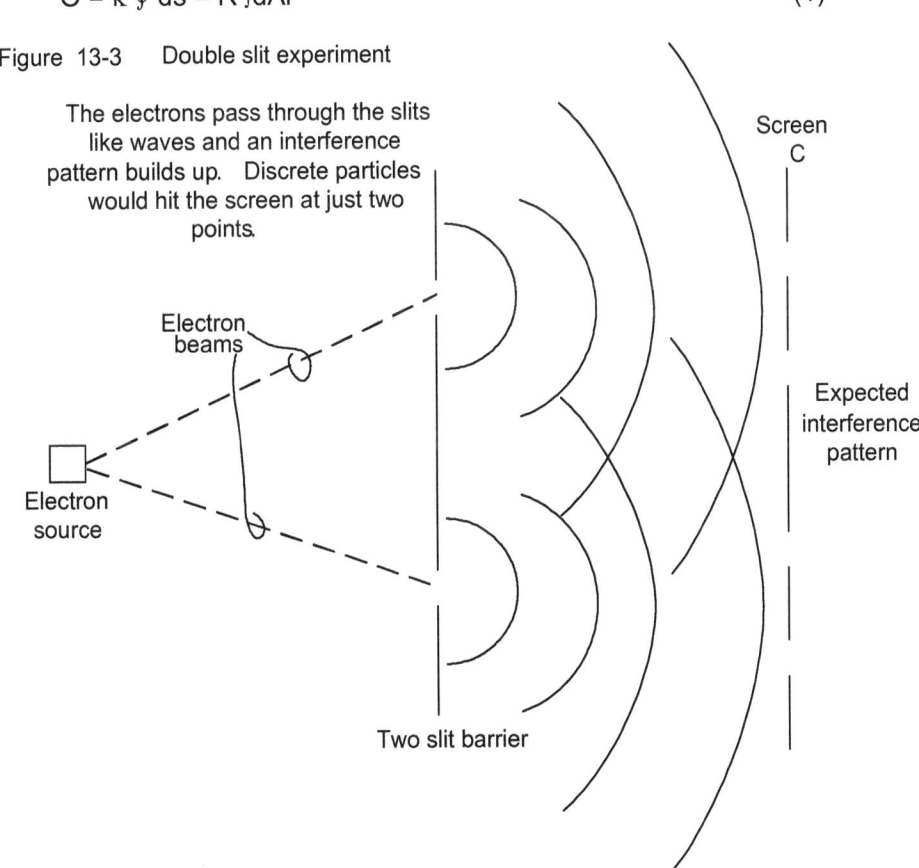

Figure 13-3   Double slit experiment

The electrons pass through the slits like waves and an interference pattern builds up. Discrete particles would hit the screen at just two points.

Electron beams

Electron source

Two slit barrier

Screen C

Expected interference pattern

Here κ is wave number, ds is the invariant distance, R is curvature, and Ar is the area enclosed by the beam.

The expected phase pattern is shown in Figure 13-3.

The Aharonov-Bohm Effect

The Aharonov-Bohm Effect is a shift in the interference pattern of two electron beams in a Young interferometer. An explanation is shown in Figures 13-3 and 13-4. Electrons pass through a double slit and an interference pattern builds up. Classical quantum mechanics predicts a certain pattern which is found in experiments and designated C in the figures.

When an enclosed magnetic field, indicated ⊙, is placed between slits, classical quantum mechanics predicts no change in the pattern. The magnetic field is totally enclosed in a metal case and cannot influence the electrons. However, the pattern shifts as designated AB in Figure 13-4. (It is greatly overemphasized.)

The only explanation given in the past was that the vacuum is a multiply connected topology and complex mathematics was necessary to show the reason for the AB effect. See Figure 13-5. This is an overly complex and is not a provable solution. The simple answer is that the field does extend beyond the barrier.

A multiply connected topology is not the smooth differentiable manifold of general relativity nor of the vacuum of special relativity. It is rather a complicated arrangement of tunnels and loops of spacetime. It is not necessary to explain the AB effect nor the other similar phase effects with this convoluted solution.

General relativity defines spacetime to be a simply connected differentiable manifold. Using O(3) electrodynamics and the Evans $\mathbf{B}^{(3)}$ field, a simpler explanation can be defined in general relativity.

Chapter 13 – The Aharonov-Bohm (AB) Effect

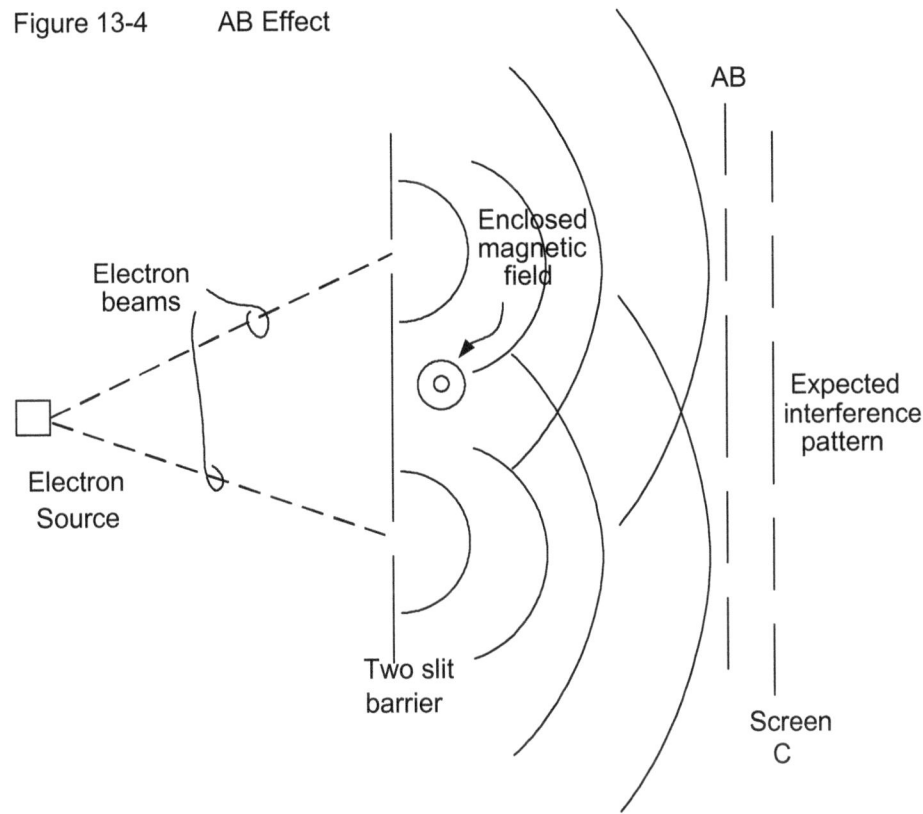

Figure 13-4   AB Effect

The previous view of the electromagnetic field is that it is something imposed on or inserted into spacetime. The correct viewpoint is that the magnetic field is spacetime itself spinning like a whirlpool.

The AB effect occurs because the magnetic field is spacetime spinning and its potential extends beyond the barrier of the solenoid coil case. The equations using torsion clearly and simply explain the AB effect. Spacetime is continuous and effects extend beyond the barrier.

The Evans unified field theory gives solutions for a number of topological phase effects that are similar or essentially the same as the AB effect. These include the electromagnetic Aharonov-Bohm effect (EMAB), Sagnac effect, Tomita-Chiao effect, and Berry phase factor. The Tomita-Chiao effect is a shift in phase brought about by rotating a beam of light around a helical optical fiber. This is the same as the Sagnac effect with several loops, and is a shift in the

Cartan tetrad of the Evans unified field theory. Similarly, the Berry phase of matter wave theory is a shift in the tetrad of the Evans unified field theory.

Figure 13-5   Multiply and simply connected topologies

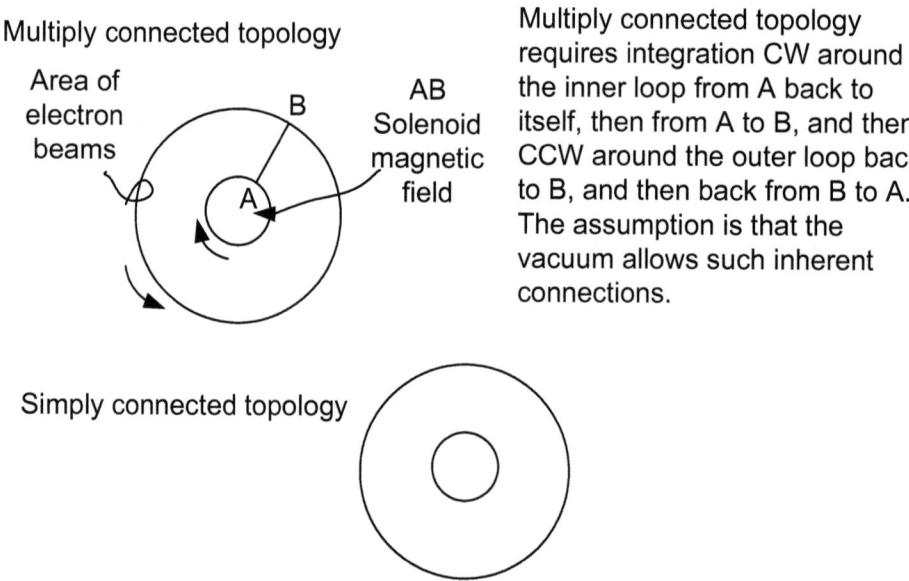

A simple fully differentiable spacetime envisioned in general relativity.

In the Evans theory the effects are simple to understand and describe. They are all related to the Inverse Faraday Effect and use $\mathbf{B}^{(3)}$ electromagnetics and O(3) electrodynamics. $\mathbf{B}^{(3)}$ introduces the conjugate product naturally into physics. On the other hand, the U(1) Maxwell-Heaviside theory cannot explain these effects except by torturous logical inversions. In particular, the conjugate product is introduced, but empirically without understanding where it originates.

Chapter 13 – The Aharonov-Bohm (AB) Effect

The Helix versus the Circle

Evans' equations give more than just an explanation of these effects. He shows that using differential geometry that the received explanations make incorrect assertions. Multiply connected spacetimes do not obey the rules that physicists have given them to explain the AB effect.

In his papers Evans often starts with a physics equation, restates it in differentials, finds a truth in geometry, and then restates the physics equation giving new insights. In this book we have avoided differential geometry since it is not simple mathematics. The following paragraph is paraphrased and simplified from Evans:

The Stokes theorem shows that a certain function of $x = 0$. However, the conventional description of the AB effect relies on the incorrect assertion that $f(x) \neq 0$. This violates the Poincaré Lemma. In differential geometry, the lemma is true for multiply-connected as well as simply-connected regions. In ordinary vector notation the lemma states that, for any function, ∇ times $\nabla x := 0$. The Stokes theorem and the Green theorem are both true for multiply-connected regions as well as for simply-connected regions. There is therefore no correct explanation of the AB effect in Maxwell-Heaviside theory and special relativity.

The difference between the generally covariant Stokes theorem and the ordinary Stokes theorem is the same as the difference between generally covariant electrodynamics and the older Maxwell-Heaviside electrodynamics. In Maxwell-Heaviside there is no longitudinal component of the photon. It describes a circle. In Evans' $B^{(3)}$ formulation the mathematics is that of a helix. See Figure 13-6. Maxwell-Heaviside special relativistic explanation is three dimensional while Evans is four dimensional.

Figure 13-6

Ordinary Stokes is a circle. The generally covariant Stokes is a helix.

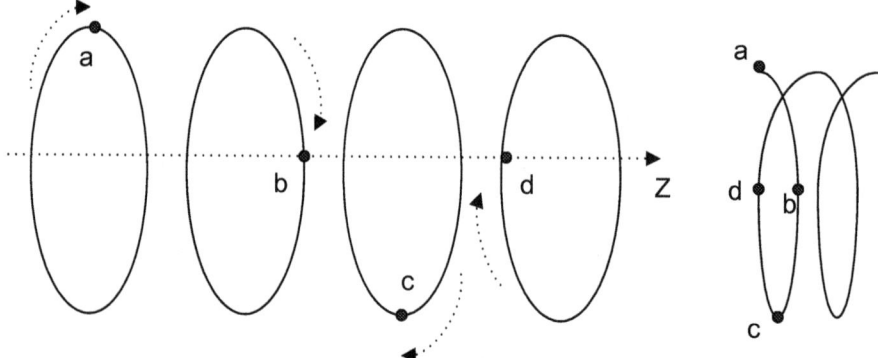

The photon exists in four dimensions. The U(1) geometry only describes three. $B^{(3)}$ gives three spatial plus time dimensions.

The diameter of a circle equals $2\pi r$. The distance that this describes when moving forward is not a full circle. Rather the arc length of the helix is drawn.

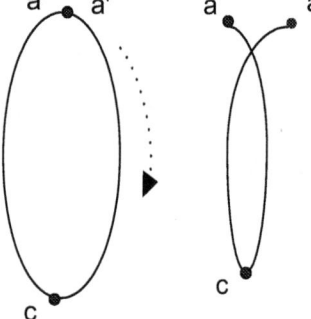

As the circle rotates and moves forward, a helix is drawn.

If the length Z were to equal $2\pi r$, then the shape would be a straight line. As it is, the implication is simply that the U(1) Maxwell Heaviside cannot describe the physical process correctly and the O(3) formulation must be used and has given correct answers to the AB and other effects.

Chapter 13 – The Aharonov-Bohm (AB) Effect

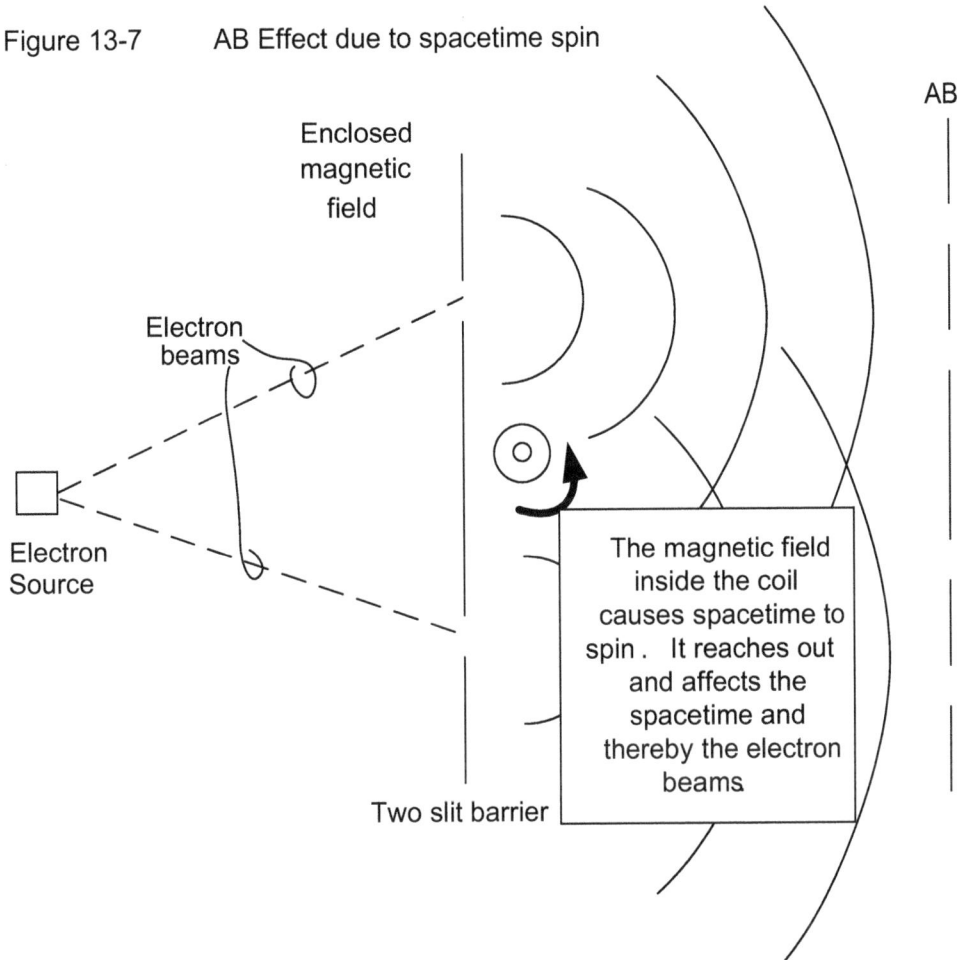

Figure 13-7   AB Effect due to spacetime spin

Summary

Phase shifts in experiments like the AB effect require awkward explanations in the standard model. These shifts received sophisticated, but simpler explanations in the Evans formulation of general relativity. There is unification of electromagnetic effects with curvature and torsion using the unified field theory. Spacetime obviously is continuous and the electromagnetic field extends beyond the enclosed coil in the AB effect. This simple explanation of the AB effect is quite elegant.

Spacetime spins outside the coil of the magnetic field.

Chapter 14 Geometric Concepts

> We believe in the possibility of a theory which is able to give a complete description of reality, the laws of which establish relations between the things themselves and not merely between their probabilities...God does not play dice with the universe.
> Albert Einstein

Introduction

There are developments that Evans makes in his papers that we have not covered in this book. Some of the more geometrical concepts are described briefly in this chapter.

The Electrogravitic Equation[54]

The Evans Wave Equation is:

$$(\Box + kT) q^a{}_\mu = 0 \qquad (1)$$

where k is Einstein's constant, T is the stress energy tensor, $q^a{}_\mu$ is the tetrad which is the gravitational potential field.

The wave equation for electromagnetism is:

$$(\Box + kT) A^a{}_\mu \qquad (2)$$

where

$$A^a{}_\mu = A^{(0)} q^a{}_\mu \qquad (3)$$

---

[54] The descriptions here are taken from "Development of the Evans Wave Equation in the Weak Field Limit: The Electrogravitic Equation" by M.W. Evans et. al. See www.aias.us.

$A^a{}_\mu$ is the electromagnetic potential field.[55] $A^{(0)}$ is the electromagnetic potential. This is the Evans Ansatz - the proposed conversion from geometry to electromagnetics using $A^{(0)}$.

If all forms of mass and energy are related through curvature, R, then an equation should exist which gives us electromagnetism as a function of gravitation. This fundamental ratio of charge e to mass m is in terms of the electrostatic potential $\varphi^{(0)}$. $\varphi^{(0)}$ would be a scalar voltage with units of volts. Then this equation should exist:

$$A^{(0)} q^a{}_\mu = \frac{\varphi^{(0)}}{c^2} q^a{}_\mu \tag{4}$$

Elsewhere it has been shown that the Evans Wave Equation becomes the Poisson equation in Newtonian gravitation in the weak field limit. The well known Poisson equation for gravitation is:

$$\nabla^2 \phi = 4\pi G \rho \tag{5}$$

where ϕ is the gravitational potential in m/s^2, G is the Newton gravitational constant, and ρ is the mass density in kg/m^3.

The acceleration due to gravity in units of m/s^2 is:

$$\mathbf{g} = \nabla \phi \tag{6}$$

Similarly equation (2) becomes the Poisson equation for electrostatics in the weak field limit:

$$\nabla^2 (\varphi^{(0)} \phi) = 4\pi G (\varphi^{(0)} \rho \varphi) \tag{7}$$

The electric field is then described as:

$$\mathbf{E} = \frac{1}{c^2} \nabla (\varphi^{(0)} \phi) \tag{8}$$

The factor $1/c^2$ is needed to adjust to S.I units.

Substituting equation (6) in equation (8) we see that:

$$\mathbf{E} = \frac{\varphi^{(0)}}{c^2} \mathbf{g} \tag{9}$$

---

[55] This is a vector valued tetrad one-form within a $\hat{C}$ negative (charge conjugation) factor $A^{(0)}$.

This gives the electric field strength in terms of **g**. The field is in units of volts per meter. It shows that there is a mutually calculable effect between gravitation and electromagnetism. There is one field, *the electrogravitic field.*

The effects will be difficult to test since they are quite small. Given two one kilogram masses one meter apart charged with one Coulomb, the gravitational force is 6.67 x 10^{-11} Newtons.

$\varphi^{(0)}$ is a fundamental voltage available from curved spacetime. There is an electric field present for each particle and it originates in scalar curvature.

The effect may be important in black hole formation as the high degree of torsion in both the mass and electrical potential in the spacetime being pulled into a hole accelerates and collapses. When we have a test black hole we will be able to confirm the effect.

The effect may be very important for extraction of energy from the curvature of spacetime near our planet. If an array of receptors can be developed, a low cost source of energy becomes available. The effect is expected to be very small.

Principle of Least Curvature

In the flat Minkowski spacetime limit we saw from Chapter 9 that R approaches $(mc/\hbar)^2$ and $|R_0| = 1/\lambda_0^2$. In this way we see that mass is expressed as scalar curvature of spacetime. This is the Principle of Least Curvature. R values here are eigenvalues of the Evans Lemma. Eigenvalues are real physical results.

While the least curvature is defined by the rest mass, the particle can have more curvature. If accelerated towards the speed of light, the particle would have greater internal curvature. If put inside a strong gravitational field, the curvature of the field and the curvature of the particle would be combined. The reference frame of the particle would be compressed.

There are several implications. One is that particle mass is curved spacetime. We have seen that the electromagnetic field is spinning spacetime. Another is that a particle never travels in a straight line since the scalar curvature of a straight line is zero. The least curvature of a particle is the least action possible, $\hbar$. In the final analysis, we find that everything in our universe is spacetime. It is not simply that everything obeys the same rules, rather we see the particle as a bit or region of highly curved spacetime. The particle presumably has torsion also since it has a frequency.

Just what equation will be found to explain the particle masses remains to be seen. There will be found some ratio or similar relationship between curvature and torsion, and the masses of the basic particles.

Non-local Effects

The Evans unified field theory also explains the violation of the Bell inequalities observed experimentally in the Aspect experiments, i.e. in regions where there are no matter or radiated fields of any kind there are still matter or radiation potentials. These give rise to the "non-locality" of quantum mechanics

The spacetime curvature R in one location extends to others. This is not action at a distance, but rather extended spacetime curvature. This offers an explanation for seemingly simultaneous actions at a distance. Rather than "entanglement" or "non-locality" as in quantum mechanics, we have a shift in spacetime curvature. No satisfactory solution has ever been found in the standard model.

The AB effect is explained by extension of spacetime torsion outside a solenoid coil's enclosure in Chapter 13. In the same exact manner the gravitational attraction experienced by masses for one another are explained by the extension of curvature outside the immediate volume of a particle. Each mass is spacetime and it extends to infinity causing curved space. Differential geometry shows that a seemingly local effect extends throughout spacetime and cause "non-local" effects.

Differential geometry has a solution giving the cause for entanglement in quantum mechanics. No satisfactory solution has been found in the existing standard model.

EMAB and RFR

Magnetic, electric, and gravitational AB effects have been observed to date. The electromagnetic AB effect has not yet been observed.

Evans proposes an experiment using what he calls the Electromagnetic Aharonov-Bohm effect (EMAB). The electromagnetic potential field was defined in equation (3) as $A^a_\mu = A^{(0)} q^a_\mu$. The field causes interaction of circularly polarized electromagnetic radiation with an electron beam.

The magnetized iron whisker of the original AB effect can be replaced by a circularly polarized radio frequency beam. The $\mathbf{B}^{(3)}$ component of the radio frequency field causes a shift in the fringe pattern of two interfering electron beams. This has not been performed to date, but if successful it could open up a new radar technology.

Similarly, radiatively induced fermion resonance (RFR) can be tested. This would lead to a new type of imaging that would be much more accurate than present MRI methods.

Calculations show that the phase shift using standard lasers at typical power densities is unobservable.

The Inverse Faraday Effect is the orbital angular momentum given to electrons in a beam by photons. This is explained by the Evans $\mathbf{B}^{(3)}$ spin field. For a beam traveling at low velocities the energy is:

$$E_{IFE} = e \hbar \mathbf{B}^{(3)} = e^2 A^{(0)2}/2m = \mathbf{p}^2/2m \tag{10}$$

where e is the charge of the electron, m is the mass of the electron, $p = e A^{(0)}$ is its linear momentum. $p^2/2m$ is the kinetic energy. The energy given to the beam is a directly proportional to the power density and inversely proportional to the square of the angular frequency.

The equations describing this can be derived from differential geometry and the $\mathbf{B}^{(3)}$ field. The result is a known equation.

Chapter 14 – Geometric Concepts

Differential Geometry

We have avoided differential geometry in this book up to now because of its complexity. Once it is understood, it simplifies physics, but getting to that level is not a common achievement.

However, even if it is not perfectly clear, a look at it is worthwhile.

The fundamentals of the unified field theory are based on these equations:

$$T^c = D \wedge q^c \qquad \text{First Maurer Cartan structure relationship.}$$

The torsion form is defined as the covariant exterior derivative of the tetrad. Here the indices of the manifolds, typically μ and v are not included since they are redundantly placed on both sides of the equation.

$$R^a{}_b = D \wedge \omega^a{}_b \qquad \text{Second Maurer Cartan structure relationship.}$$

This defines the Riemann form $R^a{}_b$ as the covariant exterior derivative of the spin connection, $\omega^a{}_b$.

$$Dq^a = 0 \qquad \text{Tetrad postulate.}$$

$$D \wedge V^a = d \wedge V^a + \omega^a{}_b \wedge V^b$$

This defines the covariant exterior derivative.

Evans simplifies these to a form with the wave number and shows further that R, scalar curvature, can be defined as the square of the wave number with units of inverse meters squared. That is:

$$R := \kappa^2 \qquad (11)$$

Then in order to derive physics from the geometry, Evans uses Einstein's postulate:

$$R = -kT \qquad (12)$$

R is geometrical curvature; kT is physics observation. Equation (3), $A^a{}_\mu = A^{(0)} q^a{}_\mu$, is used to convert from asymmetric connections to electromagnetism.

The magnetic field in differential geometry is expressed as:

$$B^a = D \wedge A^a \qquad (13)$$

The expressions in differential geometry look quite different from physics, but the logic is straightforward and all fits together. In Evans' words:

> It is now known with great clarity that the interaction of these fields takes place through differential geometry: $D \wedge F = R \wedge A$ (asymmetric Christoffel connection) …Here $D \wedge$ is the covariant exterior derivative containing the spin connection, R is the curvature or Riemann two-form, T the torsion two-form, q the tetrad one-form, F the electromagnetic field two-form and A the electromagnetic potential one-form. However, it is also known with precision that the Faraday Law of induction and the Gauss Law of magnetism hold very well. Similarly the Coulomb and Newton inverse square laws. So the cross effects indicated to exist in physics by the above equations of geometry will reveal themselves only with precise and careful experiments.

Fundamental Invariants of the Evans Field Theory

The fundamental invariants of a particle in special relativity are the spin and the mass. Regardless of the reference frame, spin and mass will be invariant and they completely define the particle. In pure mathematics these are referred to as the Casimir invariants of the Poincaré group. The first is:

$$C_1 = \mathbf{p}^\mu \mathbf{p}_\mu = (mc/\hbar)^2 = |R_0| \qquad (14)$$

which is rest curvature for the mass m.

Curvature R is then an invariant in differential geometry and in the Evans unified field theory. And since the Evans Lemma gives us R as eigenvalues, we have quantization of general relativity. The R values in the Evans Wave Equation and Einstein's postulate of general relativity are invariant observables of unified field theory.

The second is:

$$C_2 = m^2 s(s+1) \qquad (15)$$

This is a description of spin and is also a fundamental scalar invariant. C_2 is shown to be similar to a structure invariant of differential geometry. In simple terms, C_2 is the spin invariant.

The analysis has been extended from special relativity into general relativity.

Chapter 14 – Geometric Concepts

> The eigenvalues, real solutions, of the tetrad are fundamental invariants of unified field theory. Particles are solutions of the Evans Lemma.

Using differential geometry, Evans identifies two types of invariants - structure and identity. There cannot be a gravitational or electromagnetic field by itself. The two will always be present together. It is unphysical to have one disappear. However one can be very small and this still needs to be found by experimentation.

> In the real universe, there are always electromagnetic and gravitational fields present together. Neither can disappear to exactly zero.

The only quantity that enters into the essential Evans equations outside differential geometry is the fundamental potential $A^{(0)}$ which has the units of volts-s/m. $A^{(0)} = \hbar/er_0$ where e is the proton charge and r_0 is a fundamental length in meters.

$$r_0 = \lambda_c = \hbar/mc. \tag{16}$$

It is seen that the fundamental voltage is also a geometric property of spacetime.

Evans also shows that

$$mc = eA^{(0)} = e(\hbar \kappa_0 / e) \tag{17}$$

This shows that rest energy / c is the product of two C negative quantities; these are e, charge, and $A^{(0)}$ which is the potential in volt-sec/meter. For two different signs of charge, the positive and negative, the equation always gives positive mass. This is the experimental observation which has not had any theoretical foundation until now.

Evans Equations of Unified Field Theory

Origin of Wave Number

Starting with the ansatz:

$$A^a{}_\mu = A^{(0)} q^a{}_\mu \qquad (18)$$

This says the electromagnetic tetrad is electromagnetic potential times the asymmetric tetrad.

Without covering the differential geometry equations that lead to all the formulation here we find that the magnetic field B equals $A^{(0)}$ times a torsion form indicated by T. That is:

$$B^{(0)} = A^{(0)} (T^2{}_{32} + iT^1{}_{32}) \qquad (19)$$

And in the Maxwellian limit it is known that:

$$B^{(0)} = \kappa A^{(0)} \qquad (20)$$

where κ is the wave number, $\kappa = \omega/c$, ω is the angular frequency, and c the speed of light. From equations (19) and (20):

$$\kappa = T^2{}_{32} + iT^1{}_{32} \qquad (21)$$

Equation (21) shows that the origin of wave number and frequency in electrodynamics is the torsion of spacetime.

The scalar curvature, R, is defined as $R = \kappa^2$ which we can now see as a function of torsion T[56]

Processes in electrodynamics can therefore be described by the components of the torsion tensor. It is well known that the dielectric permittivity and the absorption coefficient in spectroscopy are defined in terms of a complex

---

[56] Actually Professor Evans gives us $R = \kappa \kappa^* = (T^2{}_{32})^2 - (T^1{}_{32})^2$. We are simplifying here to explain the basic ideas.

wave number, so the process of absorption and dispersion becomes understandable in terms of spacetime torsion.

Photon mass is defined by the Evans Principle of Least Curvature:

$$\kappa \longrightarrow 2\pi/\lambda_0 = 2\pi \, mc/\hbar \qquad (22)$$

Here $\lambda_0 = \hbar/mc$ is the Compton wavelength and $\hbar = h/2\pi$ is the reduced Planck or Dirac constant. If the photon mass is about 10^{-60} kg, the minimum wave number is about 10^{-18} m^{-1}, and

$$T \rightarrow 2\pi \, mc/\hbar \qquad (23)$$

The mass is the minimum value of the torsion tensor component T:

$$m = (\hbar/2\pi c)(T_{min}) \qquad (24)$$

What precise information about particles will be uncovered by these equations is unknown at the time of this writing. However it would seem that understanding of all the particles observed in experiments will be seen in terms of curvature and torsion – gravitation and electromagnetism together.

Summary

Differential geometry is physics. This is Evans' primary basis for development of any number of equations in a variety of areas. The interaction of the four fields in physics takes place through differential geometry.

A number of examples are given here in simplified form.

The electrogravitic equation shows that conversion of curvature to torsion is likely possible and may lead to a new energy source.

Particles are curvature and torsion together with mass determining the least curvature.

Entanglement in quantum mechanics is due to spacetime geometry.

The invariants of physics and the origin of mass and spin is differential geometry. The analysis has been extended into general relativity from special relativity showing another factor in unification.

The point that Evans makes is that geometry is physics.

Chapter 15   A Unified Viewpoint

> Isaac Newton used algebra to understand large scale motion, but published results as geometry. Paul Dirac used projective geometry to understand quantum mechanics, but published results as algebra. Albert Einstein used the metric of Riemann geometry to understand gravity, but published results as tensors. Myron Evans used differential geometry to explain all of physics.
> Professor John B. Hart

Introduction

As of the time of this writing, physics is still in several theoretical camps and each is certain that it has the greatest promise for success. We have general relativity including Evans' work, quantum mechanics in the standard model, and string theory and its offshoots. The evidence so far indicates that the Evans formulation is the only one that unifies gravitation and electromagnetism. It shows that quantum mechanics emerges from general relativity. It is correct in all its basic assumptions and it has been experimentally proven by a number of effects which it predicts and explains and which the standard model cannot explain.

The Evans equations complete Einstein's unification goal.

Further experimental verification will occur. That there will be corrections, modifications, and clarifications is inevitable. However, the basic concepts are now drawn out and are quite clear. There will be more discoveries in the future as physicists analyze older problems using the new methods, so this is certainly not the end of the story.

Chapter 15 – A Unified Viewpoint

This chapter reviews the material we have covered; adds more pictures, mechanical analogies, and simplified relations; and considers some implications that are a bit speculative.

The opinions here are not necessarily sanctioned by Professor Evans. Any errors in this book, but especially in this chapter, are this author's.

Review

There are problems within special relativity, general relativity, and quantum theory as they have stood in the past. Quantum theory is a theory of special relativity. Neither quantum theory nor special relativity can deal with spacetime gravitational effects. Einstein's general relativity explains spacetime, but it cannot adequately describe the other three forces within it. Each theory is well developed to explain matters within its own area, but is disjointed from the others.

The Evans Field Equation is the first insight into a plausible mechanism to obtain unification. From it, Einstein's gravitational field equation can be derived but also a new equation of spinning spacetime that describes the electromagnetic field.

The metric of spacetime must be defined as having torsion and curvature, T and R, to allow the turning of the electromagnetic field. Differential geometry already defines this metric and appears to be sufficient for the task.

Geometrically T and R are the only forms necessary to describe spacetime. The tetrad is the form that allows relations between different spacetimes. These are not new concepts, but they are used now with a new equation. Suddenly, the picture becomes much clearer.

From matrix geometry, which is well developed, the tetrad is analyzed in terms of its symmetry properties. A new concept occurs:

Symmetry indicates centralized potentials – spherical shapes. Gravitation and electric charge are symmetrical.

Antisymmetry always involves rotational potentials – the helix. Magnetism and a new form of gravitation have been found which are antisymmetric.

Asymmetry is a combination of symmetric and antisymmetric forms contained in the same shape. This is our universe's spacetime metric or manifold. Any form of mass or energy always has some of the other form however slight. Antisymmetric and symmetric curvature and antisymmetric and symmetric torsion coexist. Gravitation and electromagnetism should always be present together to some degree.

The chart at the end of Chapter 6 shows the connections between the various uses of the Evans Field Equation.

The generally covariant <u>field</u> equations of gravitation and electrodynamics are:

$$R^a_\mu - \tfrac{1}{2} R q^a_\mu = kT^a_\mu \qquad \text{Evans} \qquad (1)$$
$$R_{\mu\nu} - \tfrac{1}{2} R g_{\mu\nu} = kT_{\mu\nu} \qquad \text{Einstein} \qquad (2)$$
$$q^a_\nu \wedge (R^b_\nu - \tfrac{1}{2} R q^b_\mu) = k q^a_\nu \wedge T^b_\mu \qquad \text{Torsion (Evans)} \qquad (3)$$

In Chapter 7 Evans Wave Equation is discussed. This is an equation that is as much quantum theory as general relativity. It emerges from general relativity and it quantizes general relativity. The wave equation is $(\Box + kT)q^a_\mu = 0$.

$\Box$ is like the rate of change of the curvature. kT is Einstein's constant times the stress energy tensor. It is the energy density of a system – particle, magnetic field, electric charge, or antisymmetric gravitation.

The Evans Wave Equation $(\Box + kT)q^a_\mu = 0$ is derived from the Evans Field Equation and is the unification equation. It is as rich in quantum applications as Einstein's field equation is rich in gravitational applications.

The tetrad can be expressed in a variety of ways and this has great power to allow unification. The tetrad can be a scalar, vector, spinor, etc. That allows the same equation to describe all the forces (fields) of physics and to be an equation in special relativity, quantum mechanics, and general relativity. See Figure 15-1.

Figure 15-1 The tetrad and the forces of physics

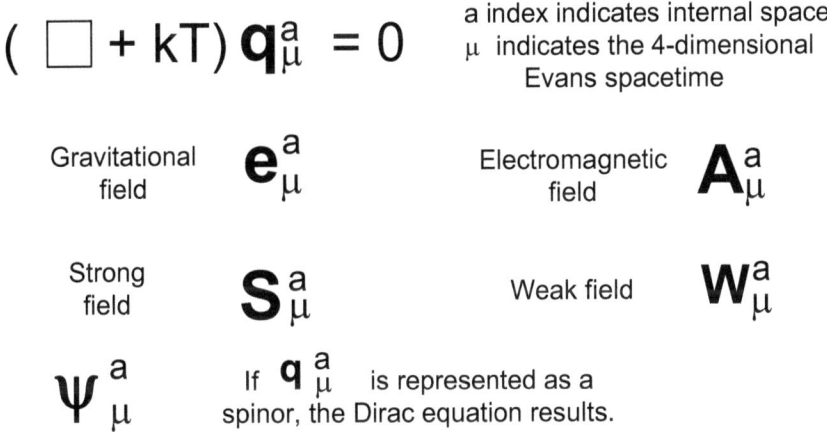

The mathematical separation of the asymmetric tetrad into symmetric and antisymmetric parts indicates more about the basic forces. The forces are representations of symmetries.

This is a wonderful example of the interaction between mathematics and physics. Since the tetrad is known to be valid in both mathematics and physics, and since the tetrad is properly described by a matrix in geometrical mathematics, it is also a proper description in physics. This follows Einstein's and Evans' contention that physics is geometry.

Subjects are discussed which in the past were seen as primarily quantum mechanics, but which can now be looked at in relativistic terms.

The Dirac and Klein-Gordon equations give us the formula for the mass-volume relationship in terms of the basic constants c, ℏ, and G.

We know energy density increase is accompanied by a decrease in the volume of any spacetime system as perceived from a low energy density reference frame. Reference frames contract in special relativity as they approach the speed of light. Wavelength decreases in electrodynamics as frequency increases with energy increase. The neutron star and black hole decrease in volume at certain mass concentrations. The black hole and the particle obey the same equation. However, so do the magnetic field, the electron, and antisymmetric gravitation. Now we see that a minimum volume exists. The black hole is not a point – it is a dot. There are no singularities. In addition, we have the formula for calculations. The volume shrinks, but never to zero.

$$mV_0 = k\hbar^2/c^2 \qquad (4)$$

Heisenberg uncertainty is not required because in unified field theory we have causal quantum mechanics. The Evans $\mathbf{B}^{(3)}$ spin field is described by and explains the Inverse Faraday Effect. O(3) electrodynamics is derived from general relativity using the Evans equations and describes spinning spacetime. It can be viewed as an intermediate theory between the standard theory and full unification theory.

Electroweak theory is discussed and one example of how Evans attacks the problem using unified field theory rather than quantum theory shows the power of the Evans equations. The Z and W bosons are described as forms of curvature. The Higgs mechanism is not needed.

Analysis of the AB effect shows the concept of spinning spacetime as the electromagnetic field. Spacetime extends beyond an immediate region and is the key concept in understanding entanglement. "Non-local" effects do not occur. Rather, entanglement is a spacetime effect.

Evans uses differential geometry applied to physics problems. Einstein developed general relativity using Riemann geometry and Evans developed the unified field theory using differential geometry. He is not the first to use it, but he succeeds in unification and uses geometry to explain it.

Only two concepts besides differential geometry are necessary to use to develop the equations of unification. One is the ansatz using conversion electromagnetic voltage $A^{(0)}$ to turn the tetrad into electromagnetics. The other is

Einstein's postulate R = -kT. This equates geometry on the left side with physics on the right side. If one reads Evans' papers in sequential order it is seen that in the beginning he started with physics, found differential geometry allowed proper development of equations, and reformulated the physics in terms of differential geometry. In the end, he clearly declares physics is geometry, just as Einstein declared.

Curvature and Torsion

Our universe requires curvature and torsion for unification of gravitation and electromagnetism. The Evans spacetime metric has both. The metric in Evans' extension of general relativity is asymmetric - it has symmetric and antisymmetric components. This is the difference between it and Einstein's general relativity, which has a symmetric Riemann metric with curvature alone. Evans has added the antisymmetric torsion to unify the previously separate components. All phenomena can be described using curvature and torsion. Differential geometry and physics are intimately related, if not identical constructs. The strongest statement is that what is true in differential geometry is true in physics. There are certain aspects that will be more physical than others will and clearly defining them will lead to more understanding.

That curvature is gravitation and torsion is electromagnetism is not controversial. Evans tells us that curvature and torsion always exist together to some degree. The one can be converted to the other. Curvature in geometry is the 1st curvature and torsion is the 2nd curvature.

Mathematics = Physics

The formation of equations with "Geometric Concept = Physics Equation" is basic in relativity and Evans' unified field theory. This is seen in Einstein's postulate R = -kT. R is geometry, -kT is physics.

$$G q^a{}_\mu = kT q^a{}_\mu \qquad (5)$$

Is one formulation of the Evans Field Equation. By breaking the tetrad, $q^a{}_\mu$, into symmetric and antisymmetric parts, a mathematical operation, a new window into physics opens. The inner structures of Einstein and Evans' field equations as well as that of curvature and torsion are revealed.

$$q^a{}_\mu = q^a{}_\mu{}^{(S)} + q^a{}_\mu{}^{(A)} \qquad (6)$$

And since the electromagnetic potential field is defined:

$$A^a{}_\mu = A^{(0)} q^a{}_\mu \qquad (7)$$

then we find symmetric and antisymmetric parts of it also:

$$A^a{}_\mu = A^{(0)} q^a{}_\mu = A^a{}_\mu{}^{(S)} + A^a{}_\mu{}^{(A)} \qquad (8)$$

We can then expand or redefine $R = -kT$ to:

$$R_1 q^a{}_\mu{}^{(S)} = -kT_1 q^a{}_\mu{}^{(S)} \qquad (9)$$

$$R_2 q^a{}_\mu{}^{(A)} = -kT_2 q^a{}_\mu{}^{(A)} \qquad (10)$$

$$R_3 A^a{}_\mu{}^{(S)} = -kT_3 A^a{}_\mu{}^{(S)} \qquad (11)$$

$$R_4 A^a{}_\mu{}^{(A)} = -kT_4 A^a{}_\mu{}^{(A)} \qquad (12)$$

where T is the energy momentum tensor. T_1 is symmetric gravitation. T_2 is antisymmetric gravitation. T_3 is symmetric torsion indicating electrostatics, and T_4 is antisymmetric torsion giving electromagnetism.

On the left side of each equation is the mathematical formulation of curvature. On the right side of each equation is the physics formulation of stress energy density.

Where Einstein had only the symmetric form of curvature of equation (7) – the gravitational field - the Evans equations indicate that there are four potential fields and four types of energy momentum.

All four potential fields are interconvertible and all forms of energy are interconvertible and:

$$T^a{}_\mu = T^a{}_\mu{}^{(S)} + T^a{}_\mu{}^{(A)} \qquad (13)$$

The stress energy can take four forms also.

Chapter 15 – A Unified Viewpoint

The Tetrad and Causality

The tetrad links two reference frames. It absorbs differences and translates vectors in one to those in the other. The base manifold is curved and torqued spacetime; the index is Euclidean spacetime.

The tetrad and the concept of moving frames as an alternate description of general relativity was invented by Elié Cartan. The tetrad is a key concept in expanding the field and wave equations into greater understanding.

The tetrad is a vector valued one-form in differential geometry. It provides the connections between the curved and torqued spacetime of our universe with the flat tangent mathematical space where we hold our vectors in order to define transformations from one reference frame to the next. When seen inside the Evans Wave Equation, $(\Box + kT)q^a{}_\mu = 0$, the values of $R = -kT$ become quantized. This is demonstrated in Figure 15-2.

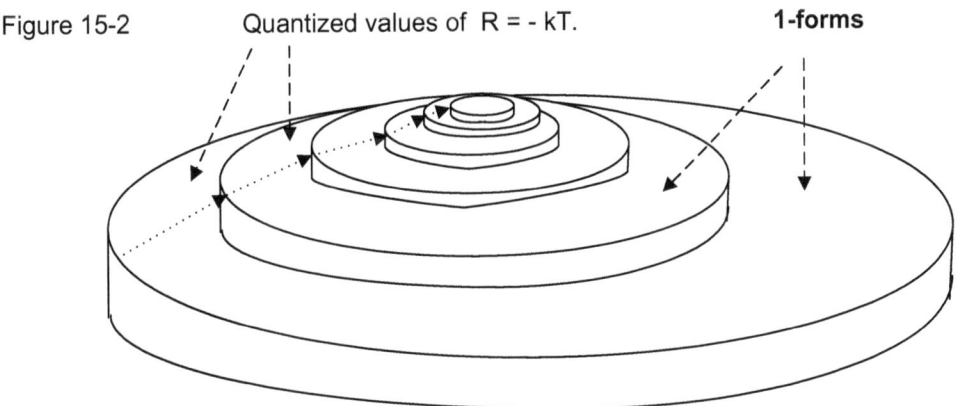

Figure 15-2   Quantized values of $R = -kT$.   1-forms

Heisenberg Uncertainty

There is no Heisenberg uncertainty in Evans' theory. The values of R that result are real and physical (eigenvalues). The result is causal wave mechanics leading to objective physics. General relativity is causal. There are no unknowable measurements. $\hbar$ is the smallest change that can occur. This does not limit measurements since there is either no change or $\hbar$ change. However, the uncertain statistical interpretation is invalid.

Statistical mechanics is not abandoned and probability still exists in say the roll of a dice. However, reality at the base is not probabilistic. We can know the position and momentum of a particle down to the accuracy of $\hbar$.

Non-locality (entanglement)

The derivation of a fundamental theorem describing "non-local" effects using the Evans unified field equations is not completed as of the time of this writing. However, it is safe to say that Evans' explanation of the AB effect shows that non-locality will have a similar explanation. Entanglement must be a geometric effect according to Einstein and general relativity. Evans' discovery of the spinning spacetime remote from the immediate cause in the AB effect indicates the origin of what we see as non-locality. Spacetime extends beyond an immediate volume and is the cause of "entangled" events.

The Aharonov-Bohm effect is a shift in R from one spacetime to another. Action at a distance and entanglement are explained using unified field theory. Curvature is the key. Since it does not exist as a basic concept in special relativity or quantum theory, they do not have the ability to explain non-local effects.

Principle of Least Curvature

Evans uses his Principle of Least Curvature to show that a particle cannot travel in a straight line. If it could, the spacetime would have zero curvature. In addition, the principle indicates that a particle always has a wave nature. From this one can derive de Broglie particle waves since the particle always has some torsion.

This means that de Broglie wave-particle duality can be derived from unified field theory and general relativity.

The Principle of Least Curvature also indicates that the phase in optics (radiated waves) and dynamics (matter waves) is derivable from unified field theory as in the AB effect.

The Nature of Spacetime

We can clearly define the vacuum as Minkowski spacetime of the standard theory. It has a special relativistic flat metric. Einstein's Riemann metric had curvature – gravitation; but it did not have torsion - electromagnetism. This is non-Minkowski or Riemann spacetime. It could explain electromagnetic effects in terms of tensors, but not describe gravitation and electromagnetism's mutual effects.

The "Evans spacetime" has both curvature and torsion. It has both a symmetric and an antisymmetric metric. Curvature R and torsion T are spacetime.[57]

It seems that differential geometry is spacetime and spacetime is differential geometry. We have evolved within this framework and are ourselves products of spacetime. Our mathematical thinking is a reflection of the basic composition of the spacetime of the universe. We are an intimate part of the process.

There are other geometries – Riemann for example, which is contained in differential geometry but has been shown to be too simple. It may be that Evans use of differential geometry will be superseded by another more complex geometry if future experiments offer problems it cannot solve. So far this expansion in scope has been the history of physics and this author certainly hopes we are not reaching the end of learning.

The universe and geometry evolved together. We are geometry. If this makes us feel a bit insignificant, keep in mind that differential geometry is just as complex as the universe. And if belief in God makes one shy away from what seems to be a simple description of existence, note that God created the geometry and one of Einstein's remarks with respect to the basic constants of the universe and their respective ratios was, "What I want to know is if God had a choice." Maybe the next generation of physicists will find a better definition, but for now spacetime is geometry is physics.

Recognition that the particle and electromagnetic wave are also forms of spacetime makes the definition a bit more concrete.

The word "spacetime" has a connotation of the nearly empty vacuum space between planets and star systems. Interestingly, or we would not be here to notice, spacetime has more aspects:

1. **Curvature.** One form of curvature is symmetric gravitation far from clumps of particles. Those are the geodesics of Riemann and Einstein. Low energy density spacetime lies between the planets and star systems. It is geometric curvature. It has some torsion also - electromagnetic waves both of recent origin and remnants of the big bang. The electromagnetic waves are not merely passing through spacetime (as special relativity and quantum mechanics have viewed it). Spacetime is the torsion and curvature. Particles are little concentrations of curvature and torsion – compressed spacetime. Particles, gravitation, and curvature are forms of symmetric spacetime. Antisymmetric curvature is a bit of a mystery still.
2. **Torsion.** The electromagnetic field, the photon, is spinning spacetime. This is antisymmetric torsion. The distance metric that Riemann geometry gives us is twisted and torqued in locations. Charge is symmetric torsion. Just what it is remains a mystery. Polarization of spacetime near charge occurs – positive (negative) symmetric torsion causes adjacent spacetime to exhibit negative (positive) torsion.
3. **Time.** Still much of a mystery. Evans' discussions and equations deal very little with time other than its translational symmetry. We may find that time is

---

[57] We do have to distinguish between T torsion and T stress energy using context.

just a spatial dimension. Time appears to be the movement of three dimensions in the 4th dimension. Time may be defined by spatial movement of the spinning field. Space is the three dimensions with which we are familiar. The time portion is the continued existence of those spaces. Time is a symmetry of the three dimensions of existence as they continue (translate) in a fourth spatial dimension. Evans gives us little new information in this area. That alone may be most telling, the implications are beyond the scope of this book.

4. **Photon, neutrino, electron, proton, (and neutron).** These are not forms *in* spacetime. They are forms *of* spacetime. More properly they are minimum curvature forms when at rest. No one has as of the time of this writing taken Evans' equation mV = k (ℏ/c)^2 or its relations and found the curvatures that relate to the masses of the basic particles. The ratios of the respective totals of each in the universe are another subject still to be explored. The mass and/or curvature ratios may lead to understanding. These four and the neutron have extended existence. The fleeting masses that have been defined in the particle zoo – some 1000 of them – are temporary transitional energy states. They are found repeatedly, but for very short times.[58] They appear to be unstable states or combinations of compression that quickly split apart into stable compressed &/or spinning states of the photon, neutrino, electron, and proton.

5. **The neutron.** The neutron's longevity[59] both inside and outside the nucleus is an interesting puzzle. The solution should help to explain the minimum curvatures and their stability. The free neutron decays into a proton, electron, and antineutrino. In terms of curvature and torsion, the neutron is composed of symmetric gravitation, symmetric torsion, and the neutrino which is yet undefined in terms of torsion and curvature. The neutron within a neutron

---

[58] For example, Σ+ has a mass of 1189 MeV and τ, an average lifespan of 10^{-10} sec or c τ = 2.4 cm. before it decays. The definition of particle is restricted here to those with more or less permanent existence.
[59] The free neutron exists for an average of 886 seconds before it decays.

star has a continuum of states and a lifespan in those states until the entire spacetime experiences the big crunch and recycles again with a big bang.

It seems inevitable that we will find that every form of spacetime has some portion or at least potential to be all four forms of field – symmetric and antisymmetric gravitation and electromagnetism.

The Particles

We need to relook at all particle physics in light of the concepts of curvature and torsion. In addition the neutrino and maybe the photon need to be added to the particle list and evaluated as forms.

Professor Evans says, "...as a working hypothesis the proton, electron, neutrino and photon are stable because they are in stable equilibrium with minimized action. The action or angular momentum occurs in units of the Planck constant. The other unstable particles have not minimized action, and transmutation occurs according to the simultaneous Evans' equations in my paper on the electro-weak field. The transmutation always tends to minimize action..."[60]

The particle is one of the forms of spacetime. It has location, mass, frequency, and sometimes charge. It is curved spacetime and it curves spacetime. The amount of curvature it causes is proportional to $1/r^2$ where r is the distance from the center of the particle. This is a symmetric property.
since defining one of the particle's aspects with the same word is inaccurate. We could use "curvature form" and "torsion form".) Curvature defines the massive aspect best and torsion defines the oscillatory aspect best. For example, the photon is an antisymmetric torsion form convertible to curvature and the proton is a symmetric curvature form with some torsion. The particles each have some aspects of both curvature and torsion as shown in Figure 15-3.

---

[60] Email communication.

Figure 15-3 The stable particles

Proton

Symmetric Spacetime
Mostly Curvature = Gravitation
Localized
+ charge

Photon

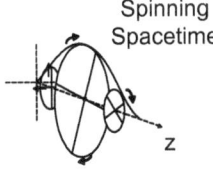

Spinning Spacetime

Antisymmetric Spacetime
Mostly Torsion = Electromagnetism
c => not local
No charge

Neutrino

Antisymmetric Spacetime?
Mostly Torsion?
= Antisymmetric gravitation?
c => not local
No charge.

Electron

Symmetric Spacetime
Mostly torsion with distinct curvature
Localized
− charge

Neutron

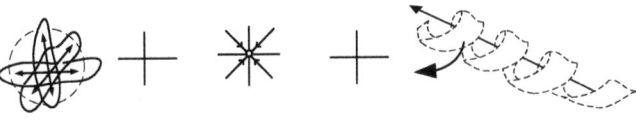

Symmetric gravitation + Symmetric Torsion +
Antisymmetric gravitation? (antineutrino)

Quantum mechanics presented us with the knowledge that the particle has a dual nature – particle and wave.[61] (We need a new or different word here

---

[61] Einstein showed the photon to be wave and particle; de Broglie showed the particle to be particle and wave.

In Chapter 9 we saw

$$\lambda_{de\,B} = \hbar / p = \hbar / mv \text{ and } \lambda_c = \hbar / mc \tag{14}$$

The frequency of a particle is a function of ℏ, its mass, and its velocity. Evans shows that this is also definable in terms of spatial curvature:

$$\Box = R_0^{1/2} = \hbar / mc \text{ [or v]} \tag{15}$$

Here the Compton or de Broglie wavelength defines energy and spatial curvature within general relativity – unified field theory when combined. Also:

$$E = \hbar c \sqrt{|R_0|} \tag{16}$$

The particle is curvature - compressed spacetime. It also oscillates and is a standing wave. Evans' new insight is that this is directly related to basic constants – ℏ and c.

This method of looking at protons and neutrons does not require the quark concept. Quarks are by definition unobservable and constrained. The quark concept may just be a mathematical tool and the particle is compressed states. The three components of the particle, if they are ever shown to be actually discrete, are not unlikely to be found to be curvature oscillations in three dimensions of spacetime. They have SU(3) symmetry and that fits into the Evans equation, but the subject is not clear.

We also see in these equations that quantum formulas have explanations in general relativity or simple geometry. Waves are curvatures.

Evans does not state that the neutrino is antisymmetric gravitation, however it seems a possibility to this author. Each of the four forces has a carrier in the standard viewpoint. Certainly it is logical to say that antisymmetric curvature is spinning gravitation. Locally it needs a force carrier so the neutrino becomes a candidate.

Chapter 15 – A Unified Viewpoint

Four forms of energy can be described within the unified field q_μ^a :

FIELD/ FORCE CARRIER	Description	POTENTIAL FIELD	DESCRIPTION	Charge
Proton[62] <Neutron>	Symmetric Centralized Gravitation	$q_\mu^{a\,(S)}$	Most massive particle with a stable curvature state. Localized.	+ <0>
Neutrino?	Antisymmetric Spinning Gravitation	$q_\mu^{a\,(A)}$	Moves continuously at c. Low mass.	0
Photon	Antisymmetric Torsion Spinning Spacetime	$A_\mu^{a\,(A)} = A^{(0)}\, q_\mu^{a\,(A)}$	Carries EM force. Moves continuously at c. Low mass.	0
Electron	Symmetric Centralized Charge	$A_\mu^{a\,(S)} = A^{(0)}\, q_\mu^{a\,(S)}$	Pulls spacetime to it. Has small mass	−
Spacetime	Hybrid Asymmetric	q_μ^a	Curvature and Torsion. Gravitation and Spin.	Irregular. The average is 0.

The Field/Force Carrier chart is as much a puzzle as an organization of the four stable forces. There is work to be done using the Evans equations to relate the internal structure of the forces.

The precise ratio or relation of the total curvature and torsion of each to one another is unknown as of the time of this writing.[63] A serious study of the relations between curvature and torsion may result in a definitive answer. Unified, there is greater power than either general relativity or quantum theory has alone.

The Electromagnetic Field – The photon

Evans gives us spinning spacetime as the electromagnetic field. Here we have one of the more definitive changes in our understanding of physics. Up

---

[62] Main curvature component of the neutron also?
[63] When this book is completed the author is going to work on the reason for the masses. Physicists better get to work on these. It would be embarrassing if a lowly mechanical engineer found the relations first and won a free trip to Stockholm.

until Evans' equations the electromagnetic field was always seen as something superimposed on spacetime, not as spacetime itself.

We can then generalize from the spinning spacetime to the other forces to say all are spacetime. While this is not strictly physics yet, it is definitely indicated by the mathematics.

$A_\mu^{a\,(A)} = A^{(0)} q_\mu^{a\,(A)}$ shows us two things. One, that electromagnetic potential (viewed as charge or potential voltage) $A^{(0)}$, is the conversion factor that takes the tetrad to the values of spinning spacetime. That charge and spin are related is established. None of the factors here or in any of the basic forms precedes the others. They all originate simultaneously. So we see that the electromagnetic field is voltage times the tetrad.

The photon is mostly wave (torsion) and a bit of particle (curvature). It has mass equivalence if not mass itself. It is a bit of spinning curved spacetime. The photon carries the electromagnetic force. The electromagnetic field outside a bar magnet can be described as standing photons.

$$(\Box + kT)\,\Psi = 0$$

The B field is the magnetic field or magnetic flux density composed of rotating, polarized photons. It can be defined by the force it exerts on a co-moving point charge. The force is at right angles to the velocity vector in the direction of travel and at a right angle to the electric field. The Evans' general relativity version is:

Just why the free photon constantly moves at c remains unexplained.

When a photon is captured in an atom and excites an electron to a higher energy orbital state, the electron occupies more space in the orbital levels. It is spread out a bit more. It is unstable and will eventually cause a photon to be ejected and it will return to its unexcited stable state.

The photon was moving at c. It suddenly stopped and some of its energy – curvature, torsion, and momentum - was converted to spacetime mass or volume. The increase in electron space may be due to a spatial component of the photon.

Research may show the photon to be the archetypical wave-particle. It is a pure field of wave when traveling. It becomes mass or space when captured in a particle. Jam enough of them together and you get a stable curvature form.

The Neutrino

The neutrino moves near c as does the photon. The free neutron becomes a proton, electron, and antineutrino. The neutrino is possibly the carrier of the antisymmetric gravitational component of spacetime. It is still a most elusive particle.

There are three neutrinos and their antiparticle versions. They are now estimated to have masses from about 3 eV to 18 MeV and a small magnetic moment. The lifespan of the electron neutrino is upwards of 10^9 seconds.

Evans states that the neutrino masses are eigenvalues of the Evans Lemma, $\Box\, q^a_m = Rq^a_{\mu,}$, and the wave function of the neutrino is a tetrad made up of the complex sum of two types of neutrino multiplied by the Evans phase. Further information can be found at www.aias.us in "Evans Field Theory of Neutrino Oscillations."

The phase factor is a torsion form and this indicates that the neutrino is a mixture of torsion and curvature since it has mass. The mixture changes with time and we see one geometric form transforming into another during the process. Presumably, the 2nd type of curvature (torsion) converts to the 1st type (mass).

It is obvious that we have a better mathematical understanding of its character than any classical description.

The Electron

It is likely intimately related to the proton or it is its exact opposite. Presumably one electron and one proton are born together. Total charge then equals zero. Curvature dominates in the proton, torsion dominates in the electron. Why? We do not know.

A naïve description would be that the electron is symmetric torsion spinning "inward" to cause charge. Then the proton is spinning "outward" to have the opposite charge. These are most likely the wrong words. Vacuum polarization is then the result of spinning in opposite directions.

The Neutron

The neutron decays into a proton, electron, and antineutrino after 10 minutes (180,000,000 kilometers.) Ignoring quarks as nothing but convenient mathematical descriptions of energy states, the neutron may be an irregular compressed state composed of minimum curvatures. But in high densities (neutron star) it is a most stable form of particle.

It is neutral in charge having a proton and electron "within" it. It obeys the $1/r^2$ inverse square rule for gravitation.

The unraveling of the neutron can indicate curvatures of the proton, electron, and neutrino. The splitting of the neutron into a positive and negative part could give insight into the torsion of charge. The weak force is electromagnetic and the neutron defines it.

The neutron is not composed of a proton, electron, and antineutrino. Rather the curvature and torsion within the neutron break apart into the curvatures and torsions that equal the proton, electron, and antineutrino

The only component the neutron seems at first sight to be lacking is the photon. No photon emerges when the neutron decays.

However, as mass builds to very large quantities, the neutronium of a neutron star is a different matter. The neutronium forms a super large nucleus without much in the way of protons. The first stage of development is for the protons of the atoms to squish with the electrons to form neutrons. Where does the antineutrino come from? One assumes it should be seen in terms of torsion in the atoms and a curvature or torsion conversion. When looking at the

Chapter 15 – A Unified Viewpoint

particles, there is much less clarity than when thinking in terms of curvature and torsion. See Figure 15-4 for speculation.

Figure 15-4   Enigmatic Neutron

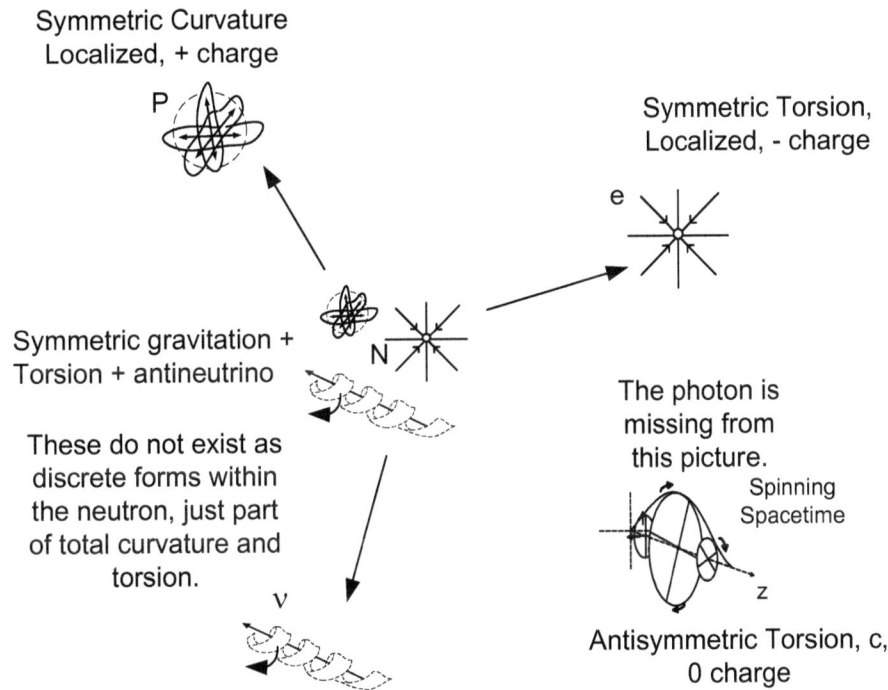

Unified Thinking

The essential point of the above descriptions is that thinking in terms of curvature and torsion becomes necessary.

Curvature, R, is centralized gravitation and antisymmetric gravitation.

Torsion, T, is electromagnetism and symmetric charge.

There is only geometry = directions of curvature and spin in four dimensions.

Note that we are still missing some vocabulary words. We refer to antisymmetric torsion as electromagnetism, symmetric torsion as charge, and symmetric curvature as gravitation. We do not have a classical term for antisymmetric gravitation.

Unified Wave Theory

There are two new fundamental equations:

1) The Evans Field Equation which is a factorization of Einstein's field equation into an equation in the tetrad:

$$G^a_\mu := R^a_\mu - \tfrac{1}{2} R q^a_\mu = k T^a_\mu \qquad (17)$$

From this equation we can obtain the well-known equations of physics.

2) The Evans Wave Equation of unified field theory:

$$(\Box + kT) q^a_\mu = 0 \qquad (18)$$

The real solutions it offers, its eigenvalues, will obey:

$$R = -kT \qquad (19)$$

This quantizes the physical values of kT that result. Its real function is the tetrad q^a_μ. The four fields can be seen to emerge directly from, and are aspects of, the tetrad itself. The wave equation was derived from the field equation and is thus an equation of general relativity. Figures 15-5 and 15-6 exemplify this.

Chapter 15 – A Unified Viewpoint

Figure 15-5   All equations of physics can be derived
from the Evans Field Equation

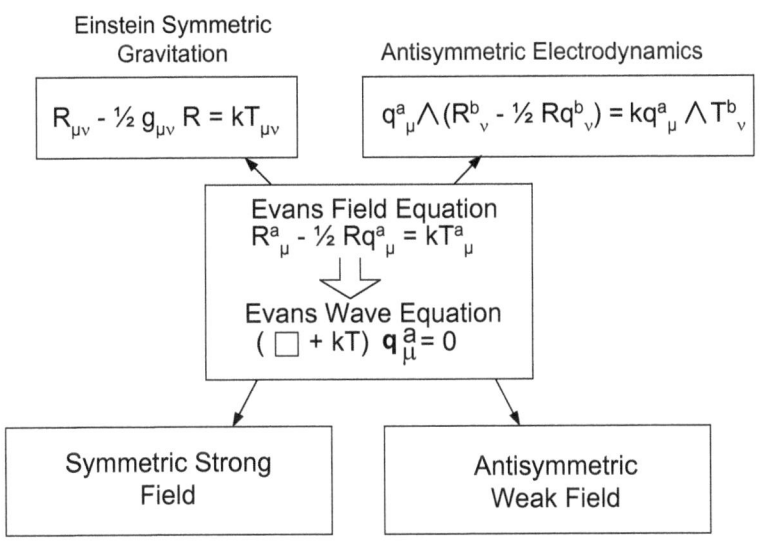

Figure 15-6

$$(\Box + kT) q^a_\mu = 0 \qquad q^a_\mu = q^a_\mu{}^{(S)} + q^a_\mu{}^{(A)}$$

Symmetric   **Antisymmetric**

Gravitational field e^a_μ Basis vectors   Electromagnetic field A^a_μ Torsion, voltage × tetrad

Strong field S^a_μ Quarks are representations of quantized gravitation.   Weak field W^a_μ Radioactivity

Single Particle Scalar ϕ^a_μ $m^2c^2/\hbar^2$ Klein-Gordon equation.   Spinor ψ^a_μ Dirac equation

$$1/\lambda_C^2 = (mc/\hbar)^2 = kT_0 = R_0 = \kappa_0^2$$
$$\lambda_C = R_0^{-1/2}$$

Scalar curvature is related to the Compton wavelength $R_0 \;\; \lambda_C$ / λ_C

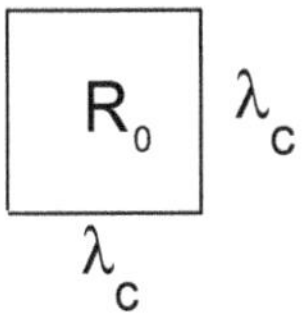

Oscillatory Universe

The equation $mV = k(\hbar/c)^2$ for the particle may be applied to the universe as a whole. The implication is that the universe is oscillatory. It can get no smaller nor larger than the equation allows. If one plugs the mass of the universe into the equation, the volume is finite, not zero. If this equation is not directly applicable, every particle (energy state) in the universe must obey it individually and the sums are the same.

Another indication is given by Evans that is not explained in this book. In a consideration of the time component of R, a derivation in R gives a cosine function that shows it bounded by ± 1. R can never be infinite and therefore the curvature must allow some volume. A singularity is impossible.

This our universe probably started at a big bang, but not from a singularity – zero volume, infinite density. It is more likely that the seed was a compressed state that had previously been a contracting universe.

If one wants to discuss the origin of universes before that, before many cycles of oscillation, it is as likely that the origin was low density. A state of stretched out flattened spacetime is similar in structure to the flat void of nothingness. Then a slight oscillation and gradual contraction to a dense kernel could occur. Then from a big bang to another flattened state with recontraction again could increase the amount of matter gradually. This was discussed in Chapter 9.

This is absurdly speculative, although fun. Mathematically, it seems the oscillations have been going on infinitely, but that too seems absurd. And who is to say that the previous contracting universes all used differential geometry. Einstein's question if God had a choice remains open.

Generally Covariant Physical Optics

We have not gone into this subject in any detail due to its complexity. It was discussed briefly in Chapter 13 on the AB effect. The bottom line is that a

Chapter 15 – A Unified Viewpoint

geometric phase factor can be used to show that the Aharonov-Bohm effect, Sagnac effect, Berry and Wu Yang phase factors, Tomita-Chiao effect, and all topological phase effects are explained in general relativity as a change in the tetrad. The generally covariant Stokes theorem, .O(3) Electrodynamics, and the $B^{(3)}$ field are involved in derivations.

General relativity and electrodynamics combined are more powerful than either alone.

Charge and Antiparticles

Figure 15-7 Charge and Antiparticle

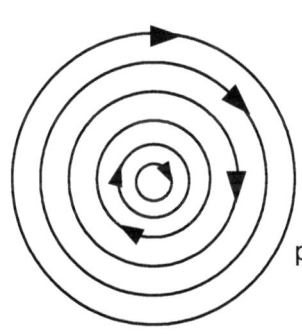

Charge?

The spin of an electron is speculatively depicted with two dimensions suppressed.
If the velocity of spin is proportional to the charge, then it is spinning faster as one approaches the center. It is similar to a whirlpool, but 4-dimensional. Spin is poorly described at this time.

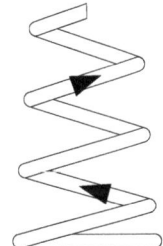

In three dimensions the velocity profile may be more like a helical spring.

Antiparticle?

The sphere rotating in time-space direction. Quite difficult to picture. Time is spatial.

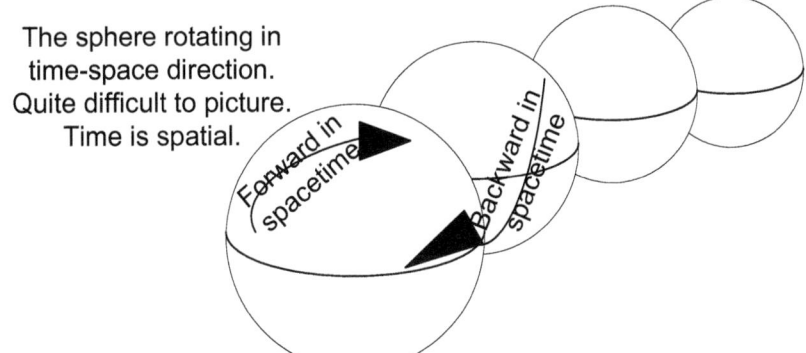

We do not have a definite mechanical picture of what charge is. Mathematically and electrically we have a better grasp. Evans adds symmetric torsion to the understanding. This is "centralized spacetime spin." Symmetric implies centralized location. Like gravity, charge is most dense towards a centralized location and it gradually dissipates away from the center. Torsion implies spin. In what direction that spin occurs is not yet clear.

We can hypothesize that spin in all four dimensions can occur. The left or right rotation could be in the 4th dimension. Then spin could occur in the time-space direction. Then we would have forward and backward spherical rotation.

One assumes that the antiparticle is spinning in a different direction from the particle. While Evans gives us the tetrad with a negative sign for the antiparticle, this author cannot draw a nice classical picture with any definitiveness. See Figure 15-7.

The Electrogravitic Field

$$\mathbf{E} = \frac{\varphi^{(0)}}{c^2} \mathbf{g} \tag{20}$$

This gives the electric field strength in terms of **g**. The field is in units of volts per meter. It shows that there is a mutually calculable effect between gravitation and electromagnetism. There is one field, *the electrogravitic field.*

The effects will be difficult to test since they are quite small.

This does show that there is a fundamental voltage $\varphi^{(0)}$ available from curved spacetime. There is an electric field present for each particle and it originates in scalar curvature.

The Very Strong Equivalence Principle

The Very Strong Equivalence Principle is the last thing we mention here. Everything is the same thing – geometric curvature and torsion. The photon, electron, proton, neutrino and distance between them are composed of gravitation and spin - curving and spinning spacetime.

From a spacetime metric with curvature and torsion of mathematics we move to physics with gravitation and spin everywhere.

There is much more to the work of Dr. Evans, and that can be found in his papers and books. The basics are presented in this book. In his words:

A Summary of Einstein-Cartan-Evans Field Theory

1) Physics is objective. There are no magic hidden forces. Geometry as developed by Riemann and Cartan is sufficient for understanding.

2) Einstein's General Relativity of 1915 was the first step in understanding. Riemann's geometry allowed curvature to be equated with gravitation. Since torsion was set to zero, spin = electromagnetism remained hidden.

3) Cartan's differential geometry, developed in the 1920's, is the next step by allowing spin (the torsion tensor) within the spacetime. It is well accepted and sufficient geometrically to explain what is known at this time.

4) The Evans $B^{(3)}$ spin field of 1991 was a start at unification. It recognized that the magnetic field had substance.

5) In 2003, Evans found a vector equation that led to Einstein's equation of General Relativity and then converted it to Wave Mechanics to develop the Unified Field Theory. From the same initial equation both General Relativity and Wave Mechanics could be realized.

6) The tetrad form of General Relativity lends itself to both curvature and spin – to gravitation and electromagnetism. The factor $A_{(0)}$ is the conversion.

7)   The ECE theory is based on Cartan geometry and is consistent within it. New physics may be discovered and more complicated geometry required, but at the present, ECE theory is sufficient.

8)   Experiments that cannot be explained in the present standard model of physics can be explained by ECE theory.

   Laurence G. Felker, Reno NV, 2007

Glossary

There are some terms here that are not used in this book, but are necessary when reading Evans' works at www.aias.us. They have been included to help the non-physicist reader.

In addition, many web sites and search engines contain much more extensive information. Among them are:

http://en.wikipedia.org/wiki/Main_Page

http://mathworld.wolfram.com/

http://scienceworld.wolfram.com/physics/

$A^{(0)}$

$A^{(0)}$ is scalar originating in $\hbar / e$, the magnetic fluxon. $A^{(0)}$ is negative under charge conjugation symmetry, C. It has the units of volts-s/m. The fluxon has units of webers = volts-seconds. e is the charge on the proton which is positive, e⁻ is the negative electron. This is a conversion factor from geometry to physics defining the electromagnetic potential field: $A^a{}_\mu = A^{(0)} q^a{}_\mu$. It is the only concept Evans uses other than differential geometry in developing the most sophisticated parts of his theory. See Tetrad and Charge Conjugation also.

Voltage or potential difference is $\phi^{(0)} = cA^{(0)}$.

Abelian

In group theory, refers to a group where the commutative law holds. If $A \times B = B \times A$, the elements commute. It is non-Abelian where $A \times B \neq B \times A$. Electrodynamics is non-Abelian. Rotation is described by non-Abelian equations.

Abstract space, Hilbert space

There are two spaces that are considered - the tangent space of general relativity and the vector space of quantum mechanics. Evans identifies these as the same space. This is a new insight. While the tangent space is a geometrical physical space, the vector space in quantum mechanics has, up until now, been considered a purely abstract mathematical tool. Hilbert space and internal space of gauge theory are other terms for the abstract space in quantum mechanics. Both are vector spaces.

In the mathematics of quantum theory the abstract spaces are needed to perform operations. The vectors that are gauged are not generally considered to be real; instead they indicate what is going on inside a particle field and give accurate numbers for interpretation.

These vectors are used to manipulate the probability equations. There are an infinite number of abstract vector spaces in which we can do calculations.

The vectors are complex numbers with the probability $|\psi^2| = $ some number from 0 to 1. It is always positive since the absolute value is taken. The number is the probability of a location, momentum, energy, angular momentum, or spin.

Hilbert space extends the vector space concept to functions. The dot product of two vectors and complex functions can be considered to make another vector space.

In relativity the tangent space is real. It is progression in time or the 4th spatial dimension. For example, to find the rate at which a curve is changing, a tangent is used. In two dimensions, the tangent appears as a line at a point on the curve. That tangent is not part of the curve. We use it to get the slope. It could be said to be in a different space.

With a two dimensional space, the tangent is easily visualized. In four dimensional spacetime, there are tangents referred to as "tangent vectors." In dealing with curved spaces, those vectors are real. Consider a 4-dimensional vector. Three vectors indicate lengths and directions in the three-dimensional space we inhabit and the 4th is time.

At any point in spacetime, the tangent vectors completely describe the spacetime at that point. These are geometric objects – they exist without reference frame and can be transformed from one frame to another.

Affine transformation

A transformation that is a combination of single transformations. A rotation or reflection on an axis. Affine geometry is intermediate between projective and Euclidean. It is a geometry of vectors without origins, length, or angles. These are the connections between the vectors in one reference frame and another. They connect the base manifold to the index. See also Christoffel symbols.

Cartan's moving frames is called an affine geometry. It is a spacetime that uses differential geometry in nearly the same way as Einstein used Riemann geometry. Evans uses a Cartan differential geometry in a more complete way.

Aharonov-Bohm effect

This is an effect occurring in a double slit experiment from the presence of a vector potential produced by a magnetic field of a solenoid. The pattern of electrons hitting a screen is shifted – the electrons are influenced by a magnetic field that does not touch them. The Evans' explanation is that spacetime itself is spinning and extends outside the solenoid.

Glossary

AIAS

Alpha Institute for Advanced Studies. This is a physics discussion group with Professor Myron W. Evans serving as director. The Father of the House is Professor John B. Hart of Xavier University in Cincinnati.

Ampere's Law

The integral of a magnetic field of a current-carrying conductor in a closed loop in spacetime is proportional to the net current flowing through the loop. That is, the magnetic field is proportional to the current. $V = IR$ is Ohms law. V is voltage, I is current, R is DC circuit resistance. Electromagnetic force, caused by a magnetic field induces a voltage that pushes electrons (current in amperes) around a closed wire. A battery would also provide a voltage.

Analytical functions and analytical manifolds

In mathematics a full differentiable manifold is analytical. It is defined continuously down to the point. A line is a simple manifold.

An analytical function is any function which is the same in the entire manifold upon which it is defined within the limits that are set. An analytical function must exist on an analytical manifold.

Differential geometry as physics assumes spacetime in a fully differentiable manifold. The spacetime does not have to be fully homogeneous – the same everywhere. Each region must however, be fully analytical.

Evans and Einstein assume that physics is geometry. There are those who think that mathematical models are only approximations to reality. Both are likely true in different aspects of physics. One of the tasks of the researcher is to find which equations are physics and which are nice approximations.

It should be noted that if physics came first – the creation of the universe – and we evolved within it, then our mathematics evolved from physics and in that way they cannot be separated, except by our errors in understanding.

Angular Momentum

Rotational momentum. For a spinning object, this is $L = I \omega$ where I is the moment of inertia and ω is the angular velocity.

The moment of inertia is rotational inertia. From $F = ma$, Newton's law, $F = maR$ where a is angular acceleration and R is length of lever arm or a radius – it depends on the shape of the rotating object. If $F = maR$ above is multiplied on both sides by R, the torque is defined as $\tau = RF = maR^2$. For a uniform sphere, $I = 2/5 \, m \, r^2$ where m is the mass and r is the radius. For a hollow cylinder with inner radius of r_1 and outer radius of r_2, $I = \frac{1}{2} m(r_1^2 + r_2^2)$.

Rotational angular momentum has units of action - energy times time, or momentum multiplied by length.

293

This is particle "spin."

Antisymmetric tensor

The simplest antisymmetric tensor is $A^{\mu\nu} = -A^{\nu\mu}$ with Greek indices 0 to 4 dimensions. This can be rewritten $A^{[\mu\nu]} = \frac{1}{2}(A^{\mu\nu} - A^{\nu\mu})$. This means that each individual element in the matrix which the tensor defines is inverted. A tensor can be written as a sum of the symmetric and antisymmetric parts:

$$A^{\mu\nu} = 1/2\,(A^{\mu\nu} + A^{\nu\mu}) + 1/2\,(A^{\mu\nu} - A^{\nu\mu})$$

The antisymmetric tensors will describe turning or twisting. The symmetric tensors will typically describe distances.

Ansatz

Postulate, conjecture or hypothesis.

B$^{(0)}$

$B^{(0)}$ is magnetic flux density.

B$^{(3)}$ field

The B field is the magnetic field or magnetic flux density composed of rotating, polarized photons. It is sometimes defined by the force it exerts on a point charge traveling within it. The force is at right angles to the velocity vector (direction of travel) and also at a right angle to the magnetic field. The **B**$^{(3)}$ field is the Evans' spin field. The **B**$^{(3)}$ field is directly proportional to the spin angular momentum in the (3) or Z-axis, and for one photon it is called the photomagneton.

In Evans' work we see the magnetic field as geometry with an O(3) symmetry. The precise operations within the theory are difficult. The definition is:

$$\mathbf{B}^{(3)*} = -ig\mathbf{A}^{(1)} \times \mathbf{A}^{(2)} = (-i/B^{(0)})\,\mathbf{B}^{(1)} \times \mathbf{B}^{(2)}$$

The **B**$^{(3)}$ field is defined as the conjugate product of transverse potentials: $\mathbf{B}^{(3)} := -i\,g\,\mathbf{A}^{(1)} \times \mathbf{A}^{(2)}$

Where $\mathbf{B}^{(1)} = \text{curl }\mathbf{A}^{(1)}$, $\mathbf{B}^{(2)} = \text{curl }\mathbf{A}^{(2)}$, $g = e/\hbar = \kappa/A^{(0)}$ and $B^{(0)} = \kappa A^{(0)}$. The **B**$^{(3)}$ spin field is non linear.

This equation is now understood to be a direct consequence of general relativity, in which any magnetic field is always defined by:

$$B = D \wedge A$$

where $D \wedge$ is the covariant exterior derivative.

The electromagnetic field is therefore the spinning or torsion of spacetime itself. The field is not in or on spacetime; it is spacetime.

It is the fundamental second or spin Casimir invariant of the Einstein group where the first Casimir invariant is mass. Thus, mass and spin characterize all particles and fields. In other words the photon has mass as inferred by Einstein and later by de Broglie.

The **B**$^{(3)}$ field and O(3) electrodynamics relate to electromagnetism as a torsion form, made up of cross products. It is non-linear, spin invariant, fundamental electromagnetic field directly proportional to scalar curvature R.

Electromagnetism is thus a non-Abelian theory which is also non-linear in the same sense as gravitation.

B$^{(3)}$ gives the longitudinal component of the electromagnetic field.

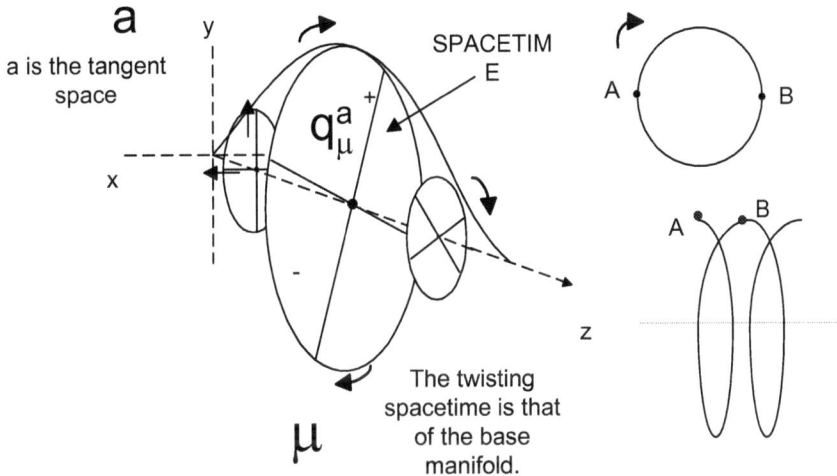

As the circle rotates and moves forward, a helix is drawn.

The photon exists in three dimensions.

The **B**$^{(3)}$ field is a physical field of curved spacetime. It disappears in flat spacetime of special relativity. O(3) electrodynamics exists in curved spacetime and defines a sphere.

Where spacetime has no curvature, it has no energy. R can therefore never be zero in our universe.

Basis vectors

Unit vectors $\mathbf{e}_{(1)}$, $\mathbf{e}_{(2)}$, $\mathbf{e}_{(3)}$, $\mathbf{e}_{(4)}$ which define a mathematical tangent space. The basis vectors of a reference frame are a group of four mutually orthogonal vectors. Basis vectors establish a unit vector length that can be used to determine the lengths of other vectors. They obey:

$$e_0^2 + e_1^2 + e_2^2 + e_3^2 = +1$$

That is, they form a four dimensional sphere. In addition:

$$\mathbf{e}_0 \cdot \mathbf{e}_0 = -1$$
$$\mathbf{e}_1 \cdot \mathbf{e}_1 = \mathbf{e}_2 \cdot \mathbf{e}_2 = \mathbf{e}_3 \cdot \mathbf{e}_3 = +1$$

$e_a \cdot e_b = 0$ if $a \neq b$; a, b are 0, 1, 2, or 3.

These come into use in curved spacetime. Orthogonal means they are at right angles to one another with respect to the multidimensional spacetime. (This is not really perpendicular, but can be imagined to be so for visualization.) Normal means they have been scaled to unit form. The basis vectors establish the unit form.

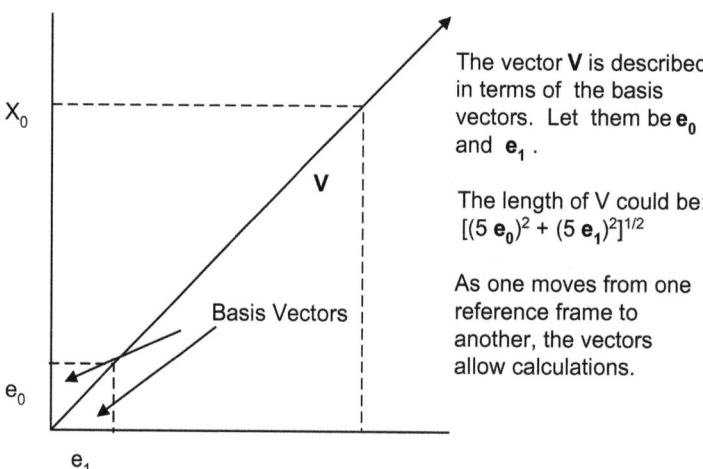

The vector **V** is described in terms of the basis vectors. Let them be e_0 and e_1.

The length of V could be:
$[(5 e_0)^2 + (5 e_1)^2]^{1/2}$

As one moves from one reference frame to another, the vectors allow calculations.

Orthonormal means they are both orthogonal and normalized. If one finds the orthonormal vectors at an event point, then one can define the vectors at any other event point in the spacetime. The spacetime is curved and some method of calculating and visualizing the spacetime is needed.

Unit vectors are tangent to a curve at a point. For example the three unit vectors of a curved coordinate system are mutually orthogonal ("perpendicular") and cyclically symmetric with O(3) symmetry. With $e_{(1,2,3)}$ the unit basis vectors and $u_{(1,2,3)}$ the coordinates at any point.

$e_{(1)} \cdot e_{(2)} = 0 \quad e_{(1)} \cdot e_{(3)} = 0 \quad$ and $e_{(2)} \cdot e_{(3)} = 0$

and

$e_{(1)} \times e_{(2)} = e_{(3)} \quad e_{(1)} \times e_{(3)} = e_{(2)} \quad$ and $e_{(2)} \times e_{(3)} = e_{(1)}$

$e_{(n)} = 1/h_i \times \partial R / \partial u_i$

and the arc length is:

$ds = |dR| = |\partial R/\partial u_1 \times du_1 + \partial R/\partial u_2 \times du_2 + \partial R/\partial u_3 \times du_3|$

The calculations to get coordinates are most tedious. The orientation of the components is shown in the drawing below courtesy of Professor Evans.

Glossary

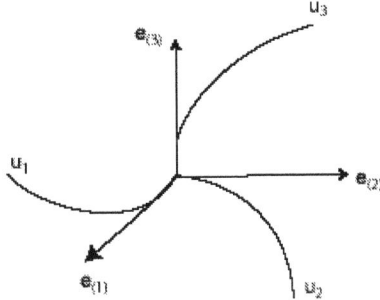

Courtesy Myron Evans

Blackbody radiation

When a sphere is in temperature equilibrium with its surroundings, the radiation given off follows a known curve. Classical physics could not explain and the formulas predicted infinite short wave radiation energy density. It was called the ultraviolet crisis. Max Planck found a method to accurately predict the curve and in so doing used h as a constant that predicted energy levels were discrete rather than continuous. This became the quantum hypothesis.

c

Speed of light in vacuum equals 2.9979×10^8 m/s. This is the speed of time also.

Canonical

If the states of a system are parameterized by time, then the approach is "canonical." Space and time are treated differently and this is not fully compatible with general relativity.

Cartan tetrad theory

A generalization of Riemann geometry. Einstein's Riemann approach to general relativity focused on geodesics – on tangents moving along the curved lines. Cartan's method used the moving frame of the tetrad along the curved lines.

Casimir Invariants

The Casimir invariants of the Poincaré group define mass and spin for any particle. The group of special relativity with the spacetime translation matrices or operators included is the ten element Poincaré group.

Causal

Real processes are causal such as the decay of the atom. General relativity sees physics as causal. Quantum theory has tended to see reality as acausal or probabilistic.

Causality implies the cause of an event occurs before the result and can be predicted, even if only statistically in some cases. Reality is inherently causal.

Charge Conjugation Invariance, C

There will be no change in some processes when antiparticles replace particles.

This is true for strong and electromagnetic interactions.

This is not true for weak interactions where only left-handed neutrinos and right-handed antineutrinos are involved. C changes a left-handed neutrino into a left-handed antineutrino (or a right-handed neutrino to a right-handed antineutrino). Processes involving right handed-neutrinos or left-handed antineutrinos have never been seen together.

P, parity, turns a left-handed neutrino or antineutrino into a right-handed one.

CP is Charge Conjugation with Parity Invariance and this turns a left handed neutrino into a right handed antineutrino (and vice versa.) This applies to weak interactions. Parity is mirror reflection.

See Symmetry and Groups.

Charge-current density vielbein

A tetrad representing the amount of electric charge per unit volume. A flowing electrical current is $J = -env$ where J is the volume of current, -e the charge on an electron, n the number of electrons, and v is the velocity of movement. J is proportional to the applied electric field, E.

Christoffel symbol (capital gamma, Γ)

These are used to map one group of vectors in one space onto another set of vectors in another space. It is a mathematics map from one physical spacetime or reference frame to another.

They are connection coefficients between spaces, manifolds, vectors, or other geometric objects. They are tensor-like objects derived from the Riemann metric used to map objects from spacetime to spacetime.

There are 64 functions which connect the basis vectors in one reference frame to those in another. Symmetry in the matrix reduces the number of functions to 40. The connections that are generated are called Levi-Civita connections.

Some of the functions can be antisymmetric and a tensor results called the torsion tensor, T.

$$\Gamma\mu\nu - \Gamma\nu\mu = T\mu\nu$$

In general relativity they are similar to gravitational force. If the symbol vanishes, goes to zero, then the reference frame is that of a body in free fall. This can always be found for any object and then the geodesic is known.

There are two types of Christoffel symbols. Γ_{ijk} is the first kind and $\Gamma_{ij}{}^k$ is the second which are the connection coefficients or affine connections seen in Evans' work. "Affine" refers to geometry intermediate between projective and Euclidean. It is a geometry of vectors without origins, length, or angles. These are the connections between the vectors in one reference frame and another. They connect the base manifold to the index.

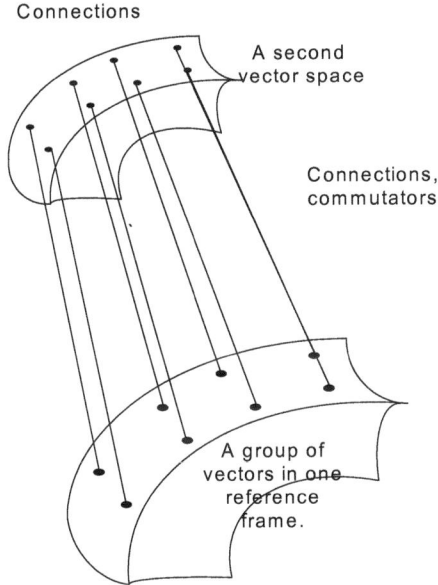

Circular Basis

The equations of electrodynamics have a simple appearance when written in circular basis. The photon, a spin-1 particle, and the spin-1 quantum operators are diagonal when written in the circular basis. This makes circular-polarized radiation a simple basis for the photon field. Evans' $\mathbf{B}^{(3)}$ spin field and O(3) electrodynamics are written in circular basis.

Classical limit

Classical refers to Newtonian physics. Our large flat spacetime macro world we experience in normal living is classical. When any relativistic quantum wave equation is reduced to flat space, we say that we have gone to the classical limit. It is easier to use the classical for mechanical explanations.

Clifford algebra

In orthogonal groups, a linear map can be defined which associates that group space with a linear vector subspace. A matrix can be built from vectors and spinors using Clifford algebra.

Commutators

In quantum theory, the commutator is defined $[A,B] = AB - BA$. These operators are antisymmetric. $[u,v] \equiv [\partial^\mu \partial_\nu] \equiv \partial^\mu \partial_\nu - \partial_\nu \partial^\mu$. The commutator is also called the Lie bracket. It is zero only if A and B commute. The Heisenberg uncertainty principle is a theorem about commutators.

Complex conjugate

The complex conjugate of $a + xi$ is $a - xi$.

Complex Numbers

$i = \sqrt{-1}$. Complex numbers exist physically in the real universe in as much as they affect actions. In quantum theory they are relevant and required to predict and explain processes. The argument has been made by Steven Hawking that time is -1 such that $ds^2 = (ct)^2 + x^2 + y^2 + z^2$. The subject is not closed. If as this author thinks, time is truly a spatial dimension, then Hawking's idea may be clarified and proved to be correct.

In the complex numbers, a rotation by i is equivalent to a 90° rotation. Therefore two 90° rotations, $i \times i = -1$, is equivalent to a 180° rotation. The i rotation may be movement in the 4th spatial dimension.

Complex numbers are often used to describe oscillation or phase shifts.

Components

A vector is a real abstract geometrical entity while the components are the coefficients of the basis vectors in some convenient basis. The vector can move from reference frame to reference frame. Components are tied to just one frame. Strangely at first sight, vectors are more real than a spacetime, say the one you are in while reading this.

Compton wavelength

$\lambda_c = h/mc$

This is the wavelength of a photon containing the rest energy of a particular particle. As the particle energy goes up, the wavelength goes down as the frequency goes up. See de Broglie wavelength also.

Connection

Given a metric on a manifold – that is, given a group of points or events with definite positions and times in a space – there is a unique connection which keeps that metric. A connection transports the vectors and tensors along a path to any new manifold. In SR the transport is simple using Lorentz transformations and only 2 dimensions are affected. A box accelerated to .87c will shrink down to .5 in the time and direction of travel. The connection will be x' = .5x.

In general relativity and curved spacetime the Levi-Civita connection and a corresponding curvature tensor is one method. Christoffel symbols also provide the connections. The Jacobi is another. Any spacetime affected by mass or energy will shrink, twist, deform, or change in a regular pattern that can be found and calculated. Curvature tensors, matrices, and vectors provide the actual mathematic tools to do so.

Elie Cartan worked out an approach to the idea of connection that used his method of moving frames.

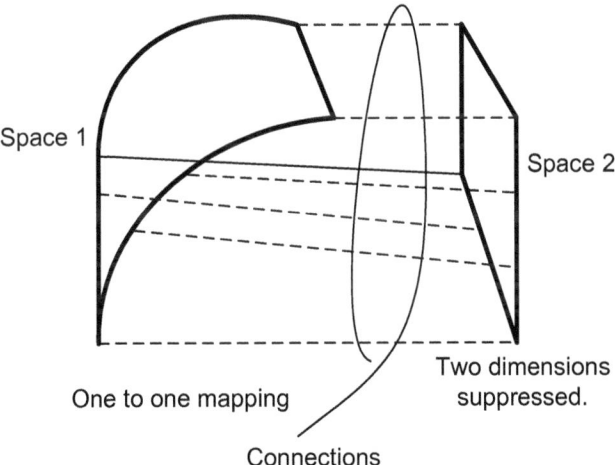

Contraction

A 1-form operating on a vector is contraction. If a de Broglie wave and a 1-form are considered together, the figure below shows how the 1-form crosses it each cycle. This is just one depiction. See Gradient and One-form also.

Contravariant

The word refers to the method and direction of projection of a vector or tensor onto a coordinate system. The coordinate system is used as the basis vectors. The origin of the coordinate axis is put at the tail of a vector. Then parallel lines are drawn to find the axes.

A contravariant vector has four perpendiculars to the base manifold of our real universe. It is in a mathematical vector space. If that vector were moved to a warped high gravitational region it would change its shape to exactly duplicate itself with reference to the new shape of the base spacetime. There are basis vectors: A=(1,0,0,0); B=(0,1,0,0); C=(0,0,1,0); D=(0,0,0,1).

The new spacetime may be twisted and scrunched by the gravitation so that the four dimensions would seem misshapen to someone far away in a more normal spacetime, but the basis vectors would be in proportion.

These vectors are tangent to the real spacetime of the base manifold. They are dual to covariant vectors.

Contravariant tensors and vectors

Contravariant vectors are tangent vectors on a manifold. A tangent space to a manifold is the real vector space containing all tangent vectors to the manifold. Contravariant tensors measure the displacement of a space – a distance.

Covariant tensors form a vector field that defines the topology of a space. A covariant vector is a 1-form, a linear real valued function on top of the contravariant vectors. These 1 forms are said to be dual or "to form a dual space" to the vector space. The full information about the curvature and distances can be drawn from either set of vectors or tensors.

Together they are invariant from one reference frame to another. Distances, volumes or areas are the same in any reference frame if found from the tensors.

If e^1, e^2, and e^3 are contravariant basis vectors of a space, then the covariant vectors are types of reciprocals. The components can be found from these vectors.

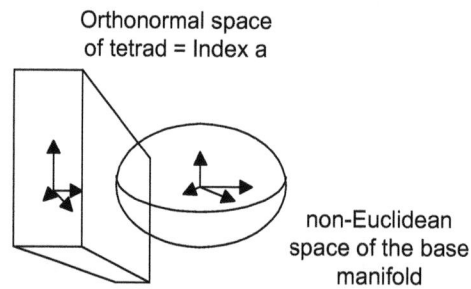

Orthonormal space of tetrad = Index a

non-Euclidean space of the base manifold

The curved spacetime depicted above is that of our real universe. The orthonormal space is the mathematical space. The tetrad relates both.

Contravariant vs. covariant

Contravariant is ordinary components / scale factors. First rank vectors have components proportional to the value of the velocity, force, acceleration, etc. which they describe.

Covariant are ordinary components x scale factors. First rank covariant vectors are proportional to the products of the elements of the other two axes.

The contravariant x covariant = 1.

Glossary

Scalars describe real physical properties.

Functions describe relations between physical properties.

Operators connect the functions.

A contravariant component is physical property / geometric function. Force = mass x acceleration is contravariant.

A covariant component is a physical quantity x a geometric function. Work = force x displacement is covariant.

Contravariant tensors (or vectors) have raised indices: a^β. These are the tangent vectors that define distances. Covariant tensors (or vectors) have lowered indices: a_β. These are used in other operations, but they give all the same information. To convert a covariant into a contravariant, the metric tensor can be used: $g^{\mu\nu} V_\nu = V^\mu$

Contravariant vectors	Tangent vectors. Mathematically, a tangent vector is a column vector, ket in quantum theory. Coordinate vectors are contravariant; they have upper indices. Their derivatives, in particular the 4-vector, are covariant; they have lower indices.
Covariant vectors	Linear forms. Dual or cotangent vectors. The dual vectors can be written in terms of their components. ω is frequently used to denote the dual vectors. One-forms are dual vectors. The action of a dual vector (field) on a vector (field) is to produce a function or scalar. The scalar is without indices and it is invariant under Lorentz transformations. Mathematically, a cotangent vector is a dual row vector, bra in quantum theory.
Cotangent space	T^*_p Any tangent vector space has a dual space. The dual space is the space with all the linear maps from the original vector space to the real numbers. The dual space can have a set of basis dual vectors. It is also called the dual vector space.

Contravariant vectors make up the vector space. They can be mapped to the linear space which has two operations - addition and multiplication by scalars.

The vector space has linearly dependent and independent vectors, and parallel vectors and parallel planes. The vector space does not have angles, perpendiculars, circles or spheres. The mapping of vectors onto scalars makes the dual linear space of covariant vectors. The dual space has the operations missing in the vector space. These are mathematical spaces, which allow us to derive covariant relationships.

The dual space contains the linear one-forms. Vectors are arrows; one-forms are planes. The gradient of a scalar function is a dual vector. This is the set of partial derivatives with respect to the spacetime. Mapping of one-forms onto vectors is called contraction.

Partial derivatives do not transform invariantly from one gravitational field to a different one. For this reason connections, tensor operators, or covariant derivatives are necessary. A covariant derivative acts like a regular partial derivative in flat space (Cartesian coordinates), but transforms like a tensor on a curved spacetime. ∂_μ is insufficient and we use the covariant derivative, del or ∇ to do this. It is independent of coordinates. ∇ is a map from one tensor field to another, from one gravitational spacetime to another. It is linear ($\nabla (A + B) = \nabla A + \nabla B$) and obeys the Leibniz product rule: $\nabla(S \times T) = (\nabla T \times S) + (\nabla S \times T)$. The x here is a tensor product.

The method to do this gets complicated. Given ∇_μ in four dimensions, the calculations require that for each direction in the spacetime, ∇_μ is the sum of ∂_μ plus a correction matrix. This is a 4 x 4 matrix called the connection coefficient or $\Gamma^\nu{}_{\mu\lambda}$. This is a long and tedious calculation and it is the concept we are most concerned with here, not the details.

Coordinate basis (Greek index)

The actual coordinates in the real spacetime of the universe. This can be a curved and twisted spacetime.

Correspondence principle

When any general theory, like Evans', is developed, it must result in the older established physics predictions. Einstein's relativity predicts Newton's equations for the orbits of planets in low gravity while adding spatial curvature in high gravity. Quantum theory predicts Maxwell's equations. Evans' equations reduce to all the known equations of physics.

Cotangent Bundle

This is a space containing all the covectors attached to a point in a manifold – a physical space. These are geometric objects that can be used to define forces, vectors, etc. and transform them to other places in the space.

The phase space is the cotangent bundle to the configuration space. This is a mathematical abstraction.

Covariant

Objective, unchanged by reference frame changes. It expresses Einstein's equivalence principle that the laws of physics have the same form regardless of system of coordinates. The coordinate system is used to find results, but the physical theory is pure geometry and needs no coordinates. The electromagnetic field tensor in terms of the four vectors is covariant or invariant under Lorentz transformations. The velocity of light is a constant regardless of the observer.

A covariant coordinate system is reciprocal to a contravariant system.

Glossary

A covariant tensor is like a vector field that defines the topology of a space; it is the base which one measures against. A contravariant vector is then a measurement of distance on this space.

$\int_{covariant}{}^{covariant}$ = invariant (e.g., mass)

From en.wikipedia.org/wiki/covariant.

Covariant derivative operator (∇ or D)

The covariant derivative replaces the partial derivative in multidimensional spacetime.

The standard partial derivative is not a tensor operator; it depends on the coordinates of the space. The covariant operator acts like the partial derivative in a flat space (the orthonormal tangent space or the flat coordinate space), but transforms as a tensor. It takes ∂_μ and generalizes it to curved space. ∇ is the covariant derivative operator. It is a map from one tensor field to another. It is the partial derivative plus a linear transformation correction matrix, the matrices are called connection coefficients, $\Gamma^\rho_{\mu\sigma}$. This is written as $\nabla_\mu V^\nu = \partial_\mu V^\nu + \Gamma^\nu_{\mu\lambda} V^\lambda$ where V is a vector.

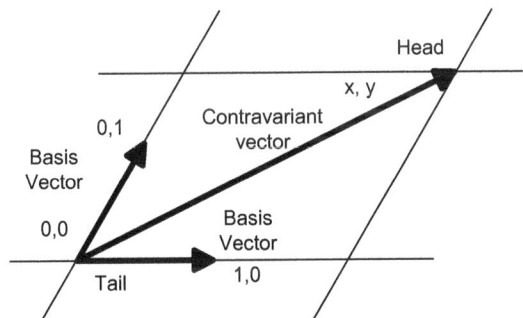

The covariant derivative is the derivative of a vector field as two points become closer together. It has components that are partial derivatives of its basis vectors. These all become zero at a point. The calculations using the covariant derivative and parallel transport allow one to analyze the curvature of the spacetime.

Upper indices are called contravariant because they transform contrary to basis vectors. Lower indices are called covariant. An (M, N) tensor is M times contra and N times covariant.

A symmetric covariant derivative is a rule producing a new vector field from an old one. The new vector field obeys several rules. ∇ = f (1 form, vector field, a vector) = new vector field. It summarizes the properties of all the geodesics that go through a point and provides parallel transport to allow comparison of the values of tensor fields and vector fields at two adjacent points.

The covariant derivative operator is D.

$DT^\lambda = dT^\lambda + \delta T^\lambda$ keeps DT^λ a tensor during transformations.

∇ is the operator on a Riemannian manifold. It was devised by Christoffel after Riemann devised the curvature tensor.

Covariant exterior derivative, D∧

An exterior derivative that is covariant. It provides a method for calculating Riemann curvature tensor and gauge fields, which are intimately related in Evans' equations. See exterior derivative. D∧ is the symbol for a covariant exterior derivative.

Covariant derivatives and covariant exterior derivatives are different. The covariant derivative is defined by parallel transport; it is not a real derivative. The covariant exterior derivative is an actual derivative. Covariant exterior derivatives are derived with respect to exterior products of vector valued forms with scalar valued forms.

Torsion never enters the exterior derivative and for that reason the covariant exterior derivative must be used for generally covariant physics.

The covariant exterior derivative takes a tensor-valued form and takes the ordinary exterior derivative and then adds one term for each index with the spin connection.

The covariant exterior derivative gives forms that look like one-forms or gradients.

Curl

$\nabla \times \mathbf{F}$ = curl (**F**). The curl of a vector field is the angular momentum or rotation of a small area at each point perpendicular to the plane of rotation.

Curvature

There are a variety of types of curvature. Keep in mind that the calculation itself is unnecessary for the non-physicist. The most startling thing in Evans' work on curvature is that wave number of quantum theory and curvature of general relativity are related. See Wave Number also.

Note that in Evans' usage, wave number is also defined $1/r$.

κ, lower case kappa, is used to indicate the magnitude of curvature. Curvature is measured in $1/m$ or $1/m^2$. $\kappa = 1/r$ and the scalar curvature is then $R = \kappa^2 = 1/r^2$.

A straight line has 0 curvature. A circle with radius r has curvature $1/r$.

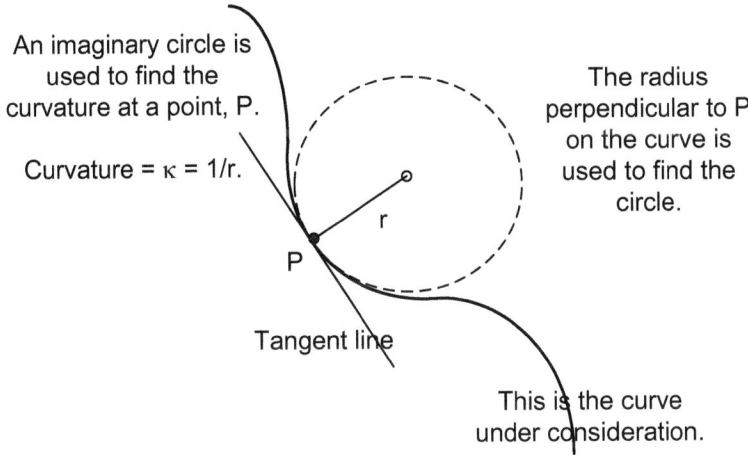

Curvature at a given point P has a value equal to the reciprocal of the radius of a circle that closely touches the curve at the point P; it is a vector pointing in the direction of that circle's center. The smaller the radius of the circle, the larger the magnitude of the curvature $1/r$ will be. When a curve is almost straight, the curvature will be close to zero. A circle of radius r has curvature of $1/r$ (or $1/r^2$ depending on the definition used).

In higher dimensions curvature is a tensor that depends on the Levi-Civita connection. This connection gives a way to parallel transport vectors and tensors while preserving the metric.

Extrinsic curvature and intrinsic curvature both exist. The extrinsic curvature is described by the Frenet formulas. These describe a spacetime curve in terms of its curvature, torsion, the initial starting point, and initial direction.

Gaussian curvature is detectable to two-dimensional inhabitants of a surface, where other types require knowledge of the three-dimensional space surrounding the surface on which they reside. Gaussian curvature describes the surface area of spheres from within.

Types of curvature include Riemann curvature, sectional curvature, scalar curvature, the Riemann curvature tensor, Ricci curvature, and others including maps, groups, and tensor fields.

The curvature tensor defines transport of a vector around a small loop that can be approximated by a parallelogram formed by two tangent vectors. Transporting vectors around this loop results in a linear transformation of the vector pairs. There is a matrix which defines change in a tangent space that result from the parallel transport along this parallelogram. The curvature tensor is antisymmetric.

Einstein equations of general relativity are given in terms of scalar curvature.

Mass, energy, and pressure cause spatial curvature. Exactly what is happening to the vacuum or as yet unknown substructure, we do not yet know. We can refer to all as mass or energy since they are equivalent.

Tensors are bold in the following. The density of mass is measured to be **T** by any observer. It is a geometric object and is measured in density/m^3 (or grams/m^3 or m/m^3 using geometric unit conversions). **T** is a tensor. The four velocity is used to arrive at the curvature. **T** = density / volume in the classical limit.

G is the Einstein tensor. It gives the average curvature in four dimensions at a point. k = 8π is a constant that is found from making the equation predict Newton's orbital laws.

G = 8π**T** is the Einstein equation that shows how curvature is produced from mass. Plug mass density into **T** and the curvature comes out after a bit of math. The units are inverse meters.

There are rules that govern how the tensor **G** is found from Riemann geometry. **T** is mass or the physical effect and **G** is mathematical curvature in spacetime or of the vacuum.

See http://mathworld.wolfram.com/Curvature.html for a good explanation.

Curved spacetime

Gravitation and velocity are curved space and time respectively or spacetime collectively. Spinning, twisted, or torqued spacetime is electromagnetism. This is also a form of curvature and is called torsion by Cartan. Torsion is sometimes called the "second curvature."

d'Alembertian

The symbol □ indicates the d'Alembertian which is the 4 dimensional Laplacian. $□^2$ is used by mathematicians to mean the same thing. □ is used by physicists. The partial derivative notation gives

$$□^2 = \nabla^2 - \frac{1}{c^2}\frac{\partial^2}{\partial t^2}$$

here the $1/c^2$ converts units to S.I. and can also be seen as converting time into an area or distance where c is the velocity of electromagnetic waves.

The d'Alembertian calculates the difference between the value of a scalar at a point and its average value in an infinitesimal region near the point.

Depending on the specific equation, the difference between the value of a function at a point and the average value in the region of the point is proportional to other changes such as acceleration of the wave. See Laplacian.

de Broglie wave

A matter wave. The quantum wave related to a particle. An electron has a standing de Broglie wave at the Bohr orbit distance. These are real and also mathematical waves.

de Broglie wavelength of a particle

$\lambda_{de\ B} = h/p = h/mv$. The equation says that the wavelength of a particle is equal to h divided by its linear momentum. Interestingly, $\lambda_{de\ B}$ can be found applying a one-form to the velocity vector – see a picture at Contraction.

Derivative

This gives the slope of a curve at any given point. A simple mathematical operation that allows one to see the direction in which one is moving. Since curves continuously change direction, the slope changes at each point. Slope is a decent physical interpretation. Strict mathematical definitions involve the size of the region or point and the directional derivative.

Differential forms

These are totally antisymmetric tensors used in exterior calculus. Electromagnetism is described well by differential forms. These are the wedge products of two one-forms that produce tubes, egg crate, or honeycomb structures that twist or turn through spacetime. One forms are families of surfaces as in a topological map.

Dimension

While mathematically well defined, physically it is not well defined. We are familiar with the three dimensions we live in and with the time dimension that has firm psychological meaning. In physics, time is a bit vague as a dimension. In general relativity, time and space are the same thing. There is no school that firmly says there is no such thing as time, but rather a change in position in space that allows evolution. In quantum theory time is ill defined. We are able to use it in equations to predict results, but just what it is remains vague.

Dirac equation

A 3+1 dimension special relativistic version of the Schrodinger equation. It predicts the antielectron. It was not usable in general relativity until Evans derived it from his wave equation.

$$i\hbar \frac{\partial \psi}{\partial t} = \frac{\hbar c}{i} \left(\gamma_1 \frac{\partial \psi}{\partial x^1} + \gamma_2 \frac{\partial \psi}{\partial x^2} + \gamma_3 \frac{\partial \psi}{\partial x^3} \right) + \gamma_4 mc^2 \psi$$

γ_n are the Dirac spinor matrices.

$i = \sqrt{-1}$, $\hbar = h/2\pi$, ψ is the wave function, t is time, m is mass of a particle

See http://mathworld.wolfram.com/DiracMatrices.html

The Evans' general relativity version is:

$$(\Box + kT)\psi = 0$$

where $kT = (mc/\hbar)^2$ and ψ is the dimensionless metric four spinor.

Directional derivative operator

This is the same as a tangent vector. $U = \partial u = (d/d\lambda)$

Dot product
Called the inner product in four dimensions. It is the length of projection of one vector upon another when the two vectors are placed so that their tails coincide.

Dual
Dual vectors = one-forms = covectors = covariant vectors. See one-forms also.

A set of basis 1-forms, ω, that are dual to a set of basis vectors denoted **e**. *J is dual to J is indicated by the asterisk. The dual space of V, the vector space, contains continuous linear functions mapped to real numbers. In tensors, elements of V are contravariant vectors, and elements of V* are covariant vectors or one-forms. Contravariant vectors are row vectors and dual vectors are column vectors.

The maps formed by the dual vectors are themselves a vector space.

Dual vectors have lower indices; vectors have upper indices.

e
The charge on an electron. $e = 1.60 \times 10^{-19}$ Coulomb.

ε_o
The permittivity of free space vacuum = 8.854×10^{-12} C^2 N^{-1} m^{-2} or C^2/J-m = per volt.

Eigen equations, eigenfunctions, eigenvalues, eigen vectors
"Eigen" can be read to mean "discrete, real, and proper."

An eigenfunction is a well behaved linear equation that has a real solution. They produce eigenvalues which are scalars. Matrixes are often used to express the set of equations with the diagonal values being eigenvalues.

When the Schrodinger wave equation is solved for a particle influenced by a force and a has a known energy as a function of its position, the permitted energies are called eigenvalues. ψ, the wave functions are called eigenfunctions. The boundary conditions must be known.

The only real solutions in a bound system like an electron are standing waves. The wave-particle must satisfy the boundary conditions that allow quantized energies obeying $\lambda = \hbar/p$. The wave must connect back onto itself at all points; this limits the size or wavelength to certain values.

The eigenvalues define the set of solutions to the Evans wave equation and the eigenfunctions are the corresponding functions. In the present standard theory, the electromagnetic field is Abelian, while the other three are non-Abelian. This becomes clear as soon as we start to use a metric four vector. In existing theory the gravitational field is the only generally covariant sector; the other three fields are not generally covariant. All four fields must be generally covariant to be the real description of physics. Evans has achieved this.

An eigenvalue is proper or real value that will result in a correct solution.

An eigen equation is a correct or proper equation.

Eigenvectors are solutions that result from use of the eigenvalues for an equation.

For example, let **Ax = y** where **x** and **y** are vectors and **A** is a matrix. This is a linear equation. The solution involves letting A**x** = λ**y** and rearranging to get (A -λ**I**)**x** = 0. The λ that satisfy the equations are eigenvalues of the matrix **A**. The eigenvectors are the solutions.

The scalar λ is called an eigenvalue of matrix A if a vector exists that solves it.

Every vector that solves the equation is an eigenvector of A associated with eigenvalue λ. The equation is called an eigenequation. Eigenvectors corresponding to the eigenvalues are linearly independent. An eigenspace is an invariant subspace.

Einstein-Cartan Theory

Cartan proposed including affine torsion to general relativity. Einstein's original concept was a metric theory. Evans' theory agrees completely and adds details that allow unification. See http://en.wikipedia.org/wiki/Einstein-Cartan_theory.

Einstein's constant

$k = 8\pi G/c^2$. The 8π is a constant that makes the Newtonian formulas result. The Einstein constant times the stress energy tensor results in curvature in mathematics as in R = -kT.

Einstein Field Equation

The equations generated from $G_{\mu\nu} = 8\pi T_{\mu\nu}$ are 16 nonlinear partial differential equations. G is curved spacetime. $T_{\mu\nu}$ is the stress energy tensor.

$G_{\mu\nu} = R_{\mu\nu} - \frac{1}{2} g_{\mu\nu}R$ is the Einstein tensor with $R_{\mu\nu}$ the Ricci tensor and $g_{\mu\nu}$ the metric tensor and R is the scalar curvature. The 10 symmetric elements define gravitation.

There are initially 256 equations but symmetries give terms that subtract each other so that the number of terms that need be evaluated drops. The Ricci curvature controls the growth rate of the volume of metric spheres in a manifold.

The torsion tensor = 0 in this formulation. By going a step backwards, using the metric vectors, and coming forward again, Evans shows that the 6 antisymmetric elements define electromagnetism.

Einstein group

The general linear group $GL(4.R)$ is called the Einstein group in physics. These are the spacetime transformations of general relativity. It is equivalent to tensor representations and half-integral spins. It allows continuous linear transformations of 4-dimensional spacetime. The rotations of the Lorentz group SO(3,1) are subgroups of the Einstein group.

Einstein tensor

Tensors are bold in this paragraph. **G** is the Einstein tensor. It is derived from the Riemann tensor and is an average curvature over a region. In the general theory of relativity, the gravitational tensor **G** = $8\pi GT/c^2$. **G** and **T** are tensors. With G and c = 1, **G** = 8π**T**. The bold **G** above is the Einstein tensor, typically not bold in modern notation. The G is the Newton gravitational constant. **T** is the stress energy density, that is the energy density.

Electric Field

The electric field is not well defined classically – we cannot define a proper picture of what it looks like. It is associated with vacuum polarization and may be symmetric spacetime spin. If spinning in one direction it is positive and in the other it would be negative. Since spin could occur in four dimensions, where the spinning occurs is not yet defined.

It is a force field which accelerates an electric charge in spacetime. Electric charges cause electric fields around them. The electric field, **E**, is a vector with a magnitude and a direction at each point in spacetime. The force is described by **F**= q**E** where q is the numerical value of the charge.

The field is a region of spacetime curved and twisted by the presence of charge and mass – there can be no force without the potential presence of mass. There is a system comprised of two coupled components - the mass and an interacting gradient in a potential flux. dv/dt involves a gradient in that flux interaction with the mass. Mass is a necessary component of force, and without mass present there cannot be a force present.

Mathematically we can deal with the field well, pictorially or classically we are still insufficient. Maybe this is because there is no picture, but this author thinks it is because we still have insufficient knowledge.

Electromagnetic field

Asymmetric spacetime is a correct physics definition.

Professor Evans does not completely agree with the following description. The engineer in the author sees standing photons and twisted spacetime as the electromagnetic field. Electromagnetic interactions occur between any particles that have electric charge and involve production and exchange of photons. Photons are the carrier particles of electromagnetic interactions and can be seen as bits of twisted curvature. This definition of photon's as electromagnetic field lines needs full examination. Inside a bar magnet for example, the presence of an electromagnetic field is not necessarily photons. Electrical engineers envision the magnetic field as photons.

Elementary Particles or Fields

Electron, nucleon (proton and neutron), neutrino, and photon.

Quarks and gluons are probably forces or fields, not actual particles. This is indicated by Evans and others' work, but is not fully accepted yet. They will be less than 10^{-19} meter in radius and exist alone for almost no time-distance.

Equivalence principles

There are weak, strong, and very strong equivalence principles and a few variations in between.

The weak equivalence principle equates gravitational and inertial mass. The strong or Einstein equivalence principle states that the laws of physics are invariant in any reference frame. The invariant interval is one result. This is the principle of relativity or "the general principle of covariance."

The very strong equivalence principle states that all mass, energy, pressure, time, spacetime, and vacuum are identical at some level. A unification theory assumes this. Reference frames are covariant. Tensors are the method of making the transformations from one reference frame to another invariant.

Evans states that all forms of curvature are interchangeable. This author states that the Very Strong Equivalence Principle exists and that all forms of existence are interconvertible and are conserved. Professor Evans has not disagreed.

Euclidean spacetime

Flat geometric spacetime. Real spacetime is curved by presence of energy. Mathematics is simpler in Euclidean geometry. The abstract index of quantum mechanics = the tangent spacetime of general relativity and is Euclidean.

Evans Spacetime

Curved and spinning spacetime. Riemann curvature is gravitation and Cartan torsion is electromagnetism. The Evans spacetime has a metric with both curvature and torsion. This is the unified spacetime. Riemann spacetime is symmetric and curved, but cannot support a description of the electromagnetic field. An antisymmetric metric is a two dimensional space called spinor space. This is needed to describe electromagnetism.

Event

Four-dimensional point.

Exterior derivative

"d" differentiates forms. It is a normalized antisymmetric partial derivative. d is a tensor unlike the partial derivative. The gradient is an exterior derivative of a one-form. d (dA) = 0

This is a type of vector. Rather than the arrow type, it is the *contour* of equal vectors. See one-form and gradient for picture. A topological map is typical. The bands of equal elevation are equivalent to the exterior derivative boundaries. If one works in three dimensions, a concentric onion shell shape would be similar to those produced by an exterior derivative. The exterior derivative is mathematical; it exists in the tangent space, not in real spacetime.

Special cases of exterior differentiation correspond to various differential operators of vector calculus. In 3-dimensional Euclidean space, the exterior derivative of a 1-form corresponds to curl and the exterior derivative of a 2-form corresponds to divergence.

Fiber bundle

In differential geometry a bundle is a group of "fibers" or mathematical connections that map vectors or products from a manifold to a space. It is a generalization of product spaces.

The tangent bundle is the group formed by all the tangent vectors. The cotangent bundle is all the fibers that are dual to them. The vector bundle is the vector space fibers. Operations like tensor products can be performed as though the spaces were linear.

In mathematics a fiber bundle is a continuous map from a topological space A (reference frame in physics) to another topological reference frame space B. Another condition is that it have a simple form. In the case of a vector bundle, it is a vector space to the real numbers. A vector bundle must be linear from the one space to the next. One of the main uses of fiber bundles is in gauge theory.

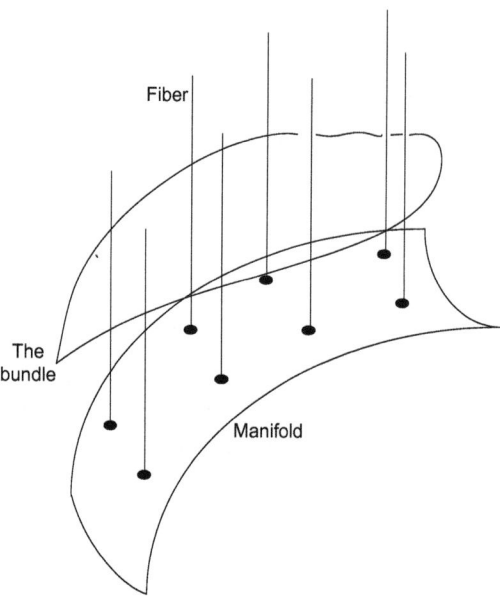

Fiber bundle of gauge theory

Given two topological spaces, A and B, a fiber bundle is a continuous map from one to the other. B is like a projection. For example if A is a human body and B is a shadow, then B would be the fiber bundle. The ground would be another topological space, C. The space C could be a vector space if the shadow is a vector bundle. It is the mapping.

Gauge theory uses fiber bundles. See tetrad for the mapping.

Field equations

Gravitation and electromagnetism can be described as fields rather than curvature and torsion. Poisson's equations are discussed in Chapter 3. Einstein's field equation gives curvature in terms of the stress energy via the Ricci and Riemann tenors. See Chapter 4.

Fluxon

The elementary unit of magnetic flux.

Force

$F = ma$ is the most common force equation in physics. A force is exerted by curvature or torsion. In one view, each of the four forces has a carrier particle – gravitation by the graviton, magnetic fields by the photon, the strong force by the gluon, and the weak force by W^+, W^-, and Z. Interactions are the results of force.

Fields explain "action at a distance." The force carrier moves into other regions.

Force	Gauge boson	Gauge group	Comment
Gravitation	Graviton	Evans	Theoretical
Electromagnetism	Photon	O(3)	Quantum Electrodynamics (QED)
Weak nuclear	W^+, W^-, Z.	SU(2)	Short range electrical interaction
Strong nuclear	8 gluons	SU(3)	Theoretical, QCD

The SU(2) and (3) symmetries will probably be dropped from the standard model in the future. The predictions of the quantum theories are mostly accurate. The explanations "why" are inaccurate. Note that the graviton and gluons have never been found.

Forms

The form number is the number of indices. The common ones used are:

Scalars are 0-forms

A one-form in Euclidean space is a vector,

The tetrad is a one-form,

The spin connection is a one-form but not a tensor.

Dual vectors are one-forms,

The Hodge dual of the wedge product of two one-forms is another one-form. In vector notation
    this is a cross product.

The antisymmetric tensor is a two-form,

The torsion tensor is a vector valued two-form.

The Riemann form is a (1, 1) tensor valued two-form.

$q_{\mu\nu}$ is a scalar valued two-form. These are two-forms due to the two subscripts and scalar valued with no superscript.

$q^c{}_{\mu\nu}$ is a vector valued two-form as seen by the subscripts and vector valued with one superscript.

$q^{ab}{}_{\mu\nu}$ is a tensor valued two-form. The two superscripts indicate tensors.

Current density is a three form

The Levi Civita tensor in four dimensions is a four-form.

A differential p form is an antisymmetric tensor.

Forms can be differentiated and integrated without additional geometrical information.

The wedge product of a p form and a q form is a p + q form.

The torsion and Riemann forms characterize any given connection.

Four-momentum

The energy momentum four vector is: $\rho^\mu = mU^\mu$. See four velocity below.

Four-vectors

In the spacetime of our universe four dimensional vectors are needed to describe physics. At each point in spacetime there is a set of all possible vectors at that point. They are located in the tangent space, T_p. These are not point to point vectors.

Four-velocity

A particle moves through spacetime with proper (its own) time τ. The tangent vector is the four-velocity or $U^\mu = dx^\mu/d\tau$. It is negative since timelike vectors are always negative. The four-velocity is separated into spatial (dx) and time (dτ) components. Then the spatial change is divided by the time change.

The four-acceleration vector is an invariant quantity. The 4-acceleration of a particle is always orthogonal to its 4-velocity.

Frequency

$$f = \nu = \omega/2\pi\nu = 1/T = v/\lambda$$

Glossary

The Fundamental Forces of Nature

Force & Symbol	Stanrd Model	Evans Model
Gravitation, **e**	Graviton (purely theoretical)	Spacetime Curvature (Symmetric Curvature)
Electromagnetism, **A**	Photons and charge	Spacetime Torsion (Antisymmetric spin)
Strong force, **S**	Gluons, color force, pions?	Form of gravitation (Symmetry needs exploration)
Weak force, **W**	W&Z bosons	Form of torsion (Symmetry needs exploration)

The Cartan tetrad, q^a_μ, within the Evans wave equation, $(\Box + kT) q^a_\mu = 0$, allows representing all the forces in the same basic equation:

$(\Box + kT) e^a_\mu = 0$ **e** is a basis vectors of gravitation

$(\Box + kT) W^a_\mu = 0$ **W** can be an SU(2) 4-vector

$(\Box + kT) S^a_\mu = 0$ **S** can be an SU(3) 3-spinor

$(\Box + kT) A^a_\mu = 0$ **A** is the electromagnetic potential.

Fundamental Particle

A particle with no internal substructures. In the Standard Model, the quarks, gluons, electrons, neutrinos, photons, W-boson, and Z-bosons are considered fundamental. All other objects are made from these particles. It may be that some of these are energy states, not particles. See particles. Evans indicates that the quarks and gluons are not fundamental.

G

G = Newton's gravitational constant = 6.672×10^{-11} Nm² kg⁻²

Gauge theory

Physics is invariant under small local changes. Symmetry transformations are local. If you rotate a particle in one region, it does not change spacetime in another region. A gauge transformation is a transformation to the allowable degree of freedom which does not modify any physical observable properties.

Gauge is a small rotation or change which is used to measure a particle's state.

Gauge refers to freedom within a theory or mathematical structure or an object. An action like a rotation that causes no effect has "gauge freedom." A circle may be rotated and there is no observable effect. Same for a sphere.

A gauge field G^a_μ is invariant under the gauge transformation. Differential geometry is used to analyze gauge theories.

If a continuous, differentiable base spacetime bundle is a Lie group, then the bundle itself is a group of gauge transformations. The spacetime is such that an exterior derivative - Dv, the covariant derivative, 1-forms, and a wedge product exist and are valid operations.

Movement from one reference frame to another in relativity is gauge invariant.

A gauge transformation is defined as $A_m = A_m + \partial_m I$. Thus a small change in a value is inconsequential. One can change the potentials of a system so that the system remains symmetrical in its field.

Electromagnetics was the first gauge theory. Potentials are not gauge invariant.

The phase angle of a charged particle has no significance and can be rotated. Coordinates can be re-scaled without effect. Gravitational gauge transformations do not affect the form of scalars, vectors, tensors, or any observables. The Riemann, Einstein, and stress energy tensor T are gauge invariant.

Physically meaningful statements are relational and are "gauge invariant." Geometry is the same regardless of the reference frame in which it is viewed. This was Einstein's principle of relativity.

Gauge invariance may become a side issue as Evans' equations are developed.

Gauss' laws

The electric flux through a closed surface is proportional to the algebraic sum of electric charges contained within that closed surface; in differential form, div $E = \rho$ where ρ is the charge density.

Gauss' law for magnetic fields shows that the magnetic flux through a closed surface is zero; no magnetic charges exist; in differential form, div $B = 0$.

Generally covariant electrodynamics

Group structure of generally covariant electrodynamics is non-Abelian and generally covariant electrodynamics must be a gauge field theory with an internal gauge gravitational field is described through the symmetric metric tensor.

Geodesic

Line of shortest distance between two points. On a sphere this is a great circle route. In a complex gravitational field, this is a straight line as observed from within the reference frame, but a curve as seen from a large enough distance away. An inertial reference frame is one that has no external forces affecting it; it moves along a geodesic in whatever spacetime it finds itself. E.g., a particle "attracted" by a gravitational field actually follows a straight line in its viewpoint'. That line is called a geodesic and can be a curve as viewed by an outside observer.

Geometric units

Conversion of mass or time to units of distance. Setting $G = c = 1$ allows equivalence to be established. For example, with c, the speed of light = about 300,000,000 m/s, we can know that $c = 3 \times 10^8$ m. Another example, $G / c^2 = 7.4 \times 10^{-27}$ gm/cm which gives a conversion factor to change mass into distance.

Use of geometric units removes conversion factors which are meaningless in the abstract. They are only needed for exact calculations. For example, instead of Newtons law being $F = Gm_1m_2/r^2$, it becomes $F = m_1m_2/r^2$. (And Coulombs law becomes $F = q_1q_2/r^2$.) This more clearly says the attractive gravitational (electrical) force is proportional to the masses (charges) and inversely proportional to the distance squared.

Planck units are related to geometric units. Use of the Planck units is similar by setting h = 1 also. Then instead of $E = mc^2 = h\omega$, we would have $E = \omega = m$ tells us energy is frequency is mass.

Gluon

The hypothetical particle that intermediates the strong interaction. Given Evans' equations, it is likely that the gluon is a force field, not an actual particle. It is likely a curvature component. This is also supported by the fact that the masses of the quarks and gluons do not add properly to the particle total.

Gradient, ∇

A gradient is a 1-form, **df**, of a function, also called a differential exterior derivative. It exists in the tangent space, not the curved spacetime of the universe. It is the rate of change, the slope, at a point. *Gradient* refers to the change in a potential field. See also one-forms.

$$\text{Gradient of a function} = \text{grad } f = \nabla f = \frac{\partial f}{\partial x} \mathbf{i} + \frac{\partial f}{\partial y} \mathbf{j} + \frac{\partial f}{\partial z} \mathbf{k}$$

∇^2 is the Laplacian operator in 3 dimensions $= \frac{\partial^2}{\partial x^2} + \frac{\partial^2}{\partial y^2} + \frac{\partial^2}{\partial z^2}$

The Gradient refers to a gradual change in a potential field. A potential gravity gradient is regular and steadily changes from strong to weak as one moves away from the gravity source.

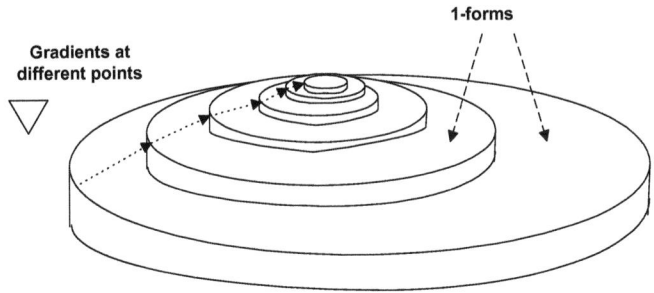

Gravitational field
In Newtonian physics the gravitational force is described as a gradient of a potential field. See Poisson's equations.

Greek alphabet

alpha	A α	nu	N ν
beta	B β	xi	Ξ ξ
gamma	Γ γ	omicron	O o
delta	Δ δ	pi	Π π
epsilon	E ε	rho	P ρ
zeta	Z ζ	sigma	Σ σ
eta	H η	tau	T τ
theta	Θ θ	upsilon	Y υ
iota	I ι	phi	Φ φ
kappa	K κ	chi	X χ
lambda	Λ λ	psi	Ψ ψ
mu	M μ	omega	Ω ω

Groups

The oversimplified basics are covered here without mathematical precision.

A group G is a collection or set of elements such as "a, b, c". The number of elements in the group is its order. a ∈ G means that a is an element of G. A group, G, is a set of at least one element with a rule for combining elements. The rule is called a product, a b, even if it is addition, rotation, etc. Any product must also be a member of the group. Identity exists such that I a + a I = a. a^{-1} exists so that a a^{-1} = a^{-1}a = I. Associative law holds: (a b) c = a (b c). Operations among the elements are called composition or multiplication, but not in the normal sense. There is only one operation defined for any given group. Group structure refers to a table of all the results of operations. This is like a multiplication table.

If G = (1, -1, i, -i) and ordinary multiplication is the product, then the product table is:

Element	1	-1	i	-i
1	1	-1	i	-i
-1	-1	1	-i	i
i	i	-i	-1	1
-i	-i	i	1	-1

SO(3) is part of the Special Orthogonal group. This is the group of matrices with a Euclidean metric. The groups Evans deals with are U(1), a circle, and O(3), the sphere. O(3) is a group of rotations in spacetime and is the simplest group that could be found that correctly describes electrodynamics. It is not impossible that in the future SO(3) would be found to apply better.

An example of how the group operates: assume A is a rotating helical line corresponding to a rotation of an angle α about the z-axis.

$$A(\alpha) = \begin{vmatrix} \cos \alpha & -\sin \alpha & 0 \\ \sin \alpha & \cos \alpha & 0 \\ 0 & 0 & 1 \end{vmatrix} = \text{the tangent vector}$$

Then at $\alpha = 0$, the tangent vector is calculated by finding the sine and cosine of 0 and substituting for A($\alpha = 0$) = tangent vector. Three vectors are found, one for x, y, and z. The process for calculation is tedious and must be repeated for every angle to find the line.

SU(2) is contained in the Special Unitary group of matrices and describes the electroweak force. Two spinors are described.

SU(3) describes the strong force as a gauge group. SU(3) derives from the three axes of rotation. The three unit particles of matter in the quark triplet substructure are as likely the mathematical recognition of the three axis of dimensionality. The existence of quarks themselves is suspect.

Gauge theory and symmetry are closely related with Lie (pronounced "lee") groups. The mathematics is beyond the scope of this book.

To meet the definition of a group there are five requirements:
1. A product operation is defined among elements.
2. Closure. All the elements must be members of the group.
3. Associativity. The operation a (bc) = (ab)c.
4. Identity element exists. There is some element, e, such that ae = ea = a.
5. Inverses exist. For a there is a^{-1}, an inverse such that a a^{-1} = e. (a^{-1} does not mean a reciprocal 1/a as in standard algebra.)

The group is Abelian if it commutes, but non-Abelian groups are more common. Commutation exists if ab = ba.

The most common group symmetry in physics is that of rotations about an axis.

Mapping M from A to B is denoted by M: A→ B. A homomorphism exists if the structure is preserved by a mapping. An isomorphism is a one to one mapping that is invertible.

See also Rotations and Symmetry

Hamiltonian
The energy of a system is found using the Hamiltonian equations.

Higher symmetry electromagnetism
Maxwell-Heaviside U(1) symmetry has been accepted for 100 years as the description of electromagnetism. It is Abelian and one dimensional which limits its use to flat space. This is perfectly adequate in normal life. However very close or inside a particle or in a strong gravitational field, the equations are inaccurate.

O(3) symmetry is needed to explain some phenomena and U(1) is only valid in the Minkowski spacetime of special relativity. O(3) electrodynamics is frame invariant in Evans' work.

Hilbert Space

A Hilbert space is a specific type of vector space. They are defined with certain characteristics such as the complex numbers with dot product and complex conjugates. Typically used in quantum theory. Hilbert space is purely abstract and infinite. See abstract space.

Hodge dual

Duality operation in three dimensions is a plane interchanged with a orthogonal vector in the right handed sense. Evans shows that electromagnetism and gravitation are dual to one another. For simple understanding substitute "perpendicular" for "dual."

The Hodge star operator can be defined as a wedge product, $\mu \wedge \nu$. This is an operation on an four-dimensional manifold giving a map between forms. The Hodge dual is like an orthogonal map from the dual to μ and ν which are one-forms.

[I asked Dr. Evans for a simple description of the Hodge dual. He replied in an email: "Thanks, this will be very useful. The Hodge dual is best described by Carroll in his lecture notes. I am not sure about pictures in this context because the concept applies to n dimensional spaces. In a well defined special case however the Hodge dual is the simple well known duality between an axial vector and an antisymmetric tensor (e.g. Landau and Lifhitz), so it applies in the context of spin to differential forms, which are antisymmetric by definition." So I have no simple description of the Hodge dual.]

Homogeneous

Homogeneous equations have typically defined radiation in free spacetime. Inhomogeneous equations refer to interaction of fields and matter. If energy from curved spacetime is possible, it is the inhomogeneous equations that will describe the process.

Homeomorphism

One to one correspondence between two objects or spaces. It is continuous in both directions. Isometric homeomorphism also has distances that correspond.

Homomorphism

Refers to group operations which maintain functions in two or more groups.

Hydrogen

A colorless, odorless gas which, given enough time, turns into humans.

Index

Typically Greek letters are used for four dimensions and Latin indices are used for Euclidean three spaces. Both upper and lower indices are used to indicate vector dimensions, tensor dimensions, etc.

Internal gauge spaces are indicated with an upper index.

Index contraction

This is a method of simplifying tensors by summing over one upper and one lower index. An (m, n) tensor becomes a (m-1, n-1) tensor. The Ricci tensor is a contraction of the Riemann tensor.

Internal index, internal gauge space

Gauge theory uses internal indices which Evans has identified with the tangent space of general relativity. O(3) electrodynamics has an internal index.

Invariance

Does not change. An invariant does not change under transformations due to velocity (curvature of time) or gravity (curvature of space). Mass is an example in general relativity.

Movement (space translation), time translation (T), and rotations are invariant.

In quantum mechanics, particle to antiparticle and the reverse is CP. It is true for all interactions except for a small group of weak interactions. Mirror reflection is parity (P). It is not true for weak interactions.

CPT is believed to be absolutely true.

When reference frames change due to say high gravitation or high velocity, there are some quantities that remain the same. Regardless of the reference frame, they will be measured to be the same. Among these are the invariant length of special relativity and the electromagnetic quantities $E \cdot B$ and $B^2 - E^2$. The value does not depend on the reference frame in which it was calculated.

The speed of electromagnetic waves, mass, the proper time between events, and charge are all invariant.

Acceleration is invariant under Galilean transformations. The speed of light is invariant under Lorentz transformations. Energy has time invariance, momentum has translational invariance, and angular momentum has rotational invariance. The invariant distance is the foundation of Minkowski spacetime. These are also conserved quantities.

Rest mass is an invariant and it is conserved. One could apply the Lorentz transformation to accelerated mass and calculate a higher value than the rest mass. This is an incorrect way of looking at it. The momentum changes, the total energy changes, but the rest mass is invariant and conserved. Imagine that if the mass actually increased at relativistic speeds, then at some velocity the particle would become a black hole. This does not occur.

It is known that mass and spin of a particle are invariants in special relativity. These are known as the Casimir invariants of the Poincaré group and are covered in Chapter 14. The spin of a particle is another invariant.

$|R_0|$ is rest curvature for the mass m and is invariant.

Certain mathematical process are invariant. Some are the trace and determinant of a square matrix under changes of basis, the singular values of a matrix under orthogonal transformations, and the cross-ratios of particles or objects.

In general, scalar quantities are invariant.

The total energy of a particle is conserved, but not invariant.

Consider a book on a scale weighing 1 kg on earth. We transport it to just outside the horizon of a black hole, recalibrate the scale and weigh it again. Depending on the mass of the hole, the weight increases. Weight is not invariant. But the mass of the book is the same. The laws of physics follow the equivalence principle. The mass must be the same regardless of the gravity or speed at which we measure it.

Conservation of energy, mass, momentum, angular momentum, electric charge, and color-charge have been identified. These quantities are invariant and are more "real" than quantities that vary.

Parity, direction, and acceleration are not invariant. The individual electric and magnetic potential field strengths are not invariant. However the combined electromagnetic field is invariant – it is the real entity.

A conserved quantity is never destroyed or created. Existence is conserved. The Evans formulation is that curvature is conserved. Total curvature may be spread out or combined in different patterns, but the total is constant.

Noether's theorem shows that invariants are usually conserved by symmetries of a physical system. Conservation laws show a correspondence between symmetries and conserved quantities. While many conserved quantities relate to a symmetry, mass had not yet been connected to symmetry until the Evans equations showed the relationship.

Conservation laws state that a property does not change as the system translates.

Inverse Faraday Effect (IFE).

Circularly or elliptically polarized light acts like a magnet upon interaction with matter. This is the Inverse Faraday Effect. Unpolarized light does not exhibit IFE.

Electrons are magnetized by the polarized light beam

Irreducible

Most basic or general form. The Einstein field equation can be further reduced to a simpler form – the Evans Field Equation. Evans is as far as we know, irreducible; the Einstein is not irreducible. An irreducible equation cannot be made any simpler and still explain all that it does.

Isomorphism

A 1:1 correspondence.

Jacobi matrix

This is the matrix of all first-order partial derivatives of a vector-valued function. It is the linear approximation of a differentiable function near a given point. $J^{\mu}_{\nu\alpha\beta} = J^{\mu}_{\nu\beta\alpha} := \frac{1}{2}(R^{\mu}_{\alpha\nu\beta} + R^{\mu}_{\beta\nu\alpha})$ where R is the Riemann tensor.

Klein-Gordon equation

$$E^2 \psi = \left(i\hbar \frac{\partial}{\partial t}\right)^2 \psi = (m^2c^4 + p^2c^2)\psi$$

The Klein-Gordon equation is Lorentz invariant but has negative energy solutions which did not have a known meaning until Evans reinterpreted. See Chapter 9. The probability density varies in time and space.

Kronecker delta

In mathematics, the Kronecker delta is a function of two variables. The answer is 1 if they are equal, and 0 otherwise. It is written as the symbol δ_{ij}, and treated as a notational shorthand rather than a function. $(\delta_{ij}) = 1$ if j=i and 0 if j not = i. δ_{ij} is the identity matrix.

Lagrangian

A Lagrangian is a function designed to sum the total energy of a system in phase space. The Lagrangian is usually the kinetic energy of a mechanical system minus its potential energy.

Laplacian

The Laplacian ∇ for a scalar function is a differential operator.

∇^2 is the Laplacian operator in 3 dimensions = $\frac{\partial^2}{\partial x^2} + \frac{\partial^2}{\partial y^2} + \frac{\partial^2}{\partial z^2}$

The Laplacian is used in electromagnetics, wave theory, and quantum mechanics. Laplace's equation is $\nabla^2 \Phi = 0$ or $\nabla^2 \Psi = 0$. The same operator in four dimension spacetime is the d'Alembertian. It can be used to find the largest value of the directional derivative or perpendiculars to curves or surfaces. For example, the average value over a spherical surface is equal to the value at the center of the sphere.

The Laplacian knows that there is no difference between the value of some function at a point and the average value in the region of the point. It sneaks up on the point to get the slope.

It can be found in:

Poisson's equation: $\nabla^2 \Psi$ = source density

Wave equation: $\nabla^2 \Psi = \frac{1}{v^2} \frac{\partial \Psi}{\partial t^2}$

Schrodinger's Equation: $\nabla^2 \Psi_n = \frac{2m}{\hbar} (V - E) \Psi_n$

This says that the difference between the value of a function at a point and the average value in the neighborhood of the point is proportional to the acceleration of the wave. It is a measure of the difference between the value of a scalar at a point and its average value in an infinitesimal neighborhood of the point.

The four-dimensional version of the gradient is:

$$\nabla_\mu = \begin{bmatrix} \frac{1}{c}\frac{\partial}{\partial t} \\ \frac{\partial}{\partial x} \\ \frac{\partial}{\partial y} \\ \frac{\partial}{\partial z} \end{bmatrix}$$

Law of Conservation

Space, time, energy, mass may be converted into one another, but never destroyed or created. Sometimes confused with invariance.

Levi-Civita symbol

Riemann tensors are formed from symmetric Levi-Civita connections. These have no torsion. Antisymmetric Cartan connections have torsion but no curvature. Neither Einstein nor Cartan put these together although both knew torsion was electromagnetism. The Levi-Civita tensor is covariant is often called the permutation tensor.

The contorsion tensor $K_{\mu\nu}$ is the difference between the Cartan and the Levi-Civita connections: $K\mu\nu = \Gamma\mu\nu - \Gamma\nu\mu$

Contorsion = 64 Cartan – 40 Levi-Civita functions = 24 functions

Levi-Civita is defined: $\varepsilon^{ijk} = \varepsilon_{ijk}$

$\varepsilon_{123} = \varepsilon_{231} = \varepsilon_{312} = 1$ and $\varepsilon_{132} = \varepsilon_{213} = \varepsilon_{321} = -1$

It is antisymmetric in the permutation of any two indices. It can be used to define the cross product of vectors, to generate an antisymmetric rank two tensor from an axial vector.

Lie group

A Lie group is a manifold whose elements form a group. The matrices of the group change vectors into other vectors. The *adjoint* matrix has a dimensionality equal to the group manifold. The mathematical formalism is beyond the scope of this book.

If fields form a non-Abelian Lie group with antisymmetric connections, then torsion exists.
See http://en.wikipedia.org/wiki/Lie_group for more information.

Linear transformation

If A and B are vectors such that Ax = B, then they are linear. Any operation where a number, vector, matrix, or tensor is multiplied by a scalar, produces a linear transformation. A graph will give a straight line.

Local

A very small region of spacetime. One can assume the dot is local. In highly curved spacetimes, say the surface of a black hole, a local region can be considered using special relativity where any larger region must use general relativity.

Lorentz group

SO(3, 1).

μ_o

Mu sub zero. Vacuum permeability = 1.26×10^{-6} m kg C^{-2}

Manifold

A manifold is a mathematical abstraction. It is a continuous or smooth surface or volume of spacetime. In general relativity it is fully differentiable, that is it has distances that are clearly distinguishable. The surface of the earth is a two dimension manifold; it has two internal or intrinsic curved dimensions. It has three extrinsic dimensions when viewed from outside. The spacetime vacuum of our universe is a manifold with four dimensions.

Mathematical operations can be performed on the abstract manifold and as far as we know, these are then valid in considering the real universe spacetime. The manifold in Evans' work is simply connected – see AB effect.

See Analytical functions.

Matrix

A matrix is a group as shown here. A matrix can be made from other matrices. For example, let M be a matrix as shown below. It can be composed of the four matrices as shown.

$$M = \begin{vmatrix} A1 & B1 & C1 & D1 \\ A2 & B2 & C2 & D2 \\ A3 & B3 & C3 & D3 \\ A4 & B4 & C4 & D4 \end{vmatrix}$$

$$M_{11} = \begin{vmatrix} A1 & B1 \\ A2 & B2 \end{vmatrix} \qquad M_{12} = \begin{vmatrix} C1 & D1 \\ C2 & D2 \end{vmatrix}$$

$$M_{21} = \begin{vmatrix} A3 & B3 \\ A4 & B4 \end{vmatrix} \qquad M_{22} = \begin{vmatrix} C3 & D3 \\ C4 & D4 \end{vmatrix}$$

A symmetric matrix has the same elements symmetric with respect to the diagonal. The trace of a square matrix is the sum of its diagonal elements. The trace of matrix M(s) below is 4.

$$M(s) = \begin{vmatrix} 1 & B & C & D \\ B & 1 & E & F \\ C & E & 1 & G \\ D & F & G & 1 \end{vmatrix}$$

A transpose of a matrix would have elements which are transposed symmetrically across the diagonal. In M here, swap A2 and B1, swap A3 and C1, etc. to make the transpose.

$$\begin{vmatrix} A1 & B1 & C1 & D1 \\ A2 & B2 & C2 & D2 \\ A3 & B3 & C3 & D3 \\ A4 & B4 & C4 & D4 \end{vmatrix} = \begin{vmatrix} A1 & 0 & 0 & 0 \\ 0 & B2 & 0 & 0 \\ 0 & 0 & C3 & 0 \\ 0 & 0 & 0 & D4 \end{vmatrix} + \begin{vmatrix} 0 & B1 & C1 & D1 \\ A2 & 0 & C2 & D2 \\ A3 & B3 & 0 & D3 \\ A4 & B4 & C4 & 0 \end{vmatrix}$$

$$M(A) \qquad = \qquad M(S) \qquad + \qquad M(A-S)$$

Asymmetric equals Symmetric plus Anti-Symmetric

Matrix multiplication

In particular, when tetrads are multiplied, we see that each element of the first tetrad affects each element of the second tetrad. Multiplication is defined as below.

For M1 M2 = M3, let

$$M1 = \begin{vmatrix} A1 & B1 & C1 & D1 \\ A2 & B2 & C2 & D2 \\ A3 & B3 & C3 & D3 \\ A4 & B4 & C4 & D4 \end{vmatrix} \quad M2 = \begin{vmatrix} A1' & B1' & C1' & D1' \\ A2' & B2' & C2' & D2' \\ A3' & B3' & C3' & D3' \\ A4' & B4' & C4' & D4' \end{vmatrix}$$

$$M3 = \begin{vmatrix} X1 & X2 & X3 & X4 \\ & \cdots & & \\ X13 & X14 & X15 & X16 \end{vmatrix}$$

The point is the resulting matrix is built of the rows of the first times the columns of the second. The result is that each element of M1 affects each element of M2.

X1 = A1 A1' + B1 A2' + C1 A3' + D1 A4'

X2 = A1 B1' + B1 B2' + C1 B3' + D1 B4'

.....

X16 = A4 D1' + B4 D2' + C4 D3' + D4 D4'

Another use is with the Dirac and other matrices is operator equations. For example, formulation of spin ½ operations.

$$S = \hbar/2 \; \sigma \text{ where } \sigma \text{ is } \begin{pmatrix} 1 & 0 \\ 0 & -1 \end{pmatrix}$$

and therfore = $S = \hbar/2 \begin{pmatrix} 1 & 0 \\ 0 & -1 \end{pmatrix}$

This is used to arrive at spin up, indicated by + ½, and spin down, -½.

Maxwell's equations

The equations that describe electric (E) and magnetic (B) fields in spacetime.

Maxwells Equations
Differential forms   MKS System   Integral form
cgs   cgs

$\nabla \cdot \mathbf{E} = 4\pi\rho$ $\nabla \cdot \mathbf{E} = \rho/\varepsilon_0$ $\int_V \nabla \cdot \mathbf{E}\, dV = \oint_S \mathbf{E} \cdot d\mathbf{a} = 4\pi \int \rho\, dV$

$\nabla \times \mathbf{E} = -\dfrac{1}{c}\dfrac{\partial \mathbf{B}}{\partial t}$ $\nabla \times \mathbf{E} = -\dfrac{\partial \mathbf{B}}{\partial t}$ $\int_S (\nabla \times \mathbf{E}) \cdot d\mathbf{a} = \oint_C \mathbf{E} \cdot d\mathbf{s} = -\dfrac{1}{c}\dfrac{\partial}{\partial t} \oint_S \mathbf{B} \cdot d\mathbf{a}$

$\nabla \cdot \mathbf{B} = 0$ $\nabla \cdot \mathbf{B} = 0$ $\int_V \nabla \cdot \mathbf{B}\, dV = \oint_S \mathbf{B} \cdot d\mathbf{a} = 0$

$\nabla \cdot \mathbf{B} = \dfrac{4\pi}{c} \mathbf{J} + \dfrac{1}{c}\dfrac{\partial \mathbf{E}}{\partial t}$ $\nabla \cdot \mathbf{B} = \mu_0 \mathbf{J} + \mu_0 \varepsilon_0 \dfrac{\partial \mathbf{E}}{\partial t}$ $\int_S (\nabla \times \mathbf{B}) \cdot d\mathbf{a} = \oint_C \mathbf{B} \cdot d\mathbf{s}$

where $\nabla \cdot$ **E, B** is the divergence, ρ is the charge density, $\nabla \times$ **E, B** is curl, **E** is the electric field, **B** is the magnetic field and J is the vector current density. μ_0 is the permeability of free space and ε_0 is the permittivity of free space.

These equations are correct in flat spacetime, but in general relativity – unified field theory – modifications are needed. The spin connection is needed to describe spacetime effects as in the AB effect.

Metric

A formal mapping of a space. Minkowski metric with flat spacetime is that of special relativity.

The metric is a map on a plane, some volume or surface like a sphere, or on a 4-space. $ds^2 = -dct^2 + dx^2 + dy^2 + dz^2$ is the Minkowski metric.

$ds^2 = \eta_{\mu\nu} dx^\mu dx^\nu$ The two vectors and the (1, -1,-1,-1) are multiplied to get the distance squared. Differential geometry as we deal with it in relativity, relates the relative differences between distances in one dimension with respect to differences in distance in other dimensions.

The metric is a mapping of distances between vectors and the surfaces of the dimensions. This is an oversimplification, but we wish to avoid going into nothing but mathematics and deal with the physical meaning of these things.

The main role of the metric in relativity is to map vectors and one forms onto each other. This allows going back and forth as the reference frame changes and calculating the degree of curvature in the spacetime. The mapping by **g** between vectors and one-forms is one to one and invertible.

Metric tensor

A symmetric tensor, also called a Riemannian metric, which is always positive. The metric tensor g_{ij} computes the distance between any two points in a four dimensional spacetime. Its components are multiplication factors which must be placed in front of the differential displacements dx_i in a generalized Pythagorean theorem.

$$ds^2 = g_{\mu\nu}x^\mu x^\nu = q^\mu q^\nu$$

The metric tensor was developed by Einstein to find a linear formula which would describe spacetime. Linear implies that the formula could take coordinates and vectors in one spacetime and move them to another spacetime – say far from a high gravitational source to very close.

The metric tensor is defined as the outer product of two metric vectors. Form the outer product of two four vectors, i.e. multiply a column four vector by a row four vector, and you have a 4 x 4 matrix, with sixteen components. If the matrix is symmetric then these reduce to ten independent equations that define gravitation. Four are adjustable and six are determined.

There are 6 antisymmetric components that are Maxwell's equations which will not fit in the Einstein tensor formulation of gravitational curvature.

The symmetric metric is always defined as the tensor or outer product of two vectors.

g is the metric tensor. It is a function which takes two vectors and calculates a real number which will be the same in any reference frame – frame invariant.

The metric tensor gives the rule for associating two vectors with a single number, the scalar product.

The metric tensor gives a function used to measure distances between points in a four dimensional curved spacetime. A distance, ds, can be calculated from the Pythagorean theorem. It comes from Riemannian curved spacetime geometry. This is:

$$ds^2 = g^2 \cdot x1^2 + g^2 \cdot x2^2 + g^2 \cdot x3^2 + g^4 \cdot x4^2$$

We want a real number to define the distance. The metric tensor takes two vectors which define the curvature that occurs and calculates the real distance between them. We cannot have a negative distance so the metric must be "positive definite" – it must be real.

Metric Vectors

The metric four-vector is: $q^\mu = (h^0, h^2, h^3, h^4)$ where h^i are the scale factors of the general covariant curvilinear coordinate system that defines the spacetime.

On a unit hypersphere:

$$h_0^2 - h_1^2 - h_2^2 - h_3^2 = 1$$

each is calculated using $h_i = \partial r$

Evans Equations of Unified Field Theory

$$|\partial u_i|$$

and the unit vectors are $e_i = \dfrac{1}{h_i} \dfrac{\partial r}{\partial u_i}$

and $R = g^{\mu\nu\,(S)} R_{\mu\nu}^{\,(S)} = (h_0^2 - h_1^2 - h_2^2 - h_3^2)\, q^\mu R_\mu$

The simple notation involves a lot of calculations. We will not go into the actual calculations, but let us envision what they require. $q^{\mu\nu\,(S)}$ is a matrix with each of the h_i used to define the elements. These allow us to scale the geometry as we move from reference frame to reference frame.

The symmetric metric tensor is:

$$q^{\mu\nu(S)} = \begin{bmatrix} h_0^2 & h_0 h_1 & h_0 h_2 & h_0 h_3 \\ h_1 h_0 & h_1^2 & h_1 h_2 & h_1 h_3 \\ h_2 h_0 & h_2 h_1 & h_2^2 & h_2 h_3 \\ h_3 h_0 & h_3 h_1 & h_3 h_2 & h_3^2 \end{bmatrix}$$

The antisymmetric metric tensor is:

$$q^{\mu\nu(A)} = \begin{bmatrix} 0 & -h_0 h_1 & -h_0 h_2 & -h_0 h_3 \\ h_1 h_0 & 0 & -h_1 h_2 & h_1 h_3 \\ h_2 h_0 & h_2 h_1 & 0 & -h_2 h_3 \\ h_3 h_0 & -h_3 h_1 & h_3 h_2 & 0 \end{bmatrix}$$

Evans goes into great detail on this subject in his work. Also see http://mathworld.wolfram.com/MetricTensor.html and Basis Vectors and Tensors in this Glossary for more.

Minkowski spacetime

The spacetime of special relativity and quantum theory. The 4 dimensional nature of Minkowski spacetime is shown in the invariant distance:

$$ds^2 = -dt^2 + dx^2 + dy^2 + dz^2$$

This is often indicated as (-1, 1, 1, 1) with the t, x, y, and z understood. This is a version of the Pythagorean theorem in time and 3 spatial dimensions. Regardless of velocity, all measurements in any reference frame agree on ds, although the components can be different. We define a special reference frame that is not being accelerated as "inertial observers." We are inertial observers.

The invariant distance, ds, is the important factor here, not the squared distance.

Minkowski spacetime does not have a spin connection. It cannot spin.

Modulus

|r| is the absolute value of a real or complex number or length of a vector. It is a scalar. If **r** = xi + yj + zk where i, j and k are orthonormal vectors, then |r| = $(x^2 + y^2 + z^2)^{1/2}$. The length of the vector r.

Momentum

p = mv for a particle in Newtonian physics and p = γmv in relativity.

For a photon p = E/c = hf/c = h/λ = $\hbar$k. The photon has no mass in the standard model but does have momentum. Evans indicates that there is mass equivalence, that is, curvature, present.

Newtonian physics

This is the physics of our normal existence. It is also called classical physics. It is valid in most situations we encounter. Relativity and quantum physics see reality at the very small and large scales differently.

Noether's theorem

Noether's theorem states laws of invariance of form with respect to transformations in spatial and temporal coordinates. In general the Noether theorem applies to formulas described by a Lagrangian or a Hamiltonian. *For every symmetry there is a conservation law and vice versa.*

> Translation invariance (or symmetry under spatial translation) means the laws of physics do not vary with location in spacetime. This is the law of conservation of linear momentum.
>
> Rotation invariance (or symmetry under rotations) is the law of conservation of angular momentum.
>
> Time invariance (or symmetry under time translation) is the law of conservation of energy.
>
> Invariance with respect to general gauge transformations is the law of conservation of electric charge.
>
> See symmetry and groups also.

non-Euclidean spacetime

One of a number of spacetime manifolds upon which physics can be considered. Minkowski spacetime is non-Euclidean, but not Riemann since it is flat, not curved. The spacetime of our universe used in Evans' work is "Evans spacetime." This is the manifold of differential geometry allowing curvature and spin.

Normal

Normal refers to the use of basis vectors (or basis one forms.) Where normalized basis vectors have been used to calculate the coordinate vectors, the system or vector is "normal." The normal vector is a multiple of some base unit vector.

O(3) electrodynamics

O(3) refers to the sphere where $r^2 = (x^2 + y^2 + z^2)^{1/2}$. The existence of the Evans spin field, $\mathbf{B}^{(3)}$ is inferred from O(3) symmetry. The structure is that of curved spacetime using covariant O(3) derivatives, not flat special relativistic spacetime where the field disappears. It is a Yang-Mills theory with a physical internal gauge space based on circular polarization.

O(3) electrodynamics has been successfully derived from torsion in general relativity in the Evans Unified Field Theory.

The electromagnetic field is the spinning of spacetime itself, not an object superimposed on a flat spacetime. The field does not travel through spacetime, it is a disturbance in and of spacetime.

O(3) is a Lie group and it is a continuous, analytical group describing the properties of three dimensional space. Elements of this group are defined by the Cartesian unit vectors $\mathbf{i}$, $\mathbf{j}$, and $\mathbf{k}$. Other members include the complex circular unit vectors, $\mathbf{e}_{(1)}$, $\mathbf{e}_{(2)}$ and $\mathbf{e}_{(3)}$. The cross product of two of these members form another member, in cyclic permutation, so in mathematics this is a "non-Abelian" property. The position vector can be defined in O(3) by $\mathbf{r} = X\mathbf{i} + Y\mathbf{j} + Z\mathbf{k}$ and the square, $\mathbf{r} \bullet \mathbf{r}$, is an invariant under O(3) transformation rotations.

The O(3) gauge allows for circular polarization about the z-axis and as simple a development of higher-order electromagnetism as possible. There are other symmetry groups that might have worked, such as SU(3). These are more complicated and allow for more freedom than necessary.

The O(3) electrodynamics is now known to be an intermediate step where the field is still pure electromagnetism. O(3) is not generally covariant. The symmetry of the Evans unified field theory is that of the complete Einstein group which is a higher symmetry than O(3) which in turn is a higher symmetry than U(1). Symmetry building is not covered in this book. The algebra is developed *The Enigmatic Photon*, (Volume Five) and in publications on www.aias.us. O(3) electrodynamics is also reviewed in Advances in Chemical Physics, volume 119 (2001).

One-forms

The one- form is an exterior derivative. Vector values at every point in a spacetime increase (or decrease) as the potential field changes distance from the gravity or electromagnetic source. The one-form taken on a vector field gives the boundary where every vector is equal. If mapped, it looks like a topological map in two dimensions or an onion in three dimensions.

See Gradient and Covariant Exterior Derivative also.

Tensors of (0, 1) type are called one-forms, covectors or covariant vectors.

The one-form defines the touching of each plane as the vector passes through.

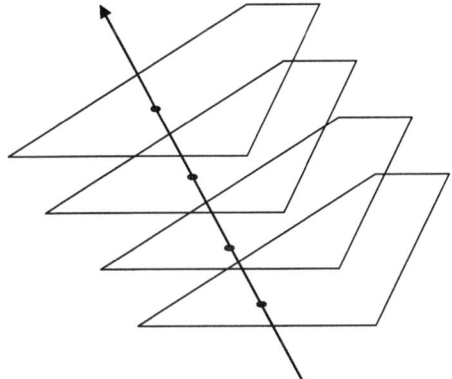

See Gradient for another example of one-forms.

A set of one-forms gives a "dual vector spacetime" to the vector spacetime. Vectors and one forms are dual to each other. They carry the same information, but are reciprocal to one another. Likewise, basis one forms are dual to basis vectors.

Complex conjugation is a process that acts like a metric by changing a vector in vector spacetime into a one form.

Vector one form $= A^0 \, y_0 + A^1 \, y_1 + A^2 \, y_2 + A^3 \, y_3$

This is called contraction and can be done between any one form and vector without reference to other vectors.

One-forms define the number of surfaces that are penetrated by a vector. They are frame independent. Their components transform as basis vectors - opposite of components of vectors. This gives the term covector. A covector transforms as a basis vector. Components of ordinary vectors transform opposite to basis vectors and are called "contravariant vectors."

Dual vectors = one forms = covectors = covariant vectors.

Any set of four linearly independent one-forms is a basis. We can use them to define a one-form basis or dual basis. One-forms map vectors onto real numbers. When a one-form is mapped onto a four dimensional vector, it gives the number of three dimensional spaces that are pierced or crossed by the vector.

To go from one-form to a vector change the time component sign.

The set of all one-forms is a covector spacetime.

The tetrad of unified field theory is a vector valued one-form.

Operators

These are any mathematical function which leaves the original formula essentially unchanged or reversible, but in a more workable form.

Evans Equations of Unified Field Theory

Orthogonal

Vectors which intersect at "right" angles and have a dot (= scalar) product of 0 are referred to as orthogonal.

Orthonormal

"Normal" refers to the use of basis vectors (or basis one-forms.) Where normalized basis vectors have been used to calculate the coordinate vectors, the system or vector is "normal." Where the system is orthogonal and normal, it is said to be orthonormal. In Evans' tetrad, the Latin spacetime indices use an orthonormal basis.

Outer product

Tensor product. For example a column vector (a, b, c) times a row vector (x, y, z) gives:

$$\begin{vmatrix} a \\ b \\ c \end{vmatrix} \otimes (x, y, z) = \begin{vmatrix} ax & ay & az \\ bx & by & bz \\ cx & cy & cz \end{vmatrix}$$

If two vectors x and y are linearly independent, the tensor or outer product generates a bivector. An arrow vector is like a directed straight line. A bivector is an oriented partial plane. A bivector is the plane surface generated when the vector x slides along y in the direction of y. The surface area is its magnitude. The product of a bivector and a vector is a trivector, an oriented volume.

Parallel transport

If an arrow points parallel to a surface in the direction of its motion, as it moves on a curved surface, it arrives back at it's origin perpendicular to its initial orientation. It has traveled in three dimensions although it sees its travel as two dimensional.

Riemannian spacetime curvature is defined in terms of the parallel transport of four vectors.

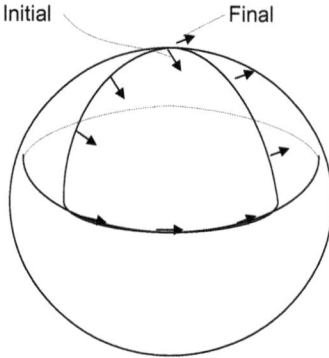

336

Particle

Imperfectly understood, a particle is like a field of energy with loose boundaries. It is a standing wave, not like a billiard ball, but rather a ring or sphere of energy or spacetime. At absolute zero, it still vibrates with some harmonic or circular motion. See Fundamental particle.

Given the Evans equations and their implications, it would appear that particles are forms of spacetime and "contain" both curvature and torsion. It is possibly adequately described a compressed spherical standing frequency wave like a four dimensional version of the string. This is associated with the Very Strong Equivalence Principle.

Alpha Particle - The helium nucleus, two protons and two neutrons.

Baryon - A half integral spin nucleon or hyperon interacting with the strong field and with mass heavier than the mesons.

Beta Particle - A high speed electron or positron emitted in radioactive decay.

Boson - A particle of zero or integral spin such as the graviton, photon, weak field bosons, gluons, pions or alpha particle,

Fermion - A particle of half integral spin, such as the proton, electron and neutron, obeying Fermi Dirac statistics, only one isolated fermion may occupy a given quantum state.

Gamma Particle - A high frequency, high energy photon.

Hadron - Any elementary particle that participates in the strong interaction.

Hyperon - Mass greater than a nucleon, decaying into a nucleon, hyperon and lighter particles.

Lepton - Any of a family of subatomic particles, including the electron, muon and their associated neutrinos, having spin 1/2 and mass less than those of the mesons.

Meson - Any of a family of subatomic particles with spin one and mass intermediate between lepton and baryon.

Muon - A lepton with mass 207 times that of the electron, negatively charged, and half life of 2.2 microseconds.

Neutrino - three distinct, stable, electronically neutral, subatomic leptons with a very small mass.

Nucleon - A proton or neutron

Neutron - An electrically neutral subatomic particle in the baryon family, 1839 times heavier than the electron, stable within the nucleus, otherwise a half life of 16.6 minutes, decaying into a proton, beta particle and antineutrino. It and the proton combine to form nearly the entire nuclear mass of the atom in the elements.

Pi Meson (or Pion) - There are two types of meson: a Pi zero (264 times heavier than an electron, zero electric charge, half life of 10E-18 seconds); and the Pi plus (273 times heavier than the electron, positive electric charge, half life of 26 nanoseconds.)

Note that it is this author's opinion that such transitory states are not "real" particles, but rather transitional curvature or energy states during processes. Unstable curvature states unravel to become stable curvature states.

Photon – not normally considered a particle as such. It converts

Proton - A stable baryon with a very long half life, spin of ½, and mass 1836 that of the electron.

Particle collisions

Inelastic collisions in particle accelerators can create more or different particles. The standard method of calculating results uses two equations: $E^2 = p^2 + (mc^2)^2$ and $E_1^2 - p_1^2 = E_2^2 - p_2^2$. (In actual calculations p^2c^2 is used instead of p.)

$$E^2 = p^2 + (mc^2)^2$$

This is the Pythagorean relationship between Total energy is the hypotenuse and momentum and rest energy are the perpendicular sides. mc^2 is invariant. Therefore with sub 1 and sub 2 referring to different reference frames due to different velocities:

$$E_1^2 + p_1^2 = E_2^2 + p_2^2$$

The spacetime that is a high velocity (near c) particle is compressed. "Explosive decompression" occurs in collisions with conservation of momentum and total energy (curvature). If high enough, the compression converts to particle pairs – that is to a number of smaller compressed regions. The regions will have stable frequencies necessary to continue to exist. If not, they will break apart again to become one of the stable particles. It has not yet been shown, but logic dictates that the stable particles will have combined curvature forms and wave frequencies similar to the electron orbits.

An additional area of research that the Evans equations open up is the affects of particle collisions in stronger gravitational fields. One assumes that in a different field, say near a black hole or even near a neutron star, that the results will be a bit different. We may find particles unstable in our low energy density reference frames become stable in strong gravitational fields.

Pauli matrices

See Group also. Pauli matrices are Hermitian or self-adjoint. This means $A = A^*$ where A^* indicates the complex conjugate. (If $A = a + bi$, the adjoint $A^* = a - bi$ for every element.)

Any non Hermitian matrix is the sum of a Hermitian and the antisymmetric Hermitian using: $A = ½ (A + A^*) + ½ (A - A^*)$.

These spin matrices are an algebra that is used in quantum mechanics of spin ½ particles. These are a geometrical product of orthonormal vectors. They are a matrix representation of 3-dimensional geometrical algebra.

$$\sigma_1 = \begin{vmatrix} 0 & -1 \\ 1 & 0 \end{vmatrix} \quad \sigma_4 = \begin{vmatrix} 0 & 1 \\ -1 & 0 \end{vmatrix} \quad \sigma_7 = \begin{vmatrix} 0 & i \\ i & 0 \end{vmatrix}$$

$$\sigma_2 = \begin{vmatrix} 0 & -i \\ -i & 0 \end{vmatrix} \quad \sigma_5 = \begin{vmatrix} -1 & 0 \\ 0 & -1 \end{vmatrix} \quad \sigma_8 = \begin{vmatrix} -i & 0 \\ 0 & -i \end{vmatrix}$$

$$\sigma_3 = \begin{vmatrix} 1 & 0 \\ 0 & 1 \end{vmatrix} \quad \sigma_6 = \begin{vmatrix} i & 0 \\ 0 & -i \end{vmatrix}$$

Pauli matrices can be developed for any spin particle. One can convert the general energy eigenvalue problem for the spin of a particle, where the Hamiltonian energy is some function of position and spin operators, into coupled partial differential equations involving the spin wave functions. These systems of equations are very complicated to solve.

The Pauli matrices are generators of SU(2) rotations. They give infinitesimal rotations in three dimensional spacetime in non-relativistic ½ spin particles. The state of a particle is given as a two component spinors. Particles must be rotated 4π to return to original state. In our normal macro world 2π is a full rotation of 360 degrees.

$\sigma_1\sigma_2 = i\sigma_3 \quad \sigma_2\sigma_3 = i\sigma_1 \quad \sigma_1\sigma_3 = i\sigma_2 \quad \sigma_1^2 \sigma_2^2 \sigma_3^2 = 1$

Hamilton's formula concerning rotations is similarly: $i^2 = j^2 = k^2 = -1$.

Phase

For an in depth study see "DERIVATION OF THE GEOMETRICAL PHASE FROM THE EVANS PHASE LAW OF GENERALLY COVARIANT UNIFIED FIELD THEORY, Myron W. Evans, January2004 at www.aias.us.

The Evans phase law is:

$$\Phi = \exp\left(ig \oint \mathbf{A}^{(3)} \cdot d\mathbf{r}\right) = \exp\left(ig \oint \mathbf{B}^{(3)} \cdot kdAr\right) := \exp(i\Phi_E)$$

Equation courtesy Myron Evans

This is applied in the AB and other spacetime electromagnetic effects.

Photon

Electromagnetic waves in packet form. Radio, infrared heat waves, light waves, and hard hitting gamma rays are photons. A photon can excite an electron to a higher orbital – it has a spatial or mass component. When it enters an atom, mass increases.

A photon is an elementary torsion-curvature form (wave-particle) with a zero charge. It has a finite non-zero rest mass found from its momentum. The photon is the quantum of the electromagnetic field.

A photon has a real longitudinal magnetic field whose quantum equivalent is the Evans' photomagneton. Circularly or elliptically polarized light acts as a magnet upon interaction with matter. This is the Inverse Faraday Effect (IFE). Unpolarized light does not exhibit IFE. This magnetization is proportional to the light intensity and the light intensity is proportional to the photon flux density. The real longitudinal magnetic field of the photon was discovered in 1992 by Myron Evans. This has far reaching implications in our understanding of the physical nature of the quantum world. The longitudinal electric field of a photon is imaginary – that it is of a different phase. Imaginary here refers to i and does not mean unreal. The transverse orthogonal magnetic and electric fields of a photon are real.

A photon is its own antiparticle. The direction of the real longitudinal magnetic field is opposite for photon and antiphoton. The photon, a quantum of electromagnetic radiation, is a magnetic dipole. An antiphoton is a photon with its magnetic polarity reversed, that is, from NS to SN or vice versa.

Photons propagate through spacetime like a wave and have particle-like behavior during emission and absorption. The wavelength of gamma rays and x-rays photons is very small so they behave like particles moving in as straight a line as the spatial curvature allows. For microwaves and radio waves the photon wavelength is quite long and so they display more wave-like attributes. The amplitude and wavelength of visible light photons is between gamma and radio waves so they exhibit more dual wave-particle behavior.

Planck constant; h

The quantum of action. $h = 6.625 \times 10^{-34}$ joule-seconds, $\hbar = h/\pi = 1.05 \times 10^{-34}$ joule-second $= 1.05 \times 10^{-34}$ N-m-s. A watt-second is a Newton meter second. This is the minimum amount of action or energy possible. $\hbar = h/2\pi$ is the fundamental constant equal to the ratio of the energy of a quantum of energy to its frequency. The units of h are mass x distance x time.

Planck quantum hypothesis

The energy of a photon E relates to its frequency according to $E = nhf$ or $E = nh\nu$. The value h is the smallest amount of energy that can possibly exist. All energy change must obey this equation.

Poisson equations

Poisson's equation for electrical fields is $\nabla^2 \phi = -4\pi\rho$ where ϕ is the electric potential in volts or joules per coulomb, ρ is the charge density in coulombs.

For gravitational fields, Poisson's equation is $\nabla^2 \phi = 4\pi G\rho$, where ϕ is the gravitational potential, G is the gravitational constant and ρ is the mass density.

Principal of General Relativity

In order to be valid, any theory or measurement in physics must be objective to all observers.

Products

For strict definitions see math references or web sites. This is a simplified set of definitions to explain what is going on in the physics applications.

Commutator product is same as matrix cross product.

Cross product. Defined in three dimensions. An example torque calculation is found in Figure 1-5. The resulting vector is perpendicular to those multiplied together.

Dot product. Pythagorean distances can be found within a manifold.

Inner product is the dot product in four dimensions. It also has more structure allowing geometry such as angles, lengths, vectors and scalars.

Matrix cross product is defined for square matrices as A X B = AB - BA.

Outer product or exterior product is same as the wedge product.

Tensor product, $\otimes$, as most products here may be applied to vectors, matrices, tensors, and spaces themselves. (The $a_x b_y$ are typically added together to find a scalar that is invariant.)

$$[a_1\ a_2] \otimes [b_1\ b_2] = \begin{bmatrix} a_1 b_1 & a_2 b_1 \\ a_1 b_2 & a_2 b_2 \end{bmatrix}$$

Tetrad product. The same format as the tensor product however the a_x would come from the index a and the b_y would come from the base manifold, labeled μ.

Wedge product is a vector product defined in a vector space. It is the 4 or more dimensional version of the cross product. Operations on matrices produces results that form parallelopipeds of electromagnetic lines and volumes of force.

Quantum gravity

Quantum gravity is the study of unification of quantum theory and general relativity starting from quantum theory. The methods considered possible in the past were: 1) starting with general relativity, develop a quantum version; 2) start with quantum theory and form a relativistic version similar to Dirac's equation in special relativity; 3) develop a totally new mathematics like string or M theory. Evan's equations show that quantum theory can be developed from general relativity.

Quantum mechanics
A physical theory that describes the behavior of matter at very small scales. The quantum theory attempts to explain phenomena that classical mechanics and electrodynamics were unable to explain. It has provided highly accurate descriptions of various processes.

Quantum number
A label on a quantum state. It can give the number of quanta of a particular type that the state contains. If the electric charge on a particle were given as an integer multiple of the electron's charge then that would be a quantum number.

Quarks
Mathematically quarks make sense, but physically they are unlikely to exist as discrete particles. The implication is that they are curvature or energy states. Quarks have fractional charge, but are supposed to be fundamental. By definition a fundamental charge cannot be divided. It then seems that quarks are either divisible and not fundamental or they are not there at all. Rather they are representations of curvature states.

They were found by bouncing energy off nuclei and curve fitting the scattering patterns. They cannot exist outside a triplet. Most likely they are a substructure that is pure curvature and torsion rather than any discrete particle like object.

The three curvatures may reflect existence in three dimensions.

R
The Ricci scalar - a measure of curvature. Curvature in four dimensions is described by the Riemann curvature tensor. Two dimensional curvature can be described by the Ricci scalar.

In the limit of special relativity, $\kappa^2 = R$ where κ (or κ in some texts) is 1/r of an osculating circle drawn at a point or dot. See curvature. The Evans equations indicate that curvature and wave number are related; a new understanding.

Rank
Number of indices. A zero rank tensor is a scalar – a simple number. A 4-vector is a rank one tensor. A four dimensional tensor is a rank two tensor.

Reference Frame
A reference frame is a system, particle, spaceship, planet, a region in spacetime say near a black hole, etc. The reference frame has identical gravity and velocity throughout its defined region. Everything in it has the same energy density. Sometimes it is necessary to shrink the size of the reference frame to a size near a point. If small enough of a region, then the laws of special relativity (Lorentz) apply. This is what is meant by all points are Lorentzian.

Renormalization

The sum of the probability of all calculated quantum outcomes must be positive one – 100%. We cannot have a negative chance or 120% or 80% chance of an occurrence.

Quantum mechanics had problems with infinite quantities resulting in some calculations where volume of a particle was assumed to be zero. To avoid these mathematical renormalization techniques were developed in quantum electrodynamics. A finite volume is assumed without proof.

Imagine a volume going to zero while the amount of mass is constant. The density goes to infinity if the volume goes to zero which is obviously wrong in reality although it is correct mathematically. The calculation: $\rho = m/V$ (density = mass / volume). When $V = 0$, then $\rho = m / V = \infty$. By assuming that V cannot get smaller than say, h^3, the density will approach a very high value, but not infinity; the equation then produces a reasonable result.

Evans has found that $mV = k(h/2\pi c)^2$. V defined here is limited. Renormalization by assuming some arbitrary small but positive volume is no longer necessary. The causal Evans volume can be used instead..

Ricci tensor

$R_{\mu\nu}$ is the Ricci tensor. It is the trace of the Riemann curvature tensor. It is a mathematical object that controls the growth rate of volumes in a manifold. It is a part of the Einstein tensor and is used to find the scalar curvature.

It can be thought of as the sum of the various curvatures spanned by a tetrad's orthonormal vector and the corresponding vector in the base manifold.

Any spacetime manifold can have a negative curvature, but a positive Ricci curvature is more meaningful.

Riemann geometry

Non-Euclidean geometry of curved spaces. Einstein gravitation is described in terms of Riemann geometry.

Spacetime curvature refers to Riemannian spacetime curvature, $R_{\lambda\mu\nu\kappa}$; there are other types of curvature. If any of the elements of $R_{\lambda\mu\nu\kappa}$ are not zero then Riemannian spacetime curvature is not zero.

Riemann tensor

Riemann calculates the relative acceleration of world lines through a point in spacetime. Its unit of measurement is $1/m^2$ which is curvature.

In four dimensions, there are 256 components – 4 x 4 x 4 x4. Making use of the symmetry relations, 20 independent components are left. $R_{\lambda\mu\nu\kappa} = (1/2) (g_{\lambda\nu} R_{\mu\kappa} - g_{\lambda\kappa} R_{\mu\nu} - g_{\mu\nu}$

$R_{\lambda\kappa} + g_{\mu\kappa} R_{\lambda\nu}) - (R/(3)(2))(g_{\lambda\nu}g_{\mu\kappa} - g_{\lambda\kappa}g_{\mu\nu}) + C_{\lambda\mu\nu\kappa}$. It is sometimes called the Riemann-Christoffel tensor. Another form is: $R^c{}_{\beta\gamma\delta} = \Gamma^c{}_{\beta\gamma\delta} - \Gamma^c{}_{\beta\delta\gamma} + \Gamma^\mu{}_{\beta\delta}\Gamma^c{}_{\mu\gamma} - \Gamma^\mu{}_{\beta\delta}\Gamma^c{}_{\mu\gamma} - \Gamma^\mu{}_{\beta\gamma}\Gamma^c{}_{\mu\delta}$

Note that while precise calculations are incredibly tedious, the basic concept of curved spacetime is simple. The Riemann tensor is complicated, but it is just a machine to calculate curvature.

Riemann gives a way to measure how much different vectors are moving toward or away from each other along geodesics. By measuring the rate of change of geodesics, one can see the curvature of spacetime. Parallel lines at a great distance move towards each other in a gravitational field.

Rotations and Symmetry

Symmetry refers to mirror reflections and similar operations.

Symmetry transformations map every state or element to an image that is equivalent to the original. In theory, the resulting effect is as symmetric as the cause – the first state. This is the principle of symmetry. The resulting object after a symmetry transformation is the same as the original.

A rotation is a movement of an entire reference frame. Regardless of direction or orientation, the system is the same physically. Rotations are orientation preserving motions in two and three dimensions. The dimension n = 2 or 3 orthogonal group is O(n) and the rotation group is SO(n). These can be written as matrices. For example, the subgroup of symmetries of the equilateral triangle is in O(2). Orthogonal transformations in three dimensions are O(3) symmetries.

Rotations in two dimensions are all of the form:

$$\begin{pmatrix} \cos a & -\sin a \\ \sin a & \cos a \end{pmatrix}$$

Left multiplication of a column vector by this matrix rotates it through an angle X degrees counterclockwise. Right multiplication gives clockwise rotation.

Other important orthogonal transformations are reflections in the y and x axes. An example is:

$$\begin{pmatrix} -1 & 0 \\ 0 & 1 \end{pmatrix} \text{ and } \begin{pmatrix} 1 & 0 \\ 0 & 1 \end{pmatrix}$$

Rotations

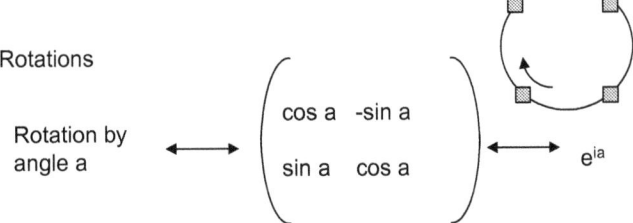

Rotation by angle a ⟷ $\begin{pmatrix} \cos a & -\sin a \\ \sin a & \cos a \end{pmatrix}$ ⟷ e^{ia}

Note that e^{ia} is 1 when angle a is 0 or $2n\pi$ since 2π is 0 in radians and e to the 0 power is 1.

Rotations about a common axis are isomorphic and Abelian. The particle appears to be described by Abelian mathematics while the electromagnetic field by non-Abelian.

It is likely that particle interactions are determined by the internal curvature and torsion structure of the particles. Mathematically, symmetry shows patterns. The energy bound in the proposed quark triplet unit particle of matter substructure may be the source of the SU(3) symmetry exhibited in the strong force interactions. Color charge has 3 axes of complex rotation and exhibits SU(3) symmetry - rotations in 8 dimensional color space. The wave function of a particle is a mathematical description of the rotation of energy bound within the particle structure.

In any event, quarks may not exist as individual particles, but there is an energy structure in the proton or neutron that has an SU(3) symmetry.

Scalar curvature (R)

The scalar curvature is R defined as $g^{\mu\upsilon} R_{\mu\upsilon}$ where $g^{\mu\upsilon}$ is the metric tensor and $R_{\mu\upsilon}$ is the Ricci curvature tensor. Scalar curvature, R in equations, is an invariant connecting general relativity to quantum theory. It is an essential concept in unified theory.

At each point on a Riemann manifold there is one real number indicating the intrinsic curvature at that point. Scalar curvature is the trace of the Ricci curvature with the metric applied. See Wave Number also.

Scalars

These are numbers with magnitude only. Mass is a scalar. Velocity is a vector – it has a direction and a magnitude. Scalars are invariant in different reference frames. They are more "real" than vectors.

Schrodinger's Equation

$$\nabla^2 \Psi_n = \frac{2m}{\hbar} (V - E) \Psi_n$$

SI Units

Quantity	Name	Unit
acceleration	meter/sec squared	m/s^2
current density	ampere per square meter	A/m^2
electric current	ampere	A
magnetic field strength	ampere per meter	A/m
mass density	kilogram/ cubic meter	kg/m^3
specific volume	cubic meter per kilogram	m^3/kg
speed, velocity	meter per second	m/s
temperature	degree Kelvin	K
wave number	reciprocal meter	m^{-1}

Derived Units

Quantity	Name	Symbol	Base units
plane angle	radian	rad	m·m^{-1} = 1
frequency	hertz	Hz	s^{-1}
force	newton	N	m·kg·s^{-2}
energy, work, heat	joule	J	m^{-1}·kg·s^{-2}
power	watt	W	J/s = m^2·kg·s^{-3}
pressure,	pascal	Pa	N/m^2
inductance	henry	H	Wb/A = m^2·kg·s^{-2}·A^{-2}
electric charge	coulomb	C	A·s
electric potential	volt	V	m^2·kg·s^{-3}·A^{-1}
magnetic flux	weber	Wb	V·s = m^2·kg·s^{-2}·A^{-1}
magnetic flux density	tesla	T	Wb/m^2 = kg·s^{-2}·A^{-1}

SI derived unit

moment of force	newton meter	N·m
angular velocity	radian per second	rad/s
angular acceleration	radian per second squared	rad/s^2
energy density	joule per cubic meter	J/m^3
electric field strength	volt per meter	V/m
electric charge density	coulomb per cubic meter	C/m^3
entropy	joule per Kelvin	J/K
electric flux density	coulomb per square meter	C/m^2
permittivity	farad per meter	F/m
permeability	henry per meter	H/m

Spacetime

This is the physics term for the universe we live in. The exact nature in mechanical terms is unknown. One belief is that spacetime is mathematical and the vacuum is the underlying mechanical reality. The vacuum in this concept is not nothing, but rather a tenuous potential field. Spacetime in classical general relativity was considered to be completely differentiable and smooth. In geometrodynamics (Misner et. al.), spacetime is more associated with the vacuum and is considered to be full of strong fluctuations at the Planck scale lengths, $L_p = (\hbar G/c^3)^{.5}$.

Evans spacetime has a metric with curvature and torsion.

Riemann spacetime has a metric with curvature only.

Minkowski spacetime has a flat metric with distance only.

Evans defines Minkowski spacetime as that in the abstract bundle and non-Riemann or Evans spacetime as the curved and torqued spacetime of the real universe. This author called it Evans spacetime which is a more appropriate term.

Evans spacetime has curvature and torsion. It is the real metric of the universe.

Spin

Spin is an abstract version of turning motion. It is the angular momentum of a particle. Spin is an intrinsic and inherently quantum property, not real motion internal to the object. Spin is quantized and must be expressed in integer multiples of Planck's constant divided by 2π. $\hbar$ is "h-bar" = $h/2\pi$. Spin 1 is then $\hbar$ and spin ½ is $\hbar/2$.

A scalar can be described as spin 0, that is, it has no spin; a vector as spin 1; and a symmetric tensor as spin 2. The electron, proton, neutrino's, and neutron have spin ½. Photon's have spin 1. The hypothetical graviton is spin 2. Negative spin is spin in the direction opposite positive.

Half spin particles are antisymmetric and are subject to the Pauli exclusion principle. Integer spin particles are symmetric and are referred to as bosons.

A spinning charged particle like an electron has magnetic moment, due to the circulation of charge in the spinning. The spin of the nucleus of an atom is the result of the spins of the nucleons themselves and their motion around one another. Spin occurs in four dimensions – the three dimensions we experience and time or the complex plane. Spin is included in conservation of angular momentum.

Curved spacetime is the "first curvature" and spin is the "second curvature."

Spin is described best by spinors as it is not a vector quantity.

Spin connection, $\omega^a{}_{\mu b}$

Any description of spacetime must be able to describe gravitational curvature and electromagnetic curvature which is spinning of the spacetime. Mathematically, the spin connection is a correct geometric definition of how to describe the spin; this is the spin connection.

The spin connection may be best described in the electromagnetic limit by being equivalent to a helix. The helix in this limit gives a picture of the spin connection. It is the geometrical object that transforms a straight baseline into a helical baseline.

The spin connection is a type of one-form that does not obey the tensor transformation law. It is used to take the covariant derivative of a spinor. In deriving the covariant exterior derivative of a tensor valued form, one takes the ordinary exterior derivative and adds one spin connection for each Latin index.

Spinors

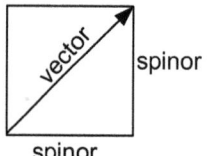

The spinor changes signs under rotations and is not really a square root, but similar.

Spinors were invented by Wolfgang Pauli and Paul Dirac to describe spin. Spinors are similar to square roots of vectors but with sign changes under rotations of 2π. They help describe the spin of an electron. The Dirac spinor is a complex number with right hand and left hand versions. Spinors can be used to represent rotation groups such as the SO(N), which Evans uses in his wave equation.

Spinors can be mapped onto real space rotations. One way to envision the process is to take a two component complex numbered spinor, say A = a, b, and find a real vector from it. The real three-vector is found from:

$x = 1/2 \, (a \, b^* + a^* \, b)$

$y = i/2 \, (a \, b^* - a^* \, b)$

$z = 1/2 \, (a \, a^* - b \, b^*)$

Then the product of A and A* is $(x^2 + y^2 + z^2)^{1/2} + x \, \sigma x + y \, \sigma y + z \, \sigma z$

With σ = Pauli spin matrices, A* is a row matrix (a* b*). Both a positive and negative spinor are associated with the same vector. We could say that the spinor is mapped to the space twice over. The rotation of 2π has no effect except a sign change.

Stokes' Theorem

One of the fundamental theorems of calculus is that the two main operations of calculus, differentiation and integration, are inverses of each other. If a function is integrated and then differentiated, the original function is retrieved. As a result, one can compute integrals by using an antiderivative of the function to be integrated.

If you add up all the little changes in a curve then you get the total of the curve. If you then break the curve into little changes again, you could retrieve those.

Stokes' theorem states that the fundamental theorem is true for higher dimensions and on manifolds. Evans has developed a covariant version.

Strong Field

The strong interaction. This holds quarks and gluons together to make protons and neutrons. Strong interactions provide the nuclear binding force. This is gravitational in nature and may ultimately be more fully explained by some antisymmetric parts of Evans' equations also.

The weak force (interaction, field) is electromagnetic. The force between electrically charged particles and between neutral atoms forming molecules is an electromagnetic interaction.

SU(2), SU(3)
Special unitary groups. See Groups.

Symbols (see SI Units also)

a	acceleration
c	speed of light in vacuum
ds	distance
g	acceleration due to gravity; gravitational field strength; metric tensor
h	Planck's constant; h-bar is Dirac's constant
i	imaginary unit; Cartesian *x*-axis basis unit vector
j	Cartesian *y*-axis basis unit vector
k	Einstein's constant; Cartesian *z*-axis basis unit vector, wave number
l	length; angular momentum
m	mass
p	momentum; pressure
q	metric vector
r	radius; distance
t	time
u	four-velocity
v	velocity
B	magnetic flux density; magnetic field
E	energy; electric field
F	force
G	Einstein tensor; universal constant of gravitation
H	magnetic field strength
K	kinetic energy (also E_k)
R	curvature; Ricci tensor
T	temperature; torsion, stress energy tensor
U	potential energy
V	potential voltage
α	angular acceleration; fine structure constant
β	v/c = velocity in terms of c
Γ	Christoffel symbol
γ	Einstein Lorentz-Fitzgerald contraction, gamma

Δ	change, difference between two quantities
ε	permittivity
η	metric –1, 1, 1, 1
θ	angular displacement
κ	torsion constant, curvature, wave number
λ	wavelength
μ	mu
ν	nu
π	3.14; ratio of circumference / diameter of a circle
ρ	density
τ	torque; proper time; torsion
ω	angular velocity, spin connection
Σ	sum
Φ	field strength
Ψ	wavefunction

Symmetric tensor

A symmetric tensor is defined as one for which $A^{mn} = A^{nm}$ with Greek indices, typically mu and nu, used when working in 4 dimensions. (Latin indices are used in 3 dimensions.)

The symmetric tensors will typically describe distances. The antisymmetric tensors will describe spin - turning or twisting.

Symmetry

See Groups and Rotations also.

A symmetrical equation, group of vectors, or object has a center point, axis line, or plane of symmetry. A mirror image is symmetrical with respect to the reflected object. A symmetrical operation results in the same or a mirror figure as the original. Reflection, translations, rotations are such operations. The symmetry group for an object is the set of all the operations that leave the figure the same.

Symmetries are the foundation for the various conservation laws. Certain characteristics remain unchanged during physical processes. Symmetries are usually expressed in terms of group theory. Symmetry is one of the fundamental mathematical components of the laws of physics defining invariance.

Einstein's special relativity established that the laws of electromagnetism are the same in any inertial reference frame – one that is not undergoing acceleration or in a gravitational field. This can be described as a symmetry group – called the Poincaré group in mathematics.

General relativity extended theory to reference frames that are accelerating or experiencing gravitational "forces." All laws of nature, including electrodynamics and optics, are field laws that are mapped in curved spacetime. A field can be reshaped without losing the relationships within it. The Einstein group defines the symmetry in general relativity. Both are Lie (pronounced lee) groups in mathematics. Tensors transform vectors into different reference frames..

The invariant distance ds (not ds^2) establishes the covariance of the laws of nature. One takes the square root of ds^2 using spinor forms. The Dirac equation does this by factorizing the Klein-Gordon equation. Spin showed ramifications - extra degrees of freedom - and can be seen in Evans as related to torsion. Quaternians can also be used to define the Riemannian metric. The 4-vector quaternion fields are like the second-rank, symmetric tensor fields of the metric of the spacetime.

Symmetry Operations or Transformations

Rotations.

A spherical system -sphere, cylinder, or circle - can be rotated about any axis that passes through its center. After a rotation the appearance of the system is unchanged. There is no smallest change allowed. An infinitesimal rotation can be made and for this reason the symmetry is continuous. (Evans does show us that h, Planck's constant, and h bar, Dirac's constant, do establish minimum rotations in physics, but mathematics allows smaller changes.)

Continuous Rotation of a Circle

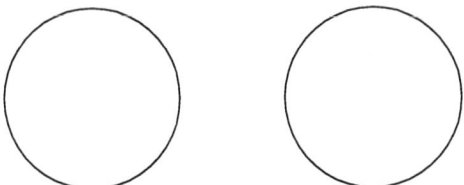

Regardless of the amount of rotation, the circle is
the same after the operation.

Discrete symmetries require unit steps. The rotation of a triangle is an example. Only complete or discrete steps of 1/3 x 360° = 120° can occur to maintain symmetry. The operation either happens in a distinct step or it doesn't happen at all.

Reflections or Parity Symmetry

The flip of the triangle is a reflection. A mirror reflects any object held up to it. Reflection in time is time reversal symmetry.

Rotation and reflection of a triangle

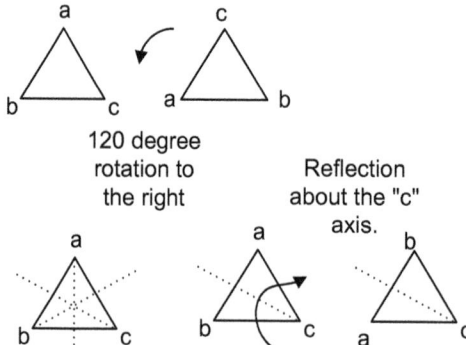

The rotation or reflection must be discrete for the triangle to appear the same.

The symmetry operations form an algebra called a symmetry group or simply group. The mathematical definition of a group includes rules:

Closure: the product of any two elements or the square of an element yields another element in the set. *Identity Element:* the unitary operation exists for each element so that one times $R = R$. *Unique inverse element:* each element has an inverse so that $R \times R^{-1} = R^{-1} \times R = 1$. *Association:* $R_1 \times (R_2 \times R_3) = (R_1 \times R_2) \times R_3$. *Commutativity:* That is $R_1 \times R_2 = R_2 \times R_1$. Group multiplication does not have to be commutative. A commutative group is Abelian (i.e., $AB = BA$ for all elements A and B). The equilateral triangle group is non-commutative, that is non-Abelian. The real universe is non-Abelian. If non-Abelian then AB-BA has a value, otherwise $AB-BA = 0$.

The continuous group of all rotations in three dimensions - symmetry group of a sphere - is SO(3). It governs the physics of angular momentum and spin. The symmetry of a sphere in N-dimensions is called SO(N). SU(N) is the symmetry of the complex number unit sphere.

Translations.

Movement is a translation in space and time. If an object is moved from one place to another, it's shape is unchanged. Spatial translations are invariant. All relativistic equations in physics are invariant under translations in space and time. The defining symmetry principle of special relativity is invariance of physical laws. This is the Strong Equivalence Principle of Einstein.

The term "proper time" is used to denote time as measured by an observer or clock within a given reference frame. It is invariant. The symbol tau, τ, is used for proper time.

Charge Conjugation.

Particle-antiparticle symmetry is called charge conjugation. A discrete symmetry of replacing all particles by antiparticles in any given reaction is called C. It is a necessary condition in quantum mechanics that the combined operations of CPT must be an exact symmetry. Experiments with neutral K-mesons indicate the violation of T-symmetry. T must be violated when CP is violated in such a way as to make CPT conserved. CPT is expected to be an exact symmetry. The violation of T means that there is fundamental information that defines a preferred direction of time.

A particle's wave function can be changed to that of its antiparticle by applying CPT operation: charge conjugation, parity, and time reversal operations. This leaves the momentum unchanged.

$C(e) = -e$.

Time reversal invariance is not a symmetry law.

Strong interacting particles (like protons, neutrons, pi-mesons) are members of the symmetry group, SU(3). One of the representations of SU(3) has eight components represented as an irreducible 8 x 8 matrix. The eight spin-0 mesons fit into one matix; the eight spin-1/2 baryons into another matrix. The SU(3) symmetry is not exact, but the pattern exists.

Given that $x^2 + y^2 + z^2 = 1$, then the set of linear maps from (x_1, y_1, z_1) to (x_2, y_2, z_2) so that $x_2^2 + y_2^2 + z_2^2$ also = 1 defines the symmetry group O(3). O(3) is the spherical symmetry group used by Evans in electrodynamics.

The group of rotations on the complex numbers is the one-dimensional unitary group U(1). O(2) is equivalent and acts on 2-dimensional vectors. Complex numbers can represent 2-d vectors.

A map between two groups which keeps the identity and the group operation is called a homomorphism. If a homomorphism has an inverse which is also a homomorphism, then it is an isomorphism. Two groups which are isomorphic to each other are considered to be "the same" when viewed as abstract groups. For example, the group of rotations of a triangle is a cyclic group

Symmetry building of the field equations

A new phenomenon may be indicated by the Evans equations. This is the inverse to spontaneous symmetry breaking, for example from SU(2) x U(1) to U(1) = SO(2). As of the time of this writing there is indication that the **B**$^{(3)}$ field could be a consequence of "symmetry restoring" or symmetry building. Evans has found that O(3) describes electrodynamics. (It could be that SO(3) is the correct group for electrodynamics.) The subtleties involve preserving motion with or without reflections and which group is more physical.

The highest symmetry is that of the unified field theory using the Einstein group.

Next is O(3) electrodynamics which is intermediate between Maxwell-Heaviside U(1). The gravitational and electromagnetic fields are decoupled in O(3). It has higher symmetry than U(1) electrodynamics.

Symmetry - Noether's Theorem

In 1905, the mathematician Emmy Noether proved a theorem showing that for every continuous symmetry of the laws of physics, there must exist a conservation law and for every conservation law, there must exist a continuous symmetry.

Rotational symmetry is the law of conservation of angular momentum.

Mirror symmetry (Parity) or particle-antiparticle symmetry is charge conjugation. Time reversal symmetry during translation is conservation of momentum. Translation symmetry is conservation of momentum. In collisions, the momentum of the resulting particles is equal to that of the initial particles.

Time translation symmetry is conservation of energy. As an object moves from time-space to time-space, existence continues.

All of the individual conservation laws amount to conservation of existence. Nothing is created or destroyed; but it can change aspects.

Evans has given us a new conservation law – conservation of curvature. That curvature can be the first kind, Riemann curvature, or the second kind, torsion.

T

Torsion. Stress energy tensor.

$T^{\mu\nu}$

The stress energy tensor or energy-momentum tensor. It describes everything about the energy of a particle or system. It contains the energy density, stress, pressure, and mass. In Evans it must also be defined by the energy within the torsion.

T is the stress energy tensor which is the total energy density. It is the source of the fields of relativity. All stress, pressure, velocity, mass, energy are contained in T. And all the components of T are the source of a gravitational field. T^{00} defines the energy density and T^i is the momentum density. In a co-moving reference frame of SR, T = p, the momentum. T is symmetric, that means that $T^{\alpha\beta} = T^{\beta\alpha}$

T is used to define the curvature or compression parameters of the spacetime due to energy density.

Tangent spaces and bundles

The tangent space is the real vector space of tangent vectors at each point of a manifold or spacetime. The bundle is the collection of all the vectors.

The tangent space in general relativity is physical, the fiber bindle space in gauge theory has heretofore been considered an abstract Yang Mills space. Evans' shows that the tangent bundle is the vector space of quantum theory.

Tangent space

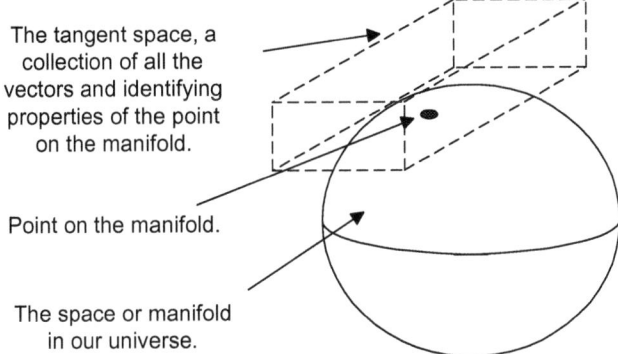

The tangent space, a collection of all the vectors and identifying properties of the point on the manifold.

Point on the manifold.

The space or manifold in our universe.

The tangent space is an infinite number of vector fields used as needed to describe the point in the real space.

Tensors

A tensor is a geometrical object which describes scalars, vectors, and linear operators independent of any chosen frame of reference.

A tensor is a function which takes one or more vectors and gives a real number which will be the same in any reference frame. Knowing the real number, a scalar, and the formula which found it allows us to move from reference frame to reference frame or from one region of a curved spacetime back into another region.

Tensors are simply formulas or math machines. Plug in some numbers, vectors or 1-forms and out comes a number or another vector or 1-form.

Tensors are linear, meaning they can be added or multiplied like real numbers. It gives the same number regardless of reference frame – high velocity or dense gravitational field.

Components in a reference frame are the values of the function when its arguments are the basis vectors, e_α, of the reference frame. Components are *frame dependent* since basis vectors are frame dependent. A rod is 1 meter long in frame A, but it will contract in a high energy density energy frame as viewed from a low energy density reference. When the entire reference frame A changes, the components will be different in different reference frames – it is the tensor which is invariant.

355

The number of dimensions of spacetime are indicated by the number of indices.

Rank is indicated by the number of different indices.

g or $g_{\mu\nu}$ is the metric tensor.

$G_{\mu\nu} = R_{\mu\nu} - \frac{1}{2} g_{\mu\nu} R$ is the Einstein tensor; sometimes called "Einstein." Note bold face letters used by some authors. Evans uses regular print which is the more modern format.

R is the scalar curvature. $R_{\mu\nu}$ is the Ricci curvature tensor. It is the only contraction of Riemann and it is symmetric. The Ricci curvature controls the growth rate of the volume of a spacetime.

$R^{\rho}_{\sigma\mu\nu}$ is the Riemann tensor

$T_{\mu\nu}$ is the stress energy tensor.

A tensor is symmetric if $f(\vec{A}, \vec{B}) = f(\vec{B}, \vec{A})$. Their components also obey $f_{\alpha\beta} = f_{\beta\alpha}$. A tensor is antisymmetric if $f(\vec{A}, \vec{B}) = -f(\vec{B}, \vec{A})$. Their components obey $f_{\alpha\beta} = -f_{\beta\alpha}$.

Spinor space is antisymmetric.

A tensor of type $\begin{pmatrix} M \\ N \end{pmatrix}$ is a linear function of M one-forms and N vectors onto the real numbers. Rank is indicated by the number of different indices, M+N above.

Zero rank tensors are scalars.

First rank tensors are vectors.

$G_{\mu\nu}$ is a second rank tensor.

If a tensor's components vanish (go to zero) in one coordinate system, they will do so in all coordinate systems.

Evaluation of a tensor, say $R^{\sigma}_{\alpha\gamma\beta}$, involves partial differentials in all four dimensions.[64]

Tensor contraction

A form of dot product for tensors. Tensors with different indices are set equal to each other and then summed. It is the operation of a 1-form on a vector.

Contraction is reduction in number of indices by applying a one form to a vector, or by summing over repeated indices of mixed tensors. For example, if a one form operates on a vector, the result gives the number of times they cross each other. It defines a point on each of the one form's planes.

---

[64] See Quick Introduction to Tensor Analysis, R.A. Sharipov, in references. He has several math texts that may be helpful.

Tensor Field

Tensors usually describe potential fields. The metric tensor $g_{\mu\nu}$ describes the gravitational field. We change local coordinates translating them to generalize, and put them in some other region of the field. For example, we can take the velocity of a particle far from a black hole and calculate a new velocity near the horizon.

Tensor product $\otimes$

Product of two tensors is another tensor. Also called the outer product.

Outer Product

$$\begin{vmatrix} q_0 \\ q_1 \\ q_2 \\ q_3 \end{vmatrix} \begin{vmatrix} q_0 & q_1 & q_2 & q_3 \end{vmatrix} = \begin{vmatrix} q_0 q_0 & q_0 q_1 & q_0 q_2 & q_0 q_3 \\ q_1 q_0 & q_1 q_1 & q_1 q_2 & q_1 q_3 \\ q_2 q_0 & q_2 q_1 & q_2 q_2 & q_2 q_3 \\ q_3 q_0 & q_3 q_1 & q_3 q_2 & q_3 q_3 \end{vmatrix}$$

Distinguished between wedge product and the inner or dot product. The inner product produces a scalar arc length. It is the Pythagorean hypotenuse in four dimensions. $|\,a\,b\,|\,|\,c\,d\,| = ac + bd$. Wedge product is defined as $|\,a\,b\,| \wedge |\,c\,d\,| = ac - bd$.

Tensor space

Tensor space is a vector space of the products between vector fields and dual vector fields **V*** composed of one forms. A linear map can be found that maps the vectors to the space.

Tetrad (vielbein)

See one-forms and basis vectors also.

A vielbein (German for many legs) is more general than tetrad and refers to any number of dimensions. The term "tetrad" means a group of four vectors which are orthonormal to one another. The tetrad defines the basis vectors for a space. The tetrad is a 4 x 4 matrix.

The advantage of using the tetrad is that it allows spinors needed for the Dirac equation to be developed in general relativity, and allows the development of the Maurer Cartan structure relations.

In the Evans equations the fields are tetrads and the gravitational and gauge fields are built from tetrads using differential geometry. Tensors were initially used in Riemann geometry to describe curved spacetime. Tetrads give an alternate but equivalent mathematical method of describing general relativity. They are four reference vector fields, e_a. With a = 0, 1, 2, and 3 such that $g(e_a, e_b) = \eta_{ab}$. The tetrad has a linear map from an internal space bundle to the tangent bundle and defines a metric on spacetime.

The tetrad formulation of general relativity is similar to a gauge theory and allows viewing general relativity in a fuller sense with electromagnetism, the strong, and the weak fields explained in the same terms. The tetrad is the frame field e = basis coordinates in the tangent bundle.

A spin structure is locally a tetrad where it has a local Lorentz index and spacetime index. A spin structure contains more information than a metric structure.

The tetrad relates orthonormal bases to coordinate bases.

The torsion tensor is defined for a connection on the tangent bundle. It is the covariant exterior derivative of the tetrad.

Given two topological spaces, A and C, a fiber bundle is a continuous map from one to the other. C is like a projection. For example if A is a human body and C is a shadow, then the mathematical lines connecting them is B and would be the fiber bundle. The ground would be another topological space, C. The space C could be a vector space if the shadow is a vector bundle. The bundle is the mapping. The tetrad is a matrix at each point that describes the connections.

The tetrad is the eigenfunction for the O(3) representation and SU(2) and SU(3) if they should continue to be used.

The tangent space to a manifold is connected by a bundle. The tetrad provides the connections. Those connections could be quite complicated. In the case of the shadow, the formulas describing the angles are contained in the tetrad.

Gauge theory uses fiber bundles. Spinor bundles are more easily described by the tetrad than by the more traditional metric tensors.

It is important to know that the tetrad is a group of four vectors that are orthonormal to one another. The tetrad defines the curvature of the spacetime. By orthonormal we mean that they are at "right" angles to one another *with respect to the curved spacetime*. They are not perpendicular since the spacetime is not Euclidean. These vectors are both orthogonal and normalized to unit magnitude allowing the basis vectors to be used. Two vectors are orthogonal if their dot product equals 0. They are not necessarily at right angles to each other in flat spacetime. The four dimensional spacetime is curved and twisted and the vectors must align.

The tetrad is defined by:

$$V^a = q^a_\mu V^\mu$$

Where V^a is a four-vector in the orthonormal space and V^μ is a four-vector in the base manifold. Then q^a_μ is a 4 x 4 matrix whose independent components correspond to the irreducible representations of the Einstein group.

Each component of the matrix is an eigen-function of the Evans Wave Equation:

$$(\Box + kT)q^a_\mu = 0$$

The gravitational field is the tetrad matrix itself.

General expression for the tetrad matrix

$$q^a_\mu = \begin{bmatrix} q^0_0 & q^0_1 & q^0_2 & q^0_3 \\ q^1_0 & q^1_1 & q^1_2 & q^1_3 \\ q^2_0 & q^2_1 & q^2_2 & q^2_3 \\ q^3_0 & q^3_1 & q^3_2 & q^3_3 \end{bmatrix}$$

The electromagnetic potential field is the matrix:

$$A^a_\mu = A^{(0)} q^a_\mu$$

where $A^{(0)}$ is volt-s/m.

In terms of basis vectors, the inner product can be expressed as $g(\hat{e}_{(a)}, \hat{e}_{(b)}) \eta_{ab}$ where $g(\,,\,)$ is the metric tensor. And $\hat{e}_{(\mu)} = \partial_\mu$ where μ indicates 0, 1, 2, 3, the dimensions of spacetime base manifold. This indicates the curvature at the point.

$$\hat{e}_{(\mu)} = q^a_\mu \hat{e}_{(a)}$$

This is most important in the mathematical development of the unified field theory, but a full explanation is impossible here. It is shown for completeness. The tetrad refers to a 4-dimensional basis-dependent index notation.[65]

In defining the tetrad ($V^a = q^a_\mu V^\mu$), two contravariant vectors are used and where V^a is any contravariant vector in the orthonormal basis indexed a and V^μ is any contravariant vector in the base manifold, μ.

It is possible to use both a contravariant and a covariant vector in the space of 'a' and in the space of μ. Carroll is an excellent source of information on the tetrad and is essential preliminary reading for Evans' works. See references.

Using the definition of q^a_μ, the tetrad in $V^a = q^a_\mu V^\mu$, V could represent position vectors, x, and the tetrad would be $x^a = q^a_\mu x^\mu$.

The tetrad could be defined in terms of Pauli or Dirac matrices or spinors, or Lorentz transformations.

The tetrad mixes two vector fields, straightens or absorbs the non-linearities, and properly relates them to each other.

---

[65] Vielbein refers to any dimensional approach with triad the 3-dimensional version and pentad the 5-dimension. Vierbein is synonymous with tetrad.

q^a_μ can be defined in terms of a scalar, vector, Pauli two-spinors, or Pauli or Dirac matrices. It can also be a generalization between Lorentz transformation to general relativity or a generally covariant transformation between gauge fields.

When metric vectors are used, the tetrad is defined:

$$q^a = q^a_\mu q^\mu$$

and the symmetric metric is defined by a dot product of tetrads:

$$q_{\mu\nu}^{(s)} = q^a_\mu q^b_\nu \eta_{ab}$$

where η_{ab} is the metric (-1, 1, 1, 1).

In the most condensed notation of differential geometry, the Greek subscript indices representing the base manifold (non-Euclidean spacetime) are not used (because they are always the same on both sides of an equation of differential geometry and become redundant). Here we keep them for basic clarity.

The tetrad is a vector valued one form, and can be thought of as a scalar valued one form (a four vector, with additional labels). It is the mapping. If a contravariant 4-vector V^a is defined in the tangent (representation) space and another, V^μ, is defined in the non-Euclidean spacetime (base manifold) of general relativity, then the tetrad is the matrix that relates these two frames:

$$V^a = q^a_\mu V^\mu$$

where the superscript a represents the three space indices and one time index and μ represents the four dimensions of the base manifold of our spacetime.

The Evans Wave Equation with tetrad as eigenfunction states:

$$D^\nu D_\nu q^a_\mu = (\Box + kT) q^a_\mu = 0$$

Tetrad postulate

Cartan's concept that the values in the curved spacetime of our universe can be connected and defined in the flat mathematical index space. Also called "moving frames." The Palatini variation of general relativity uses the tetrad.

The tetrad postulate can be stated a number of ways. $\nabla_\mu e^a_\mu = 0$ or $Dq^a_\mu = 0$ are two.

Regardless of the connection (affine, Christoffel) in three-dimensional Euclidean space, the tetrad postulate (and Evans Lemma) state that there is a tangent to every curve at some point on the curve. The tangent space is then the flat two-dimensional plane containing all possible tangents at the given point.

The basis chosen for tensors does not affect the result.

Theory of everything – Unified Field Theory

A theory that unifies the four fundamental forces – gravity, strong nuclear force, weak force and electromagnetic force. Quantum gravity is one such attempt. General relativity and quantum chromodynamics can be brought together by a unified theory. So far GUT's or Grand

Unified Theories have not brought gravity into consideration. Evans equations unify physics as they are developed since gravity and electromagnetism are now shown to be two aspects of a common geometric origin described by differential geometry.

Time

The fourth dimension. In general relativity and Evans' unified field theory it is treated as a 4th spatial dimension. In black holes the mathematics shows reversal of time and space indicating that they are different aspects of the same basic thing. In Minkowski spacetime, time and space are combined into spacetime.

The use of c in many equations indicates that time is being converted to spatial distance. For example ct in the invariant distance in special relativity or 1/c in various wave equations or the d'Alembertian.

c is the velocity of time.

Time becomes an operator in the quantum sector of Evans' equations.

Torsion

Twisting and curving spacetime. Spinning spacetime.

Torsion is the rate of change of an osculating plane. It is positive for a right hand curve and negative for a left hand curve. The radius of torsion is ϕ or $\sigma = 1/\tau$

If referring to a group, then the torsion is the set of numbers that describe the torsion of the individual elements.

Cartan torsion is the antisymmetric parts of a Christoffel symbol.

Also referred to as "dislocation density" or the "second curvature" in some texts where "first curvature" is our standard definition of curvature in three planes. Torsion is then curvature in the 4th dimension.

Torsion is zero only when a curve fits entirely inside a plane.

Torsion tensor

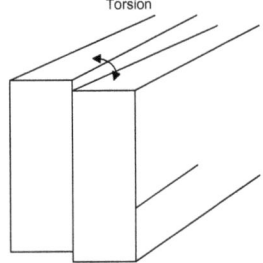

The torsion of a curve is a measurement how much the curve is turning, or of twist. The torsion tensor measures the torque of one 4-vector field with respect to another 4-vector field.

Torsion describes a connection in spacetime which is antisymmetric. The torsion tensor is a connection on the tangent bundle and can be thought of as the covariant exterior derivative of the tetrad.

Trace
Sum of the diagonal terms of a matrix.

Unified Field Equations
The Evans equations are the unified field equations. He has developed homogeneous and inhomogeneous, equations. They are reduced to the Einstein and Maxwell Heaviside limiting forms and they become the field equations of O(3) electrodynamics. The equations can be used to derive all the known equations of quantum mechanics, general relativity, and Newtonian mechanics.

Unit vectors
See Basis vector.

Vacuum
The word vacuum can be used several ways. It is the spacetime of quantum mechanics and experiences polarization and has content. We can see more clearly from Evans' work that curvature and torsion are everywhere. Since curvature and torsion can produce particles, the vacuum is more substantially understood due to Evans' clear explanation of electromagnetism as twisting or spinning spacetime.

Vacuum has been described erroneously in some texts as a total absence of anything. This could be a matter of definition, but void would be preferred for nothing.

The stochastic granular school sees the vacuum as a tenuous something. Given that quarks can be pulled out of the vacuum, it has at least a potential and that makes it something. It may be that the electromagnetic fields of the universe are the only thing present in the vacuum and the vacuum polarization is field based.

In the older viewpoint, given that an electron's location approaches infinite distance ($r \rightarrow \infty$) from the center of every atom at a low probability ($|\Psi|^2 \rightarrow 0$), then the vacuum is full of fleeting potential matter and electromagnetics. In Evans' viewpoint, curvature and torsion in one region extend into others.

False vacuum is hypothesized to have a positive energy and negative pressure and to be the force that drove the expansionary period of the universe. This is purely theoretical.

Void is the real nothing outside the universe and maybe exposed inside the ring singularity of a Kerr black hole. That void is as close as the spacetime in front of your face, but is unreachable. The vacuum interface to the void is everywhere in 4-dimensional spacetime.

Glossary

Evans gives us a definition of quantum vacuum equaling Minkowski spacetime. This is quite logical since it is the vacuum of special relativity and quantum theory. Spacetime is then curved vacuum. Vacuum is then very low mass density spacetime. It can however be severely curved or torqued near a black hole.

Vectors

Vectors are numbers with magnitude and direction. The vector is sometimes indicated as a bold letter or with an arrow → above it. In order to be added or multiplied together, they must have the same units. A negative of a vector has the same magnitude but is opposite in sign which indicates direction.

The vector is a geometric object. It does not need a reference frame to exist. It can be changed from one reference frame to another by math operations.

The velocity of a particle is a vector tangent to the path in spacetime that the particle follows. This is tangent to the 4-dimensional path.

Components of a vector are the axes of a rectangular coordinate system.

Tensors and vectors are similar in this respect.

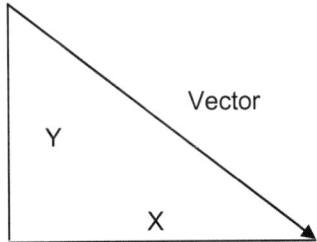

The expressions are often: x_0, x_1, x_2, and x_3. Thus x_a means x_0, x_1, x_2, and x_3.

At each point in spacetime there are a variety of vectors describing the curvature, torsion, mass, and energy.

The 4-velocity of a particle is a vector tangent to the path in spacetime that the particle follows. This is tangent to the 4-dimensional path.

Vector Field

A map which gives a real or complex vector for each point on a space. The bundle is all the vectors of the entire space. A vector field is a section of the entire bundle. The vectors exist in the mathematical tangent space. The information about each vector is within the field. The number of dimensions is defined and the bundle could be infinite. Typically, the tangent space to our real four dimensional spacetime has all the mathematical information about the mass, curvature, etc.

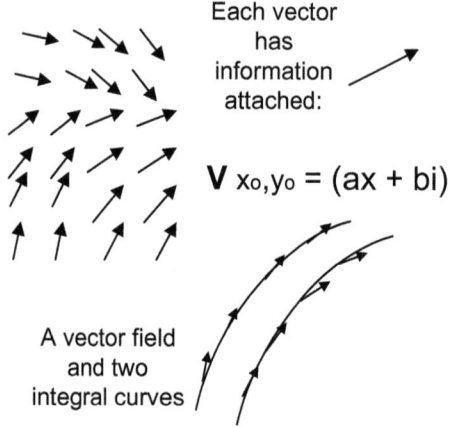

V xo,yo = (ax + bi)

A vector field and two integral curves

Vector space

A set of vectors can be defined for which addition and multiplication produce real or complex scalars. The vectors are placed in a vector space and can be moved among reference frames by tensors.

In quantum mechanics vectors describe the state of a system – mass, velocity, etc. The vectors can be mathematically manipulated in the vector space. Just why is unknown, but those darn weird vector manipulations are brought back to our universe from the mathematical space and they are very accurate in predicting results of experiments.

Vielbein

See Tetrad.

Wave equation

Wave equations are partial differential equations:

$$\nabla^2 \psi = \frac{1}{v^2} \frac{\partial^2 \psi}{\partial t^2}$$

∇^2 is the Laplacian and ψ is a probability. ($\square^2$ is the d'Alembertian, the 4-dimensional version of the Laplacian.)

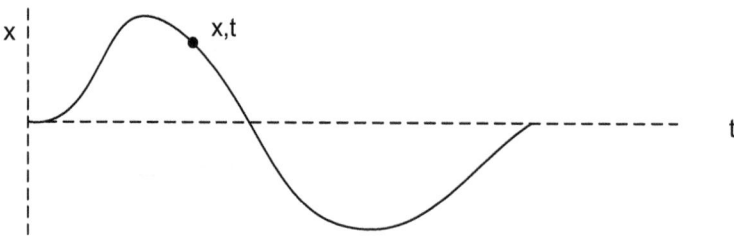

A sine wave is a function of x and t in two dimensions.

Wave equation, Evans

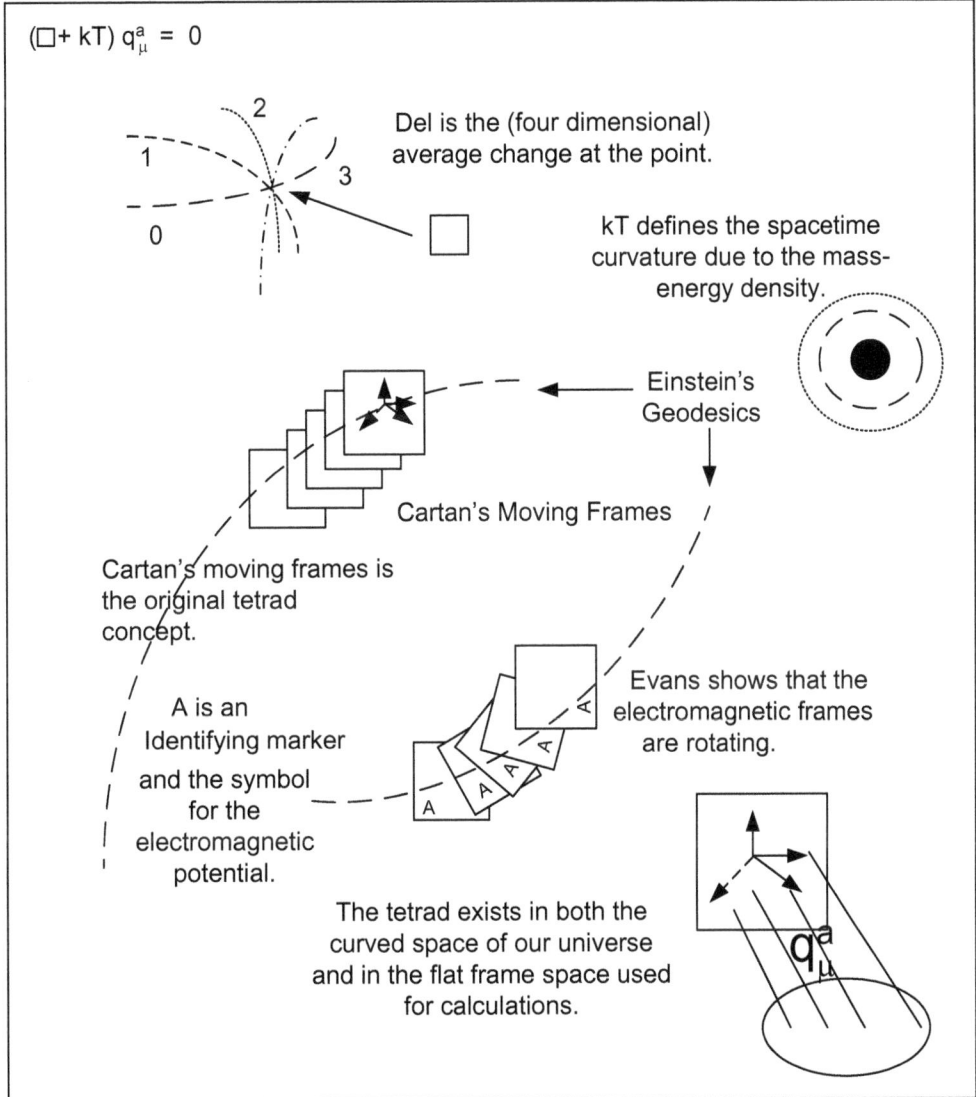

Wave or Quantum Mechanics

Study of basic physics starting with the Planck quantum hypothesis. The dual nature of existence as a wave and particle is recognized as a foundation concept.

Wave number

$\kappa = 1/\lambda$ or $k = 2\pi/\lambda$. The rest wave number is defined by Evans as $\kappa_0 = 1/\lambda_0 = |R_0|^{1/2}$. It is thus defined by scalar curvature, R. This is a unification of basic general relativity with wave mechanics.

If sinusoidal, a wave with wave length λ and frequency ω can be written in terms of wave number. For example, the electric field can be written $E = E_0 (\kappa x - \omega t)$. Here $\kappa = 2\pi/\lambda$. In rotational motion ω is angular velocity = radians per second.

The main point in Evans' work is that in the weak field limit, wave number and curvature are equated.

See http://scienceworld.wolfram.com/physics/Wavenumber.html

Wave velocity

$v = \lambda f = (2\pi/k)(\omega/2\pi) = \omega / k$

Weak field

Low curvature and torsion. Low mass, energy, velocity, electrical potential, or gravitational field.

Wedge or exterior product

The wedge product is a vector product defined in a vector space. It is the multi-dimensional version of the cross product.

References

(All web references last accessed in July, 2007.)

Advances in Physical Chemistry, Modern Nonlinear Optics, Three Parts, Evans, M. W. and Kielich, S. (Editors), Wiley Inter-Science, New York, 1995 to 1997.

Classical and Quantum Electrodynamics and the B(3) Field, Evans, M. W. and Crowell, L.B., World Scientific, Singapore, 2001.

The Enigmatic Photon Volumes 1 to 5, Evans, M. and Vigier, J.-P. et.al., Kluwer Academic Publishers (now Springer), Dordrecht, 1994 to 2002.

Generally Covariant Unified Field Theory – The Geometrization of Physics Volumes I, II and III, Evans, M. W., Arima Publishing, Bury St. Edmunds, Suffolk, England, 2005 to 2006.

The Geometry of Physics, An Introduction, Frankel, T., Cambridge University Press, 1997.

Gravitation, Misner, C. W., Thorne, K. S., Wheeler, J. A., W.H. Freeman and Company, New York, 1973.

Lecture Notes on General Relativity, Carroll, S. M. at http://pancake.uchicago.edu/~carroll/notes/ or http://xxx.lanl.gov/PS cache/gr-qc/pdf/9712/9712019v1.pdf

Methods of Mathematical Physics, Morse, P. and Feshbach, H., 2 volumes, McGraw Hill Book Company, Inc., 1953.

The Meaning of Relativity, Einstein, A., Princeton University Press, 1921, 1945.

Physics for Scientists and Engineers, Giancoli, D. C., Prentice Hall, New Jersey, 2000.

Quantum Mechanics: Symmetries, Greiner, W. and Mueller, B. Springer, New York, 1994.

Quantum Field Theory, 2nd Edition, Ryder, L. H., Cambridge University Press, Cambridge, England, 1996.

Quick Introduction to Tensor Analysis, Sharipov, R. A., http://www.physics.rutgers.edu/~sergei/507/Sharipov.pdf

Towards a Nonlinear Quantum Physics, Croca, J. R., World Scientific Pub. Singapore ; River Edge, NJ, 2003.

Index

∂, the Partial Differential, **82,** 104, 127, 129, 311, 356, 364
$A^{(0)}$, 97, 113f, 132, 146f, 154, 168, 175 184, 190, 254f, 258f, 261f, 268, 270, 279f, 289, 291, 294
AB effect, see Aharonov-Bohm effect
Abelian, 99, 225, 291, 295, 310, 318, 321, 327, 334, 345, 352
Abstract space, Fig 4-12, 291
Acceleration, 12, 19, 34, 36, 55, 82, 93, 111, 120, 127, 183, 188, 191, 255, 293, 302f, 308, 316, 323f, 326, 343, 346, 349f
Action, 64, 69, 111, 200, 218ff, 221, 230, 257, 272, 276, 293, 303, 315, 317, 340
Affine transformation, 292
Afshar, 215
Aharonov-Bohm effect, 3, 174, 230, **245-253**, 257f, 268, 272f, 286f, 292, 327, 330
AIAS, Alpha Institute for Advanced Study, 6, 102, 109, 139, 231, 242, 244, 281, 291, 334, 339
Ampere, 112ff, 116ff, Fig 5-4, 230, 293, 346
Analytical function, 293, 327
Angular Momentum, 24, 27, 56, 61, 63, 66, 69, 75f, 78, 129, 187, 217ff, 223ff, 230, 258, 276, 292ff, 305, 323f, 333, 347, 349, 352, 354.
Ansatz, **132**, 255, 262, 268, 294
Antiparticles, 14, 115, 130, 190, 298, 287
Antisymmetric tensor, 187, 294, 309, 315f, 322, 350
Asymmetric, 95, 136f, Fig 6-2, 139, Fig 6-9, 154f, 179, 184, Fig 8-3, 194, 259, 267, 269, 279, 312, **328,**
Atom, 14f, 30, 57, 59, 66, 70f, 75, 78, 159, Fig 8-6, 197, 199, 280, 282, 297, 337, 339, 347, 349, 362
B field, Fig I-4, **113ff**, Fig 5-3, 116f, Fig 5-4, Fig 5-5, 118f, **330,** see Magnetic Field also
$B^{(3)}$ Spin Field, 5f, 72, 147, 168f, Fig 7-4, 173, 183, Fig 8-3, 191, Chapter 11, 246, 248, 250ff, 258, 268, 287, 289, 294f, 299, 334, 353
Basis vectors, 28f, Fig 1-6, Fig 1-7, 49, 51f, 92, Fig 4-11, 97, Fig 4-12, 101, 103, 106f, Fig 4-16, 124, 135, 147, 150, 152, 165, Fig 12-2, Fig 12-6, Fig 15-5, 295f, 298, 300ff, 305, 310, 317, 332, 334f, 336, 355, 357ff, 362
Big Bang, 5, 18, 68, 180, 206, 274, 276, 286
Blackbody, 57
Blackbody radiation, 297
Black holes, 2, 4f, 12, 15, 22, 30, 38, 43, Fig 2-6, 46, Fig 2-9, 49, 68, Fig 3-5, 74, 91, 124, 126, 174, 200, 209f, 235, 256, 268, 323f, 327, 338, 342, 357, 361ff
Bohr complementarity, 214
Boltzman constant, 73
Boson, Fig 1-4, 14, 134, 165, 169, 233, Fig 12-1, 237ff, 268, 315, 317, 337, 347
C, charge conjugation, see Charge Conjugation

Index

c, velocity of light, 18, 20ff, Fig 1-2, Fig 1-3, 34, 37, 39, 64f, 70ff, 82, 90f, 106, 183, 197, 204f, Fig 9-3, 208, 210, 212f, 261, Fig 15-3, 278ff, Fig 15-4, Fig 15-5, 297, 319, 326, 330, 333, 349, 361
Canonical, 162, 201, 297
Cartan differential geometry, 6, 9, 97, 101, 146, 195, 292
Cartan tetrad theory, 297
Cartesian vectors, 106
Casimir Invariants, 260, 294, 297, 324
Causal, 5, 58, 61, 162, 187f, 215, 217, 219ff, 268, 271, 297f, 343
Charge Conjugation, 167, 190, 229, 255, 291, 298, 352ff
Charge, 2, 3, 14, 18f, 45f, 112ff, Fig 5-2, 117, 175, **190**, 274, 279, Fig 15-7, 291
Charge-current density vielbein, 298
Christoffel symbol, Γ, 142, Fig 6-7, 260, 292, 298f, 301, 306, 344
Circular Basis, 231, 299
Classical and Quantum, 145
Classical limit, 299, 308
Clifford algebra, 300
Closure, 321, 352
Commutators, 299f
Co-moving reference frame, 20, 354
Complex conjugate, 219, 229, 300, 322, 338
Complex numbers, 292, 300, 322, 333, 348, 352f
Components, 3, **Fig 1-6,** 36f, 47, 49, 52, 70, 84, **90f,** 124f, 138, 152, 159, 166f, 179, 190, 194, 204, 223, 226, 228ff, 236, 262, 269, 278, 296, 300, 302f, 305, 312, 316, 331f, 335, 343, 350, 353ff, 358, 363
Compression, Fig 1-1, 17f, 21, Fig 1-3, Fig 1-9, 36, **Fig 2-1,** 40f, 43, Fig 2-6, 92, 126, 147, 183, 193, 206, 275, 338, 354
Compton Wavelength and Rest Curvature, 204
Compton wavelength, **127f**, Fig 5-8, 132, 147, 173, 199, 204f, Fig 9-3, 220, 226, 234, 263, Fig 15-6, 300
Conservation Laws, 18f, 111, 239, 241, 324, 326, 350, 354, see Noether also.
Contraction, 18, 20ff, 36f, 40f, 70, 102, 155, 184, 229, 286, 301, 304, 323, 335, 356
Contravariant vectors, 302f, 310, 335, 359
Contravariant, 102, 301, 302ff, 359f
Coordinate basis (Greek index), 152, 304
Coordinate system, 3, 79, 97, 101f, 106, 296, 301, 304, 331, 361
Copenhagen School, 61, 162, 217
Correspondence Principle, 24, 304
Cotangent Bundle, 304, 314
Coulomb force, 73, 112ff, 119-122, 185, 230, 256
Covariant derivative operator, 305
Covariant Exterior Derivative, 108, 147, 259f, 294, 306, 324, 347, 358
Covariant, 3, 12f, 98f, 102, 108, 111, 121, 124, 139ff, 146ff, 153, 162f, 168, 172ff, 189, 219, 228, 231, 234, 237, 241f, 244f, 251f, 259f, 266, 286f, 294, **302-306,** 310, 313, 318, 326, 331, **334f,** 339, 347f, 358ff, 362

Croca, 65, 162, 214, 219f
Cross Product, see Vector Product
Curl, **86**, **Fig 4-7**, 108, 294, 306, 314, 330
Current, 112-119, Fig 5-1, Fig 8-2, 188, 293, 298, 316, 330, 346
Curvature and Torsion, Fig I-2, 56, 76, **136**, Fig 6-2, 152, Fig 6-9, 154, 178, Fig 8-1, 181, Fig 8-3, 189, 206, 210, 227, 235, 253, 257, 263, **269f**, 273ff, 279, 282, 288f, 313, 315, 337, 342, 345ff, 362, 366
Curvature, 6f, 9, Fig I-2, 12, 15, 18, 37, 39f, Fig 2-5, 44, Fig 2-7, see R, curvature
Curved spacetime, 2, Fig I-4, 35ff, **41,** Fig 2-4, Fig 2-9, 49f, 52, Fig 4-1, 79, 98f, 126, 135, 142, Fig 6-5, 151, 180, 188, 191, 194, Fig 10-2, 226, 228, Fig 11-5, 230, 256f, 276, 280, 288, 295f, 301f, 304, 308, 311, 319, 322, 327, 331, 334, 344, 347, 351, 355, 357f, 360
d'Alembert equation, 63, 104f, **123**, 152, 157ff, Fig 7-1, 161ff, 173, 191, 194f, Fig 8-8, 201, 308, 325, 361, 364
de Broglie guiding theorem, 72
de Broglie wavelength, 127f, 200, 215, 220, 278, 309
Del, ∇, Fig 4-7, **104,** Fig 7-2, Fig 8-8, 304
Derivation of the Boson Masses, 237
Derivatives, **81f**, Fig 4-2, 84, 89, 102, 104ff, 107f, Fig 4-17, 147, 259f, 294, 304ff, 309, 313f, 318f, 334, 347, 358, 362
Differentiable manifold, 8, 46, 79, 179, 193, 248, 250, 293
Differentiable, 318, 325, 327, 346
Differential forms, 89, 218, 309, 322, 330
Differential Geometry, 5f, 9, 15, 25, 33, 49, 51, 53f, 81, 97f, 100f, 103, 109, 121, 131, 135, 137, 146, 149ff, 160, 163, 165, 174, 177, 179, 189f, 194f, 215, 240, 244, 251, **257-265**, 268f, 271, 273, 286, 289, 291ff, 314, 318, 330, 333, 357, 360f
Dimension, 6f, Fig I-1, 9, 11f, 22, Fig 1-4, 25f, **27**, 32, 36ff, 41, Fig 2-4, 47, 49, 51ff, 69, 79ff, 84, 84ff, 89ff, 93, 102, 103, 106, 123f, 125f, 130, Fig 7-1, 165, 171, 179, 188, 190ff, 210, 225, 228, Fig 11-6, 251, Fig 13-6, Fig 15-1, 283, Fig 15-7, 292, 295, 308, **309**, 313, 320, 322, 327, 330ff, 342, 343ff, 347f, 352, 360, **361**, 363, 365
Dirac Equation, 63, 130f, 172f, 189f, 201, 203ff, 267, Fig 15-5, **309,** 351
Dirac spinor, 130, 201, 309
Directional Derivative, 86, **104f,** Fig 4-15, 309, 325, see Gradient Vector also
Divergence, **87**, Fig 4-8, 108, 314, 330
Dot product, 26, **Fig 4-4, 83ff**, 102, 184, 292, 310, 322, 341, 356ff, 360
Double slit experiments, Fig 3-2, Fig 3-3, 72, 215, Fig 13-3, 248, 292
Dual, 102f, 107, 109, 148, 227, 277, 302ff, 310, 315, 322, 335, 340, 357, 365
Duality, 56, 58, 128, 200, 204f, 273, 322
E, charge, 13, 113ff, Fig 5-2, Fig 5-4, Fig 5-5, 122, 188, 222ff, Fig 11-5255, 288, 298, 312f, 330
E, energy, 18, 24, 58, 69ff, 74, 78, 114, 181, 197, 200, 204f, 209, 212f, 244, 277, 319, 340
$e^a{}_\mu$, , 98, 317, 360
Egg crate grid, 140

Eigen equations, 78, 310
Eigen vectors, 310
Eigenfunctions, 171, 310
Eigenvalues, 29, 65, **78**, 146, 152, 157, 161f, 169, 237, 242, 256, 260f, 271, 281, 284, 310f
Einstein constant, 135, 311
Einstein Field Equation, **135,** 138, 153, 155, 159, 194, 227, **311,** 324
Einstein group, 166, 294, 311, 334, 351, 353, 358
Einstein tensor, **37,** 94, **125,** 308, 311f, 331, 343, 349, 356
Einstein, 4f, 2-9, Fig I-1, 10, 12, 16, 18, 24f, 34f, 37ff, 47, 49, 51, 53, 55, 57f, 60, 69, 71f, 77, 79, 94f, 100f, 110, 114, 120, 125f, 130ff, 134f, 138-143, 145ff, 150, 153ff, 159, Fig 7-2, 162-167, Fig 7-3, 172, 175, 177f, 180f, 183, Fig 8-3, 187f, 191, 194ff, 200ff, 204f, 208f, 214, 221f, 226ff, 231-235, 237, 241, 244, 254, 259f, 264-270, 272ff, 277, 284, Fig 15-5, 286, 289, 292, 294, 297, 304, 307f, 311ff, 315, 318, 324, 326, 331, 343, 349-353, 356, 358, 362, 365
Einstein-Cartan-Evans, see Unified Field Theory
Electric Field, Fig I-4, 19, 111ff, 117ff, 122, 166, 187f, 190, Fig 11-1, 226f, 229f, 255f, 280, 288, 298, 312, 330, 340, 346, 349, 366
Electrogravitic Equation, 123,173, 254, 263
Electromagnetic Field, Fig I-4, 14, 35, 50, 55, 79, 97, 99, Fig 5-1, 139, 141, Fig 6-6, 146, 150f, 165, 168, 175, 186, Fig 8-4, 187, Fig 8-5, 190, Fig 8-6, 193, 194, 222, 224ff, Fig 11-4, 227, 229, Fig 11-6, 247, 249, 253, 257, 260f, 265, Fig 15-1, 268, 274, 279f, Fig 15-5, 294f, 304, 310, **312f,** 324, 334, 340, 345, 353, 362
Electromagnetism, 2f, 5ff, 9, Fig I-4, 14, 37, 46, Fig 2-9, 51, 68, 95, 99, 121, 123f, **134f,** Fig 6-2, 141ff, 145ff, 150ff, 159, 164, 166ff, 173f, 178ff, 181, 183ff, Fig 8-3, 189, 193f , 195, 223, 226, 228, 229, 232, 236, 255, 256, 260, 264, 265, 267, 270, 271, 274, 277, 285, 289, 290, 296, 309, 312, 314, 316, 322, 323, 327, 335, 351, 358, 362, 363
Electron, 14f, 18, 56f, Fig 3-1, 59f, 63, 66, 69ffm 75, 78, 112f, 115, 117, 128, 134, 146, 151, Fig 7-5, 171, 175, 190, Fig 8-6, 197, 199f, **210f,** Fig 9-5, 223, 231, 233, 235, 237f, 242, 246, Fig 13-3, 248, Fig 13-4, Fig 13-5, Fig 13-7, 258, 268, 275f, Fig 15-3, **279ff,** Fig 15-7, 289, 291f, 293, 298, 308ff, 312, 317, 324, 337ff, 342, 347f, 362
Electroweak theory, 2, 68, 151, 166, Fig 8-2, 189, Fig 12-1, Chapter 12, 268, 321
Elementary Particles or Fields, 312
EMAB, 249, 258
Energy, 2, 4,14, 17-24, Fig 1-2, Fig1-3, Fig 1-7, Fig 1-9, 35-38, Fig 2-1, Fig 2-2, 41, Fig 2-4, 43ff, **56ff,** 61, 63ff, Fig 3-4, 68-79, 94, 104, 111, 123-129, 135-141, 145ff, 150f, 154-157, Fig 7-2, 172f, 175, 179-185, 189-193, 197, 199ff, 204f, Fig 9-3, 208ff, Fig 9-6, 219f, 231, 239ff, 258, 261, 267, 270, **278,** 297, 300, 313, 315, 321, 323, 325f, 333, 337f, 340, 354, 365
Entanglement, see Non-locality
Equivalence principles, 6, 91, 54f, 110, 172f, 180f, 194, 288, 313, 337

Euclidean spacetime, Fig I-1, 22, 24, 28, Fig 2-10, 89, 100, 102, 134, 136, 139, Fig 6-7, 160, 166, 180, 191, 201, 271, 302, 313ff, 323, 333, 343, 358, 360
Evans Field Equation, Chapter 6, **139 Equation (8)**, Chapter 7, 181, 187, 190f, 194f, 197, 201, 215, 217, Fig 10-2, 220, 234, 240, 254f, 260, 266, 271, 284, Fig 15-5, 310, 317, 358, 370
Evans Spacetime, 9, Fig I-2, 11, 33, 62 165, Fig 7-3, 179, Fig 8-3, 188, Fig 8-8, 205, 231, 235, Fig 15-1, **313**
Evans Wave Equation, 63, 69, 123, 130, 145f, **152,** 153 Equation (17), Chapter 7
Event, 22, 25, Fig 1-9, 52, 79, 85, **91**, 125, 272, 296, **313**
Explosive decompression, 183, 338
Exterior (Outer or Tensor) Product, Fig 4-9, 89, 341, 366
Exterior derivative, 89, 107f, Fig 4-17
Faraday's Law, 117, Fig 5-4
Fiber bundle, Fig I-3, 99, 101, 149, Fig 6-8, 167, Fig 7-3, 314f, 358
Field equations, 15, 140, 153, 160, 170, 173, 215, 266, 270, 272, 315, 353, 362
Field, 3ff, Fig I-4, Fig 2-8, Fig 5-2, Fig 5-5, Fig 5-6
Flux, 113ff, **Fig 5-3,** 116f, Fig 5-4, 119, 167, 184, 280, 291, 294, 318, 340
Fluxon, 184, 291, 315
Force, 2f, 5, 12, Fig I-4, 14, 26, Fig 1-5, 34ff, 46, 55, 68f, 83, 94, 99, 110-116, 116, 120, 134, 135, Fig 6-3, 146, 159, 165,f 169ff, 174, 179ff, Fig 8-2. 183, 185, 189, 233, Fig 15-1, 278ff, 303, 312f, 315, 317, 320f, 348ff, 360
Forms, Fig 1-8, Fig 4-9, 89ff, 102, 108, Fig 4-18, **152**, Fig 7-5, 179, Fig 9-3, 218, 265, 275, 279, Fig 15-4, 303f, **309**
Four-spinor, 172f, 202f, 309
Four-momentum, 72, 90, 94, 316
Four-Vectors, Fig 2-3, Fig 2-9, Fig 2-11, 51f, 63, 84f, 90f, Fig 4-11, 111, 137, 165, 303, 316f, 342, 351, 360f
Four-velocity, 72, 90, 94, 135, 147, 316, 349
Frames, 16, Fig 1-1, 22, 32, 34, Fig 2-1, 74, **98,** Fig 6-4, see Tetrad and Reference Frame also
Frequency, 18, 58, 69ff, 128, 197-201, 210, 213, 258, 262, 276, 278, 300, 316, 319, 337, 340, 366
Fundamental Invariants of the Evans Field Theory, 260f
Fundamental Particle, 317, 337
$g = e/\hbar$, 223, 226,
g, coupling constant 238, 294,
G, curvature, 37, 73, 94, 126, 135, 153f, 156, 175, 184, 270, 308, 311f, 349, see R also
g, gravitational acceleration, 55, 120, 188, 255ff
G, gravitational constant, 37, 64, 120f, 123, 255, 312, 317, 341
g, metric tensor, 31, 54, 91ff, 164f, 169, Fig 15-5, 331, 356f, 359
Gauge invariant fields, 141, 151, 153, 166, 318
Gauge theory, 98f, 101, 148f, 167, 187, 291, 314f, 317f, 321, 323, 355, 357f
Gauss' laws, 114, Fig 5-2, **115,** Fig 5-3, 119, 260, 318
Gell-Mann, 170, 189
General Relativity and Quantum Theory, 2ff, 145, 165, 188, 265

General Relativity, 2ff, Fig I-1, 8ff, Fig I-3, 12, 16, 19, 21, 24f, 26, 29, 31, Chapter 2, Fig 2-2, 56f, 62f, 67ff, 73f, 79, 84, 87, 90f, 98ff, 111, 124ff, 130ff, 135, 141, 143, 145ff, Fig 6-8, 151f, **154**, 156, 160, 162, 165ff, Fig 7-3, 169ff, 177ff, 185ff, 194f, Fig 8-8, 200ff, 213f, 219ff, 223, 228ff, 235, Fig 2-12, 237, 243, 245, 248, Fig 13-5, 253, 260, 263ff, 271ff, 279f, 284, 287, 289, 291, 297f, 301, 311, 327, 341, 345, 355, 357, 360ff

Generally Covariant, 3, 12, 98f, 102, 121, 139ff, 146, 153, 162f, 168, 172, 174, 189, 219, 228, 231, 234, 237, 241, Fig 12-6, **244f**, 251, Fig 13-6, 266, 286f, 306, 310, 318, 334, 339, 360

Geodesic, 34f, Fig 2-4, 43, 79, Fig 4-1, 93, Fig 6-4, 142f, Fig 6-5, Fig 7-2, 274, 297f, 305, 318, 344, 365

Geometricized and Planck units, 73f, 110, 319

Gluon, Fig I-4, 14, 56, 67, 135, 146, 170f, 189, 313, 315, 317, 319, 337, 348

Gradient Vector, 86, 89, **104f**, Fig 4-15, 309, 325, see Directional Derivative also

Gradient, 79, 86, Fig 4-7, **89**, 102, **104f**, 107f, 120f, Fig 5-6, 123f, 165, 304, 306, 312f, 319f, 326

Grand Unified Field Theory (GUFT), 68, 152

Gravitation and electromagnetism, 2, 95, 121, 123, 143, 150, 154, 166, 173, 178, 185f, 189, 193, Fig 11-5, 256f, 263f, 266, 269, 273, 276, 288f, 315

Gravitation, 2ff, 12, Fig I-4, 17, 19, 21, 31, 35ff, Fig 2-2, Fig 2-9, 55, 68, 76, 94, 97, 110,f 119ff, Fig 5-6, 125, 134ff, Fig 6-2, 138ff, 146, 150ff, 163ff, 172ff, 180ff, Fig 8-2, 190ff, 226f, 255ff, 264ff, 270ff, Fig 15-3, 283ff, Fig 15-5, 296, 308, 311, 315, 317, 323, 324, 332, 344, 350

Gravitational constant, see G

Gravitational field, 3, 5, 9, 12, 17f, 21f, 24, 34ff, Fig 2-9, 49f, 52f, 93, 97ff, 111, 120f, Fig 5-6, 123f, Fig 6-3, 141, Fig 6-6, 146, 150, 153, 165, 167, 170f, 174, 186, Fig 8-4, 188, 227, 237, 256, 261, 265, Fig 15-1, 270, Fig 15-6, 304, 310, 318, 320, 322, 338, 341, 344, 349f, 354f, 357, 359, 366

Groups, 300, 307, 311, 320ff, 333f, 348ff, 353

Hamiltonian, 321, 333, 339

Heisenberg Uncertainty Principle, 5, 56, 65, 163, 173, 187f, Chapter 10, Figure 10-1, 268, 271, 300, 272

Helix, Fig I-2, 137, Fig 6-5, Fig 7-4, Fig 8-1, Fig 8-4, Fig 11-3, 226, Fig 11-5, Fig 13-2, 247, **251**, Fig 13-6, 266, 295, 347

Higgs, 5, 171, 189, 233, **237f**, 244, 268

Higher symmetry electromagnetism, 321

Hilbert Space, see Abstract Space

Hodge dual, **Fig 4-18,** 315, 322

Homeomorphism, 322

Homogeneous, 19, 68, 180, 293, 322, 362

Homomorphism, 321f, 353

Hydrogen, 66, 322

Index, 28, 47, Fig 2-9, 49, 84, 96ff, Fig 4-12, 102, Fig 4-16, 108, 145f, 152, 162f, 165, 167, 168ff, 189, 234, Fig 12-2, Fig 12-3, Fig 12-6, 267, 271, 292, 299, 302, 304, 306, 313, 323, 341, 347, 358ff

Index contraction, 323

Inertia, 34, **55,** 110, 173, 180, 293, 313, 318, 332, 350
Inner Product, 26, 31, **84f**, 92, 141, 155, 167, 310, 341, 357, 359
Interference pattern, Fig 3-2, 246, Fig 13-3, 248, Fig 13-4
Internal gauge space, 323, 334
Internal index, 171, 179, 323
Invariance, 18, 20, 92, 100, Fig 6-3, 298, 318, 323, 326, 333, 350, 352f
Invariant distance, **22**, Fig1-4, 24, 26, 28, 31, Fig 1-9, 51f, 54, 91, 248, 323, 332, 351, 361
Inverse Faraday Effect (IFE) 147, 222f, 250, 258, 268, 324, 340
Irreducible, **125,** 166, 324f, 353, 358
Isomorphism, 321, 325, 353
Jacobi matrix, 325
Kerr black hole, 49, 209, 362
Klein Gordon Equation, 63, 131, 152, 165, 172f, 193, 201f, 214, **220**, 267, Fig 15-6, 325, 351
Kronecker delta, 325
Lagrangian, 325, 333
Laplacian, 104, 121, 123, 129, 308, 319, 325, 364
Law of Conservation, 326
Least action, 219, 257
Least curvature, 67, 171, 173, 197, 204f, 210, 212f, 219f, 234, 237, 256f, 263, 272f
Levi-Civita symbol, 134, 143, 298, 301, 307, 326
Lie group, 166, 318, 327, 334
Linear transformation, 96, 305, 311, 327
Local, 21, 35f, 59, 76, Fig 15-3, 279, Fig 15-4, 317, **327**
Lorentz group, 311, 327
Lorentz–Fitzgerald contraction, 20, Fig 1-2, 70, 209, 349
Magnetic field, 2, Fig I-4, 16, 63, 114ff, 140, Fig 6-3, 165, Fig 7-4, 188, 190, Fig 8-7, Chapter 11, 246, 248, Fig 13-4, Fig 13-7, 259ff, 266f, 280, 289, 292ff, 312, 319, 324, 330, 340, 346
Manifold, 8, 22, 46ff, Fig 2-9, 81, Fig 4-3, 89, 96ff, 99ff, Fig 4-12, Fig 4-16, 124, 156, 163f, 169ff, 193, Fig 8-7, 219,235, Fig 12-2, Fig 12-3, 248266, 271, 292f, 299-302, 314, **327**
Mass, 2,10, I-4, 18f, 35f, 40ff, Fig 2-4, Fig 2-8, 47, 55, 58ff, Fig 3-5, 70ff, 78ff, Fig 4-1, 91, 94, 110f, 120-139, Fig 5-6, Fig 6-1, 145, 147, 154, Fig 7-2, 169, 173, 177, 180, 189, 196, 199ff, 205, Fig 9-3, 208-212, Fig 11-6, 233f, 237-242, 255-258, 260-264, 275f, 278-282, 294, 305, 307ff, 312f, 319, 333, 337-341, 354, 366
Mathematics = Physics , 131, 269
Matrix , 28, 47, Fig 2-9, 89, 92, 95-98, 131, 147150, 166, 174, 184f, Fig 8-8, 201, 294, 298, 304f, 310f, 325, 327, **328f,** 331f, 338, 341, 348, 353, 357ff, 362
Maxwell's Equations, **114**, 304, 330f
Maxwell-Heaviside, 169, 173, 222, 226, 250ff, 321, 353, 362

(The) Metric, **6-12**, 24, 31, Fig I-1, Fig 1-9, 47, 51ff, 82, 85, 91-96, 98, 104, 139-143, 153f, 163, Fig 7-3, 179f, 205, 227, 228-231 Fig 11-6, 235, 265f, 269, 273f, 300, 313, 330, **346f**
Metric tensor 31, 52, 54, 91ff, 95f, 101, 125, 136f, 172, 187, 189, 228f, 303, 311, 318, 331f, 345, 347, 349ff, 356ff
Metric Vectors, 51f, Fig 2-11, 91, 137, 172, 178, 189, 231, 311, 331, 360
Minimum particle volume, 197, 213
Minkowski spacetime, 3, 7, Fig I-1, 11, 22, Fig 1-4, 24, 33, 44, 47, 51, 62, 96, Fig 6-8, 152, Fig 7-3, 180, 191, Fig 9-4, 235, Fig 12-3, 244, 256, 273, 322f, 332f, 347, 361, 363
Modulus, 333
Momentum, Fig I-4, **24,** 27, 56, 58, 65f, **69-72**, 90f, 104, 111, 125, 128f, 139, 141, 180, 183, 185, 191, 201, 210, 215-221, Fig 10-2, 239ff, Fig 12-4, 270, 280, 309, 316, 323f, **338, 354**
Moving frames, 98, Fig 6-4, Fig 7-2, 271, 292, 301, 360, 365
Multiply connected topology, 248, Fig 13-5
Nature of Spacetime, 4ff, 20, 22, 33, 61, **273**
Neutrino Oscillation Mass, 241ff
Neutrino, 14, 233, 237, 241ff, 244, 275f, Fig 15-3, 278f, 281f, 312, 337
Neutron, 14, Fig I- 1, 134f, 159, Fig 7-5, 171f, 179, Fig 8-6, 233, Fig 12-1, 275, Fig 15-3, 279, 282
Newton's law of gravitation, 119
Newton's laws of motion, 110f
Noether's Theorem, 173, 180, 324, 333, 354
Non-Euclidean Geometry, Fig I-1, 93, 343, see Riemann and Cartan differential geometry also
Non-Euclidean spacetime, Fig I-1, Fig 4-12, 136, 139, 166, 191, 302, **333**, 343, 360
Non-locality, 3, 257, 268, 272
O(3) electrodynamics, 5, 101, 146, 165, Fig 7-4, 168f, 171, 173, Fig 8-3, 187, Fig 8-6, 230, 268, 294f, 296f, 315, 320, 322f, 334, 344, 353, 358, 362
Ohm, 112
One-forms, Fig 1-8, 65, 102, Fig 4-13, 255, 260, 271, Fig 15-2, 303f, 310, 315, **334**
Operators, 89, 95, 104, **124,** 129, 152, 158, 161, **303**
Optical phase laws, 6, 174, 230, 245, 249f, 286f
Origin of charge, 3, 190
Origin of Wave Number, 262
Orthogonal, 26, 79, Fig 4-1, 84, Fig 4-11, 163, 265, **296,** 320, **336**
Orthonormal, Fig I-3, **84,** 97f, Fig 4-12, Fig 6-7, Fig 6-8, 149f, Fig 8-8, **296,** 302, **336**
Outer (Tensor or Exterior) Product, Fig 4-9, **89f,** 101, 137, 155, 166, 184, 331, 336, 341, 357
Packet, 339
Palatini variation, 51, 53, 98, 175, 189, 360, see Tetrad also
Parallel transport, 142f, Fig 6-5, 305ff, **336**

Parity Symmetry, see Reflections
Particle accelerator, 183f
Particles, 2ff, 11, Fig I-4, 14, 18f, 34, 37f, 46, 56ff, 67, 76, 115, 125, 130, 171, 173f, 179, 189f, 193, 209f, 230, 233, 237ff, 239ff, 261, 274ff, Fig 15-3, 283, 312f, 317, 337ff, 345, 347, 349, 353f
Pauli matrices or spinors, 56, 98, 165, 338f, **347f**, 359f
Phase Effects, 245, Fig 13-1, 248f, 287
Photon, 24, Fig 3-2, 63, 71f, 127f, Fig 7-4, 168, 174f, 180, 199, 206, Fig 9-3, 222, **Chapter 11,** 234, 251, Fig 13-6, 274ff, Fig 15-3, 279f, 294f, 299f, 315, 334, 339f
Planck constant, 69
Planck quantum hypothesis, see quantum hypothesis
Poisson's Equations, 120-123, 152, 255, **340**
Postulate, 3, 16, 19, 39, 47, Fig 3-4, 69, 132, **294**
Potential, 45f, **Fig 2-8**, 59, 63, 66, 97, 112ff, 117, 120ff, Fig 5-6, 135, 141, 146f, Fig 7-2, 165, 174f, 181f, 188, 190f, 249, 254ff, 258, 261f, 270, 279, 290, 292, 317, 319, 349, 357, 359, 366
Pressure, 35, 42f, 126, 135, 307, 354, 362
Principle of General Relativity, 12, 34, 228f, 231, **341**
Principle of Least Curvature, 67, 173, 197, **204f,** 210, 219f, **234,** 237, **256f,** 263, 272f
Probability distribution, Fig 3-1, 59, Fig 3-3, 63, 71, 129f, 163, 187, 219f, 272, 292, 325, 343, 362, 364
Pythagorean equation, 20, Fig 1-4, 51f, 54, Fig 3-5, 72, 84, 91f, 331f, 338, 341
$q^a{}_\mu$, 49, 52, 85, 96f, 99, Fig 4-12, 135, 137, 139f, 146f, 151ff, 156, 158,ff Fig 7-2, 162, 164ff, 184ff, 189ff, 194ff, Fig 8-8201, 203, 217, 220, 226, 234, 238, 254f, 258f, 261, 266, 270, 281, 284, Fig 15-5, 291, 317, 358ff, 365
Quantum Chromodynamics, 56, 67, 74, 360,
Quantum Electrodynamics, 56, 67, 74, 315, 343
Quantum gravity, 68, 341, 360
Quantum Hypothesis, 16, 58, Fig 3-4, 64, 69, **70f,** 78, 221, 297, 340, 365
Quantum Mechanics, Fig 1-5, 9, Fig I-3, 11, 15, 25, 39, 44, 47, Chapter 3, 74, 99, Fig 4-12, 147, 130, 149, Fig 6-8, 153, 156, 162f, Fig 7-3, 187f, 200, 203, 205, 214f, Fig 12-2, 237, 248, 257f, 263f, 313, **341,** 343, 362, 365
Quantum Numbers, 66, 74, Fig 3-6,
Quarks, 6, Fig I-4, 14, 56, 67, 135, 159, 165, Fig 7-5, 171, 189, Fig 8-6, 278, 282, Fig 15-6, 313, 317, 319, 321, 342, 345, 348
R, curvature, 9, Fig I-2, 15, 34, **37,** 39, 42, 44, 47, 69, 94, 114, 125f, 129, 135, 139, 141, 143, 145, 154,f 157, 162, 164, 184, **192ff,** 204, 226, 228, 259, 262, 269f, 272, 284f, 306, 311, 332, **342, 345**
Radiatively induced fermion resonance (RFR), 258
Rank, **94,** 102, 231, 302, 326, 342
$R^a{}_\mu$, **138ff,** 143, **145,** 148, 153f, 155, 161, 164, 259, 266, Fig 15-5,
Reference Frame, 3, 12, 17, **Fig 1-1,** Fig 1-2, 21f, Fig 1-7, Fig 2-1, Fig 2-2, 342
Reflections, 292, 323, 350-353
Renormalization, **67,** 209, 233, 244, 343

Resistance, 55, 112, 293
Rest energy, Fig 5-8, 183f, 204f, **Fig 9-3**, 220, 261, 300, 338
Ricci tensor, 94, 125, 138, 307, 311, 323, 341, **343**, 345, 356
Riemann Curved Spacetime, Fig I-1, 9, Fig I-2, Fig I-3, 31, 33, Fig 2-7, 62, 79, 101, 136, 146, 167, 178ff, Fig 8-1, Fig 8-3, 231, 235, 273, **313**, 336, 343f
Riemann geometry, 7, Fig I-1, 9, 31, 84f, 87, 91, 259, 289, 292, 297, 308, **343f**
Riemann tensor, 49, 93, 125, 312, 315, 323, 325, 331, **343f**
Rotation (physical), 63, 78f, 96, 137, 165, **Fig 9-1,** 198, 226, 266, 291, 293, 306, 317, 321
Rotations (mathematical), 96, 165, 187, 226, 291, 293, 333, 344f, **351f**, 354
Sagnac Effect, 230, 245, 249, 287
$S^a{}_\mu$, **151**, 165, 284, 317, 367
Scalar curvature, R, 94, 284, **345,** see R also
Scalar Product, 51, Fig 4-4, **83f**, 90, 141, 331, 336, see Inner Product also
Scalars, 3, **25f**, 47, 54, 62f, 79, Fig 4-4, 83ff, Fig 4-8, 89, 94, 98, 102, 104, Fig 4-18, 130, 109, 303f, 306ff, 310f, 315, 324, 336, 345
Schild's ladder, see parallel transport
Schrodinger's Equation, 58, **61**, 65, 129f, 152f, 173, 188, 201, 204, 214, 217, 309, 326, 345
Schwarzschild, 209
Self-gravitation, 35, 43, 135
SI Units, 73, 346
Singularity, **5,** 180f, 197, 206, 210, 268, 286, 362
Skew-symmetric, 95, 166
Spacetime, Introduction, 17ff, 41, Fig 4-2, 43, 47, 78, Fig 4-3, Fig 6-1, Fig 6-6, Fig 7-3, Fig 8-1, Fig 8-5, Fig 10-3, 222, 228, Fig 11-5, 247, 268, 273, Fig 15-3, 279, Fig 15-4, , 313, 317, 343, **346, 362**
Special Relativity, 2, 6f, Fig I-1, Fig I-4, 9, 12, **16,** Chapter 1, 34ff, 40ff, 49, 53, 55, 62f, 72, 76, 90, 121, 125, 136, 171f, 180, 191, 194, 201, 204, 205f, 209, 218, 221, 225f, 234, Fig 12-3, 247, 251, 260, 263, 265, 267f, 274, 295, 297, 322f, 327, 330, 332, 342, 350, 352, 360
Spin connection, 101, 107f, Fig 6-9, 259f, 306, 315, 333, **347**
Spin, 6, 9, Fig I-2, 19, 33, 56, 58, 63, 66, 76, 100f, 147, 154, 175, 182f, 190f, 209, 211 , 224, Fig 11-2, 225, Fig 13-7, 260, 280, 283, Fig 15-7, 288f, 294f, 299, 312, 317, 329, 333f, 337ff, **347f**, 352, 358, see $B^{(3)}$ spin field and spin connection also.
Spinning spacetime, 361
Spinor, 101, 285, **348**, 356, 358
Standard model, 2f, 58, 68, 99, 146, 171, 174, 189, 233f, 237, Fig 12-3, 241, 244, 253, 257f, 264, 290, 315, 317, 333
Stokes Theorem, 251, Fig 13-6, 287, 348
Stress energy, Fig 1-3, 38, Fig 2-5, **94**, 125, 159, 208, 270f
String theory, 4, 68f, 188
Strong Field, (Force), 2, 3, 5, 13f, Fig I-4, 19, 99, 113, 134f, 145f, 151, 159, 163ff, 170f, 174f, 182, 189ff, 192, 231, 234, 256, 267, 285, 315, 317, 319f, 337, 348, 357, 360

SU(2), SU(3) symmetry, 67f, 131, 146, 165f, 169f, Fig 7-5, 170f, 189, 192, 278, 315, 317, 321, 334, 339, 345, 349, 352f, 358
Symmetries of the Evans Wave Equation, 192
Symmetry, 3, 43, 67f, 113, 137, 164f, 168ff, 184, 233, 237, 265, 274f, 290, 317, 320ff, 333, 344f, 350-354
Tangent bundle, 79, 68, Fig 6-8, Fig 7-3, 235
Tangent spacetime, Fig I-3, Fig 2-3, 101, 146, 160, 166, 169ff, 175189, 313
Tensor contraction, 356
Tensor Field, see Vector Field
Tensor space, 357
Tensors and Differential Geometry, 53
Tensors, 9, 27, 31, 37, Fig 2-9, 49-54, 62, 85, 89, 91f, Fig 4-11, 93-98, 125, 137ff, 157, 185, 187ff, Fig 8-7, 201, 228ff, 262ff, 289, 294, 298, 300, 304ff, 311-318, 311-326, 331f, 336, 341-347, 350f, 354-361
Tetrad postulate, 160, **203**, 259, **360**
Tetrad, Fig 2-9, 96, Fig 4-12, 145, Fig 6-7, 195, Fig 8-8, 259-262, Fig 15-1, 269f, 271, 284, Fig 15-5, 297f, 302, 317, 335, 341, 357ff, 365
Theory of everything, see Unified Field Theory
Thermodynamics, 155, 184,
Time, 7, I-1, 9, 11f, 16f, 18, 20, Fig I-2, 22, Fig I-4, 24ff, Fig 1-9, 40, Fig 2-3, 47, 49, 61, 64, 69, 73, 90ff, 111, 113, 126ff, Fig 7-4, 180, 209, Fig 13-6, 274f, 286, Fig 15-7, 292f, 297, 300, 308, 308f, 323, 333, 352f, 361, see spacetime also
Torque, 26, Fig 1-5, 27, 32, 69, 75, 85, 126, 153, 183, 190, 228, 271, 293, 347, 361, 363
Torsion, 9, Fig I-2, 12, 47, 56, 62, 76, 8095, 97, 100f, 108, 121, 134, 136, Fig 6-2, 140ff, Fig 6-5, 144-152, 159, 166, 168, Fig 8-1, 178-189, 193f206, 210, 227, 231. 232, 235, 249, 253, 257, 262-270, 272-276, Fig 15-3,279--284, Fig15-4, Fig 15-5, 295, 298, 306f, 311, 315ff, 326f, 340, 346, 354, 361ff, 366
Trace, 94f, 166, 324, 328, 343, 345, 362
Translations, 274, 297, 323, 333, 350, 352f
U(1) symmetry, 169, 227
Unified Field Equations, 15, 139 Equation (8), 271, 362,
Unified field theory, 2, 4ff, 12, 31, 37f, 45f, 49, 51, 53, 68, 76, 102, 143, 152, 156, Fig 7-2, 164, 169, 171, 174, 177, 206, 215, 221f, Fig 11-2, 225, 227, 237, 245, 249f, 257, 259ff, 268f, 272f, 278, 284, 289, 330, 334f, 353, 359ff
Unified Wave Theory, 285
Unit vectors, Fig 4-12, 105, 295f, 332, 334, see Basis Vectors also
V_0, 180, 197, 208ff, 217, 219, 221, 244, 268
Vacuum, 2, 4, 9, 11f, 19, 44, 46f, 55ff, 66f, 99, 112, 147, 151, 180, 193, 219, 231, 235, 244, 248, Fig 13-5, 273f, 312f, 327, 346, 362f
Vector Field, 63, 86, 89, 99, 225, 302, 305f, 334, 355, 357, 359, 361, 363f
Vector Multiplication, 108, Fig 4-18
Vector Product, 26f, **86f**, **Fig 4-5**, 331, 336, 341, 366

Vector Space, 10f, Fig I-3, 27, 32, 47, 51, 62, 76, 79, Fig 4-3, 100, 102, 148, 291ff, 299, 301ff, 310, 314f, 322, 335, 341, 354f, 357f, 364, 366
Vectors, Fig 1-6, 25, 83
Very Strong Equivalence Principle, 55, 180f, 194, 288, 313, 337
Vielbein, see Tetrad
Volt, 69, 112ff, Fig 5-1, 117, Fig 5-4, 122ff, 146f, 168, 184, 246, 255f, 261, 266, 280, 285, 288, 291, 293, 310, 340, 346, 349, 359
Wave Equation, 104, 126, Fig 5-7, Chapter 7
Wave Mechanics see Quantum Mechanics
Wave number, 6, 72, 129, Fig 5-9, 197, Fig 9-1, 200, 230, 210, 215, Fig 10-2, 219, 246, 248, 259, 262f, 306, 342, 345f, 349f, 366
Wave velocity, 366
Wave-particle duality, 56, 58, 128, 200, 273,
$W^a{}_\mu$, **151**, 165, 267, 285, 317
Weak field limit, 24, 37, 94, 125, 135, 189, 202, 208, Fig 12-3, 255, 366,
Weak field, force, 2f, 5, 14, I-4, 68, 99, 113, 134f, 147, 151, 163ff, 169, 180, 189, 192, 202, 267, 284, 337, 357, 366
Weber, 114, 167, 184, 291, 346
Wedge product, 87, Fig 4-9, Fig 4-10, Fig 6-3,
Ψ, spinor, 165, 172f, 201ff, 267, 285
Ψ, wave function, 59, 61, Fig 3-3, 66, 104, Fig 4-14, 126f, 129ff, 165, 217, 292, 309f, 325f, 345, 362, 364
γ, $1/\sqrt{(1-(v/c)^2)}$, 20, 24, 70f, 183, 216, 239, 333
γ_μ, see Dirac spinor